MISSION COLLEGE
LIBRARY

59.95

Control Engineering

Series Editor
William S. Levine
University of Maryland

Editorial Advisory Board

Okko Bosgra
Delft University
The Netherlands

Graham Goodwin
University of Newcastle
Australia

Petar Kokotović
University of California
Santa Barbara
USA

Manfred Morari
ETH
Zürich, Switzerland

William Powers
Ford Motor Company
USA

Mark Spong
University of Illinois
Urbana-Champaign
USA

Iori Hashimoto
Kyoto University
Kyoto, Japan

Nonlinear Control and Analytical Mechanics

A Computational Approach

Harry G. Kwatny
Gilmer L. Blankenship

With 93 Illustrations and a CD-ROM

Birkhäuser
Boston • Basel • Berlin

Harry G. Kwatny
Department of Mechanical Engineering
and Mechanics
Drexel University
Philadelphia, PA 19104, USA
hkwatny@coe.drexel.edu

Gilmer L. Blankenship
Department of Electrical
and Computer Engineering
University of Maryland
College Park, MD 20742, USA
gilmer@eng.umd.edu

Library of Congress Cataloging-in-Publication Data
Kwatny, Harry G.
 Nonlinear control and analytical mechanics : a computational approach / Harry G.
Kwatny, Gilmer L. Blankenship.
  p. cm.
 Includes bibliographical references and index.
 ISBN 0-8176-4147-5 (alk. paper)
  1. Automatic control. 2. Nonlinear control theory. 3. Mechanics, Analytic.
I. Blankenship, G. (Gilmer), 1945– II. Title.
TJ213 .K96 2000
629.8—dc21   00-039793
                                                  CIP

Printed on acid-free paper.
© 2000 Birkhäuser Boston

Birkhäuser B®

All rights reserved. This work may not be translated or copied in whole or in part without the
written permission of the publisher (Birkhäuser Boston, c/o Springer-Verlag New York, Inc., 175
Fifth Avenue, New York, NY 10010, USA), except for brief excerpts in connection with reviews or
scholarly analysis. Use in connection with any form of information storage and retrieval, electronic
adaptation, computer software, or by similar or dissimilar methodology now known or hereafter
developed is forbidden.
The use of general descriptive names, trade names, trademarks, etc., in this publication, even if the
former are not especially identified, is not to be taken as a sign that such names, as understood by
the Trade Marks and Merchandise Marks Act, may accordingly be used freely by anyone.

ISBN 0-8176-4147-5
ISBN 3-7643-4147-5   SPIN 10734732

Production managed by Louise Farkas; manufacturing supervised by Jeffrey Taub.
Typeset by the authors in LaTeX.
Printed and bound by Hamilton Printing Co., Rensselaer, NY.
Printed in the United States of America.

9 8 7 6 5 4 3 2 1

To Lynne and Elyse

Contents

Appendix

Preface

During the past decade we have had to confront a series of control design problems – involving, primarily, multibody electro-mechanical systems – in which nonlinearity plays an essential role. Fortunately, the geometric theory of nonlinear control system analysis progressed substantially during the 1980s and 90s, providing crucial conceptual tools that addressed many of our needs. However, as any control systems engineer can attest, issues of modeling, computation, and implementation quickly become the dominant concerns in practice. The problems of interest to us present unique challenges because of the need to build and manipulate complex mathematical models for both the plant and controller. As a result, along with colleagues and students, we set out to develop computer algebra tools to facilitate model building, nonlinear control system design, and code generation, the latter for both numerical simulation and real time control implementation. This book is an outgrowth of that continuing effort. As a result, the unique features of the book includes an integrated treatment of nonlinear control and analytical mechanics and a set of symbolic computing software tools for modeling and control system design.

By simultaneously considering both mechanics and control we achieve a fuller appreciation of the underlying geometric ideas and constructions that are common to both. Control theory has had a fruitful association with analytical mechanics from its birth in the late 19th century. That historical relationship has been reaffirmed during the past two decades with the emergence of a geometric theory for nonlinear control systems closely linked to the modern geometric formulation of analytical mechanics. Not surprisingly, the shared evolution of these fields has paralleled the needs of technology. Today, mechanicians and control engineers are brought together in fields like space systems engineering, robotics, ground and sea vehicle design, and biomechanics. Consequently, our integrated approach provides a rich set of models and control design examples that are of contemporary and practical interest.

Control theory would be quite sterile without concrete connections to the natural world. The process of modeling is just as central to control engineering as is control theory itself. A control system design project does not begin when a control engineer is handed a model; it begins at the onset of model formulation.

Our main thesis is that a full appreciation of the meaning and significance of either theory benefits by developing their connection and by applying them to meaningful examples. The capability to do the latter requires supporting computational tools. In this book, we highlight and exploit the computational infrastructure common to both modern analytical mechanics and nonlinear control. To achieve the full benefits of the concepts now available, we need to exploit symbolic as well as numerical computing techniques. However fortuitous it may be, it is only during the past decade that symbolic computing technology (or computer algebra) has matured to the level of serious engineering application.

We emphasize symbolic computing because it is essential for working with nonlinear, parameter-dependent systems and it is a relatively new tool for engineers. Symbolic computing does not replace numerical computing. It supplements and enhances it. Recognizing the distinctions between symbolic and numerical computing and how best to integrate them is a significant challenge. We will use symbolic computing for several purposes:

1. to perform basic mathematical operations (like implement a coordinate transformation or compute a Lie bracket),

2. to build explicit mathematical models,

3. to simplify models (e.g., via Taylor linearization or symmetry reduction),

4. to generate numerical simulation code,

5. to implement nonlinear control constructions (such as compute an inverse system or perform feedback linearization),

6. to generate numerical code for implementing controllers.

In this work we employ examples of various levels of complexity from simple examples that illustrate a theoretical point in a transparent way to examples with detailed models for which results are too complex to exhibit in print, but can nevertheless be manipulated using a computer. We will provide examples of the latter type using electronic media, specifically, *Mathematica* notebooks. The point is that when working with engineering grade models it is not reasonable to visually examine or manually manipulate symbolic expressions by hand. However, it is possible to work effectively with such expressions using a computer.

Many of us were attracted to control systems engineering because it enables a broad exposure to numerous areas from traditional engineering disciplines to computer and information sciences and mathematics, to economics, biology, and even social sciences. Indeed, it would be quite a challenge to find an engineer in the field for more than a few years without cross-disciplinary experience. While, in recent years, the need for a multidisciplinary approach has been touted as generally necessary for technological progress, it has always been that way in the control field. Because of the extraordinary scope of control applications, control engineers have traditionally sought out the unifying principles that make it possible to function creatively in a varied and complicated environment.

From its emergence as a coherent discipline, control engineering has involved a high level of abstraction. Mathematics, perhaps the ultimate unifying principle, and certainly the most successful language invented by man to clarify and communicate complex ideas without ambiguity, has been a cornerstone of its development. In writing this book, we had to make choices to balance competing objectives. One of the most difficult was to establish a correct level of mathematical abstraction and rigor. We view ourselves as engineers, not mathematicians, and it is from that point of view that we came to a judgment. Mathematicians may decide that our arguments lack rigor and some engineers may find our discussion too formal. However, we can judiciously sacrifice rigor for accessibility, but we often need precise statements to clearly identify the range of applicability of a technique or to establish reliable machinery for computing.

We are indebted to many students and colleagues whose collaborations with us on various research and engineering projects contributed in countless ways to the writing of this book. In particular, we would like to acknowledge Dr. Reza Ghanadan of Bell Laboratories, Mr. Chris LaVigna, Dr. Carole Teolis, and Mr. Eric Salter, all of Techno-Sciences, Inc., for their contributions to the development and application of the *ProPac* software package, and to Mr. Gaurav Bajpai of Drexel University for his careful reading of the manuscript.

Philadelphia, Pennsylvania Harry G. Kwatny
College Park, Maryland Gilmer L. Blankenship

Chapter 1

Introduction

As inexpensive processors have become increasingly ubiquitous in all manner of physical devices, the opportunities and demand for using them to improve functionality and performance has pushed control design technology to new limits. While 'emergent' application areas like robotics, biomedical and micro-electro-mechanical systems bring with them their special requirements, traditional fields like the aerospace, automotive, marine and process industries have also expanded the role of automation. In many of the new control problems a direct confrontation with nonlinearity is unavoidable.

Notwithstanding the advances in our understanding of nonlinear dynamical behavior and in nonlinear control theory itself, the state of control design for nonlinear systems must be considered embryonic as compared to that of linear systems. This is in part because the possibilities of nonlinear behavior are so vast and varied, but also because of the lack of tools for working efficiently with nonlinear problems of even modest engineering scale.

The control design process, while not rigidly structured, always includes three crucial elements:

1. model building

2. control design

3. control implementation

A typical control design project might follow the process in the flow chart of Figure (1.1). Modeling is central to formulating the control design problem as well as solving it. In our view it is an integral part of control system design. The diagram also accurately suggests that model building, control design and control implementation may be repeated several times during the course of a project. Clearly, tools that facilitate and automate these processes are necessary.

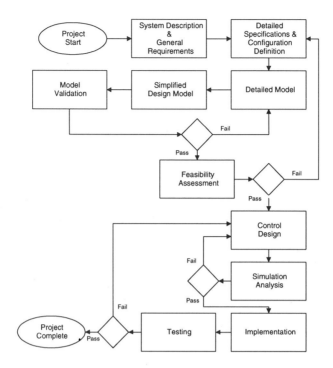

Figure 1.1: A typical design project flow chart.

Primarily for these reasons this book addresses both modeling and control and exploits symbolic computing as a means for minimizing the painful calculations, expression manipulaton and coding that would otherwise be required. A software toolbox, *ProPac* , is included with the book. It is a package to be used with the *Mathematica* [1] computer algebra system. More will be said about the software later.

1.1   Scope and Organization

This book provides an integrated treatment of geometric nonlinear control and analytical mechanics. Their common geometric foundation and the recurrent cross fertilization between the two fields is certainly justification enough for doing so. In fact, the two subjects are so well matched that in describing one it is impossible to resist drawing examples from the other. However, not the least important factors motivating us to unify this material derive from the practical considerations described above. Control system design simply can not be divorced from modeling.

---

[1]For information about *Mathematica* visit the web site www.wolfram.com

Chapters 2 and 3 deal with important preliminary material. A short summary of ordinary differential equations including basic Lyapunov stability concepts is given in Chapter 2. Our treatment of these topics is brief and focuses on those items of immediate use. There are many excellent texts for the interested reader to gather additional information. A somewhat more detailed introduction to differential geometry is provided in Chapter 3. Yet, we are still selective in our choice of material from a vast literature, including only what we think is essential background. Basic calculations using *Mathematica* and *ProPac* are introduced.

Our treatment of analytical mechanics is based on the Hamilton-Lagrange formulation. It begins with a general construction for the kinematic parameters of multibody tree structures in Chapter 4. System configuration coordinates are defined in terms of a natural and general parameterization of the individual joints. Formulas that define node inertial positions and velocities in terms of configuration coordinates and generalized quasi-velocities are derived. These calculations are implemented in the accompanying software. The dynamics of tree structures as well as systems with closed loops are developed in terms of Poincaré's form of Lagrange's equations in Chapter 5. Closed loops are treated by adding constraints to an underlying tree. Constraints may be algebraic relations among the configuraton variables and/or holonomic or nonholonomic differential constraints. Constructive procedures for deriving the equations are presented and, again, implemented in the accompanying software. Examples illustrate the assembly of models of undersea vehicles, robotic systems, ground vehicles and other systems.

Nonlinear control is the subject of Chapter 6. Here, we discuss smooth affine control systems. Basic concepts of nonlinear controllability and observabilty and local decompositions via coordinate transformation are discussed first. In terms of control system design the focus of this book is on feedback linearization and dynamic inversion. Exact (state) linearization as well as partial (input-output) linearization are fully described. The chapter closes with a discussion of nonlinear observers. Of course, computation is a key issue. *ProPac* functions that implement the required calculations are introduced and illustrated.

Feedback linearization methods are strongly model based, relying, in fact, on direct cancellation. Consequently, robustness is a major concern and we devote the next two chapters to robust control. Chapter 7 addresses smooth robustification of feedback linearizing controllers. It begins with a discussion of how uncertainty propagates through the reduction to normal form of a nominal system. The notion of matched and nonmatched uncertainty is developed. Then Lyapunov redesign for matched uncertainty and robust stabilization via backstepping for strict triangular nonmatched uncertainty are described. Adaptive control methods for systems with uncertainty that can be parameterized are then presented. Once again, software tools that implement the required calculations are described and illustrated.

Chapter 8 deals with variable structure control system design. The view of variable structure control as a nonsmooth, robust variant of input-output linearization is emphasized. The chapter begins with a general discussion of discontinuous dynamics including a formulation of Lyapunov stability analysis in that context. Methods for sliding mode and reaching control design are presented. Chattering reduction via regularizaton and other methods are described. The inherent robustness of variable structure controls with respect to matched uncertainty is established and a backstepping method is described for strict triangular nonmatched uncertainty. Supporting software is illustrated. The method also applies to a class of nonsmooth plants that includes a variety of discontinuous friction models and other so-called 'hard' nonlinearities. Examples are given.

1.2   Software

Most of the examples in this book have been devloped using the software package *ProPac* developed by Techno-Sciences, Inc., Lanham, MD. *ProPac* 2.0 is included with this book. It is a *Mathematica* package that provides a set of symbolic computing tools for modeling multibody mechanical systems as well as for linear and nonlinear control system design and analysis. There are excellent introductory books and tutorials available for *Mathematica*. Many of these are identified on the Wolfram web site.

The *ProPac* CD contains a set of tutorial and application notebooks. These include Dynamics.nb and Controls.nb which introduce the basic modeling and control tools available in *ProPac* . On-line help is available through *Mathematica's* Help Browser. For more information, notebooks and other documents visit the web site: www.technosci.com.

Using *ProPac* requires version 3 or 4 of *Mathematica*. That is all that is required to develop the equations of motion, for conducting numerical simulations within *Mathematica*, and building the C source code required for simulations in SIMULINK [2]. Use of the latter requires MATLAB/SIMULINK and a C compiler as recommended by the MathWorks for compiling MEX-files on the user's platform. Functions in *ProPac* generate C-code that compiles as SIMULINK S-functions. In this way modules for the plant and controller are easily generated for inclusion in SIMULINK simulations. Controllers, with embelishments like filters, etc., can be downloaded into DSP boards via MATLAB's Real Time Workshop. The setup is illustrated in Figure (1.2).

1.2.1   Installing *ProPac*

To install *ProPac* , follow the two step procedure:

Step 1: Put the entire ProPac directory in Mathematica's Applications direc-

---

[2]for information about MATLAB/SIMULINK visit the web site www.MathWorks.com

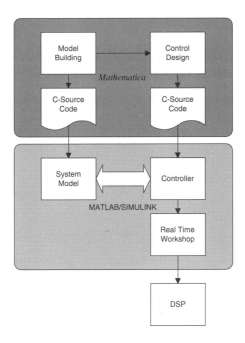

Figure 1.2: *ProPac* is a *Mathematica* package with links to MAT-LAB/SIMULINK.

tory. For the PC the full path is

```
C:\Program Files\Wolfram Research
                    \Mathematica\3.0
                              \AddOns\Applications\
```

Step 2: Start Mathematica and rebuild the Help index. The latter is accomplished with the following simple procedure. From the main menu choose: Help ⇒ Rebuild Help Index

Once this is done, on-line help is available. In the Help Browser select Add-ons and then TSi ProPac.

The Mex folder contains 3 C-source files that need to be included when compiling MATLAB/SIMULINK MEX files. These may stored in any convenient location, but must be available at the time of compilation.

1.2.2   Package Content

ProPac consists of seven packages: Dynamics, ControlL, ControlN, GeoTools, MEXTools, NDTools, and VSCTools. Once ProPac is loaded all of the functions in these packages are available for use and the appropriate packages will be automatically loaded as required. In general, a user does not have to be concerned about loading any particular package. To load *ProPac* simply enter `<<ProPac`` (most of the package functionality is available in Mathematica 2.2, enter `<<ProPac` Master``).

Dynamics contains the model building functions and ControlL and ControlN the linear and nonlinear control analysis functions, respectively. GeoTools includes basic functions used in differential geometry calculations. NDTools contains supporting functions for working with nondifferentiable nonlinearities and VSCTools contains functions for variable structure control. MEXTools includes functions for creating C-code files for both models and controllers that compile as S-functions for use with MATLAB/SIMULINK.

The Tables in the Appendix contain a brief summary of many of the available functions. More details and numerous examples can be found in the help browser and in the notebooks.

Chapter 2

Introduction to Dynamical Systems

2.1 Introduction

In this chapter we briefly review some basic material about nonlinear ordinary differential equations that is important background for later chapters. After a preliminary discussion of the basic properties of differential equations including the existence and uniqueness of solutions, we turn to a short discussion of stability in the sense of Lyapunov. In addition to stating the most important theorems on stability and instability we provide a number of illustrative examples. As part of this discussion we introduce Lagrangian systems – a topic to be treated at great length later. This chapter is concerned exclusively with dynamical systems (as opposed to control systems) and with smooth systems (as opposed to systems that contain nondifferentiable nonlinearities). Those topics will be treated in later chapters. It is presumed that the material discussed is not new to the reader and we provide only a short summary of those elements considered immediately relevant. For a more complete discussion many excellent text books are available. We reference a number of them in the sequel.

2.2 Preliminaries

A *linear vector space* , $\mathcal{V}$- over the field R is a set of elements called vectors such that:

1. for each pair $x, y \in \mathcal{V}$, the sum $x+y$ is defined, $x+y \in \mathcal{V}$ and $x+y = y+x$.

2. there is an element '0' in $\mathcal{V}$ such that for every $x \in \mathcal{V}$, $x + 0 = x$.

3. for any number $a \in R$ and vector $x \in V$ scalar multiplication is defined and $ax \in V$.

4. for any pair of numbers $a, b \in R$ and vectors $x, y \in V$: $1 \cdot x = x$, $(ab)x = a(bx)$, $(a + b)x = ax + bx$.

A linear vector space is a *normed linear space* if for each vector $x \in V$- there corresponds a real number $\|x\|$ called the *norm* of x which satisfies:

1. $\|x\| > 0$, $x \neq 0$, $\|0\| = 0$

2. $\|x + y\| \leq \|x\| + \|y\|$ (triangle inequality)

3. $\|ax\| = |a| \, \|x\| \; \forall a \in R, \; x \in V$

When confusion can arise as to which space a norm is defined in we replace $\|\bullet\|$ by $\|\bullet\|_V$.

A sequence $\{x_k\} \subset V$, V a normed linear space, *converges* to $x \in V$ if

$$\lim_{k \to \infty} \|x_k - x\| = 0$$

. A sequence $\{x_k\} \subset V$-is a *Cauchy sequence* if for every $\varepsilon > 0$ there is an integer, $N(\varepsilon) > 0$ such that $\|x_n - x_m\| < \varepsilon$ if $n, m > N(\varepsilon)$. Every convergent sequence is a Cauchy sequence but not vice versa. The space - is *complete* if every Cauchy sequence is a convergent sequence. A complete normed linear space is called a *Banach* space .

The most basic Banach space of interest herein is n-dimensional *Euclidean* space, the set of all n-tuples of real numbers, denoted R^n. The most common types of norms applied to R^n are the p-norms, defined by

$$\|x\|_p = \left(|x_1|^p + \cdots + |x_n|^p \right)^{1/p}, \quad 1 \leq p < \infty$$

and

$$\|x\|_\infty = \max_{i \in \{1, \ldots, n\}} |x_i|$$

An ε-*neighborhood* of an element x of the normed linear space V is the set $S(x, \varepsilon) = \{y \in V | \; \|y - x\| < \varepsilon\}$. A set A in V is *open* if for every $x \in A$ there exists an ε-neighborhood of x also contained in A. An element x is a *limit point* of a set $A \subset V$ if each ε-neighborhood of x contains points in A. A set A is *closed* if it contains all of its limit points. The *closure* of a set A, denoted $\bar{A}$, is the union of A and its limit points. A set A is *dense* in V if the closure of A is V.

If B is a subset of V, A is a subset of R, and $\{V_a, a \in A\}$ is a collection of open subsets of V such that $\cup_{a \in A} V_a \supset B$, then the collection V_a is called an *open covering* of B. A set B is *compact* if every open covering of B contains a finite number of subsets which is also an open covering of B. For a Banach space this is equivalent to the property that every sequence $\{x_n\}, x_n \in B$, contains a subsequence which converges to an element of B. A set B is *bounded* if there exists a number $r > 0$ such that $B \subset \{x \in V \,|\|x\| < r\}$. A set B in R^n is compact if and only if it is closed and bounded.

A function f taking a set A of a space $\mathcal{X}$ into a set B of a space $\mathcal{Y}$ is called a *mapping* of A into B and we write $f : A \to B$. A is the *domain* of the mapping and B is the *range* or *image*. The image of f is denoted $f(A)$. f is *continuous* if, given $\varepsilon > 0$, there exists $\delta > 0$ such that

$$\|x - y\| < \delta \Rightarrow \|f(x) - f(y)\| < \varepsilon$$

A function f defined on a set A is said to be *one-to-one* on A if and only if for every $x, y \in A$, $f(x) = f(y) \Rightarrow x = y$. If f is one-to-one it has an inverse denoted f^{-1}. If the one-to-one mapping f and its inverse f^{-1} are continuous, f is called a *homeomorphism* of A onto B.

Suppose $\mathcal{X}$ and $\mathcal{Y}$ are Banach spaces and $f : \mathcal{X} \to \mathcal{Y}$. f is a *linear map* if $f(a_1 x_1 + a_2 x_2) = a_1 f(x_1) + a_2 f(x_2)$ for all $x_1, x_2 \in \mathcal{X}$ and $a_1, a_2 \in R$ (or C). In general, we can write a linear mapping in the form $y = Lx$, where L is an appropriately defined 'linear operator.' A linear map f is said to be *bounded* if there is a constant K such that $\|f(x)\|_{\mathcal{Y}} \leq K \|x\|_{\mathcal{X}}$ for all $x \in \mathcal{X}$. A linear map $f : \mathcal{X} \to \mathcal{Y}$ is bounded if and only if it is continuous. A linear map from $R^n \to R^m$ is characterized by an $m \times n$ matrix of real elements, e.g., $y = Ax$. The 'size' of the matrix A can be measured by the *induced p-norm* (or gain) of A

$$\|A\|_p = \sup_{x \neq 0} \frac{\|Ax\|_p}{\|x\|_p}$$

for which we write the following special cases

$$\|A\|_1 = \max_{1 \leq j \leq n} \sum_{i=1}^{m} |a_{ij}|$$

$$\|A\|_2 = \sqrt{\lambda_{\max}(A^T A)}$$

$$\|A\|_\infty = \max_{1 \leq i \leq m} \sum_{j=1}^{n} |a_{ij}|$$

Here, $\lambda_{\max}(A^T A)$ denotes the largest eigenvalue of the nonnegative matrix $A^T A$.

f is said to be (*Frechet*) *differentiable* at a point $x \in A$ if there exists a bounded linear operator $L(x)$ mapping $\mathcal{X} \to \mathcal{Y}$ such that for every $h \in \mathcal{X}$ with $x + h \in A$

$$\|f(x + h) - f(x) - L(x)h\| \,/\, \|h\| \to 0$$

as $\|h\| \to 0$. $L(x)$ is called the derivative of f at x. If $f : R^n \to R^m$ is differentiable at x then $L(x) = \partial f(x)/\partial x$, the Jacobian of f with respect to x. If f and f^{-1} have continuous first derivatives, f is a *diffeomorphism*.

A function $f : A \rightarrow B$ is said to belong to the class C^k of functions if it has continuous derivatives up to order k. It belongs to the class C^∞ if it has continuous derivatives of any order. C^∞ functions are sometimes called *smooth*. A function f is said to be *analytic* if for each $x_0 \in A$ there is a neighborhood U of x_0 such that the Taylor series expansion of f at x_0 converges to $f(x)$ for all $x \in U$.

Consider a transformation $T : \mathcal{X} \rightarrow \mathcal{X}$, where $\mathcal{X}$ is a Banach space. $x \in \mathcal{X}$ is a *fixed point* of T if $x = T(x)$. Suppose A is a subset of Banach space $\mathcal{X}$ and T is a mapping of A into a Banach space $\mathcal{B}$. The transformation T is a *contraction* on A if there exists a number $0 \leq \lambda < 1$ such that

$$\|T(x) - T(y)\| \leq \lambda \|x - y\|, \quad \forall x, y \in A$$

Proposition 2.1 (Contraction Mapping Theorem) *Suppose A is a closed subset of a Banach space $\mathcal{X}$ and $T : A \rightarrow A$ is a contraction on A. Then*

1. *T has a unique fixed point $\bar{x} \in A$*

2. *If $x_0 \in A$ is arbitrary, then the sequence $\{x_{n+1} = T(x_n), \, n = 0, 1, \ldots\}$ converges to $\bar{x}$.*

3. *$\|x_n - \bar{x}\| \leq \lambda^n \|x_1 - x_0\| / (1 - \lambda)$, where $\lambda < 1$ is the contraction constant for T on A.*

Proof: [1], page 5.

We will make use of the following important theorem.

Proposition 2.2 (Implicit Function Theorem) *Suppose $F : R^n \times R^m \rightarrow R^n$ has continuous first partial derivatives and $F(0,0) = 0$. If the Jacobian matrix $\partial F(x, y) / \partial x$ is nonsingular, then there exists neighborhoods U, V of the origin in R^n, R^m, respectively, such that for each $y \in V$ the equation $F(x, y) = 0$ has a unique solution $x \in U$. Furthermore, this solution can be given as $x = g(y)$, i.e., $F(g(y), y) = 0$ on V, where g has continuous first derivatives and $g(0) = 0$.*

Proof: [1], page 8.

2.3   Ordinary Differential Equations

Existence and Uniqueness

Let $t \in R$, $x \in R^n$, D an open subset of R^{n+1}, $f : D \rightarrow R^n$ a map and let $\dot{x} = dx/dt$. We will consider differential equations of the type

$$\dot{x} = f(x, t), \quad x \in R^n, \, t \in R \tag{2.1}$$

When t is explicitly present in the right hand side of (2.1), then the system is said to be *nonautonomous*. Otherwise it is *autonomous*. A solution of (2.1) on a time interval $t \in [t_0, t_1]$ is a function $x(t) : [t_0, t_1] \rightarrow R^n$, such that $dx(t)/dt = f(x, t(t))$ for each $t \in [t_0, t_1]$. We can visualize an individual solution as a graph $x(t) : t \rightarrow R^n$. For autonomous systems it is convenient to think of $f(x)$ as a 'vector field' on the space R^n. $f(x)$ assigns a vector to each point $x \in R^n$. As t varies, a solution $x(t)$ traces a path through R^n. These curves are often called *trajectories* or *orbits*. At each point $x \in R^n$ the trajectory $x(t)$ is tangent to the vector $f(x)$. The collection of all trajectories in R^n is called the *flow* of the vector field $f(x)$. This point of view can be extended to nonautonomous differential equations in which case the vector field $f(x, t)$ and its flow vary with time.

Example 2.3 (Phase portraits) *For two dimensional systems the trajectories can be plotted in a plane. We will consider two systems, the Van der Pol system*

$$\begin{bmatrix} \dot{x}_1 \\ \dot{x}_2 \end{bmatrix} = \begin{bmatrix} x_2 \\ -0.8(1 - x_1^2)x_2 - x_1 \end{bmatrix}$$

and the damped pendulum

$$\begin{bmatrix} \dot{x}_1 \\ \dot{x}_2 \end{bmatrix} = \begin{bmatrix} x_2 \\ -x_2/2 - \sin x_1 \end{bmatrix}$$

Both of these systems are in so-called phase variable form (the first equation, $\dot{x}_1 = x_2$, defines velocity) so the trajectory plots are called phase portraits.

Van der Pol

```
In[1]:= f = {x2, -x1 + 0.8 (1 - x1^2)x2}; x = {x1, x2};

In[2]:= graphs3 = PhasePortrait[f, x, 15, {{-6, 6, 0.5}, {-5, 5, 5}}];

In[3]:= Show[graphs3, AxesLabel → {x1, x2}, PlotRange → {{-4, 4}, {-4, 4}},
            DisplayFunction → $DisplayFunction]
```

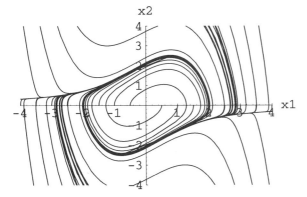

Damped Pendulum

In[4]:= $f = \{x2, -\sin[x1] - x2/2\}; x = \{x1, x2\};$

In[5]:= $graphs2 = PhasePortrait[f, x, 15, \{\{-20, 20, 0.5\}, \{-3, 3, 3\}\}];$

In[6]:= $Show[graphs2, AxesLabel \rightarrow \{x1, x2\}, PlotRange \rightarrow \{\{-10, 10\}, \{-3, 3\}\},$
          $DisplayFunction \rightarrow \$DisplayFunction]$

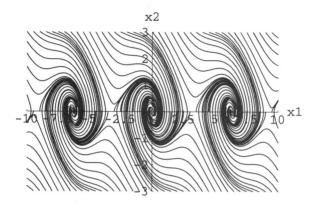

The above examples illustrate several important properties of nonlinear dynamical systems. In both cases the flow directions are to the right in the upper half plane and to the left in the lower half plane (recall $\dot{x}_1 = x_2$). Thus, it is easily seen that trajectories of the pendulum ultimately converge to rest points corresponding to the pendulum hanging straight down. These have the property that $f(x) = 0$. Any point $x \in R^n$ satisfying the condition $f(x) = 0$ is called an *equilibrium point*. The pendulum has an infinite number of equilibria spaced π radians apart. Some of these are attracting (the pendulum points straight down) and some repelling (straight up).

In contrast, all trajectories of the Van der Pol equation approach a periodic trajectory. Such an isolated periodic trajectory is called a *limit cycle*. Some systems can exhibit multiple limit cycles and they can be repelling as well as attracting. Equilibria and limit cycles are two types of 'limit sets' that are associated with differential equations. We will define limit sets precisely below. As a matter of fact, these are the only type of limit sets exhibited by two-dimensional systems. More exotic ones, like 'strange attractors' require at least three dimensional state spaces.

The existence and uniqueness of solutions to (2.1) depend on properties of the function f. In many applications $f(x, t)$ is continuous in the variables t and x. We will impose a somewhat less restrictive characterization of f. We say that a function $f : R^n \to R^n$ is *locally Lipschitz* on an open and connected subset

$D \subset R^n$, if each point $x_0 \in D$ has a neighborhood U_0 such that

$$\|f(x) - f(x_0)\| \le L \|x - x_0\| \tag{2.2}$$

for some constant L and all $x \in U_0$. The function $f(x)$ is said to be *Lipschitz* on the set D if it satisfies the (local) Lipschitz condition uniformly (with the same constant L) at all points $x_0 \in D$. It is *globally Lipschitz* if it is Lipschitz on $D = R^n$. We apply the terminology 'Lipschitz in x' to functions $f(x, t)$ provided the Lipschitz condition holds uniformly for each t in a given interval of R.

Note that C^0 functions need not be Lipschitz but C^1 functions always are. The following theorems relate the notion of Lipshitz with the property of continuity.

Lemma 2.4 *Let $f(x, t)$ be continuous on $D \times [a, b]$, for some domain $D \subset R^n$. If $\partial f/\partial x$ exists and is continuous on $D \times [a, b]$, then f is locally Lipschitz in x on $D \times [a, b]$.*

Proof: (following Khalil [2], p. 77) For $x_0 \in D$ there is an r sufficiently small that

$$D_0 = \{x \in R^n \mid \|x - x_0\| < r\} \subset D$$

The set D_0 is convex and compact. Since f is C^1, $\partial f/\partial x$ is bounded on $[a, b] \times D_0$. Let L_0 denote such a bound. If $x, y \in D_0$, then by the mean value theorem there is a point z on the line segment joining x, y such that

$$\|f(x, t) - f(t, y)\| = \left\| \frac{\partial f(t, z)}{\partial x} (x - y) \right\| \le L_0 \|x - y\|$$

■

The proof of this Lemma is easily adapted to prove the following.

Proposition 2.5 *Let $f(x, t)$ be continuous on $[a, b] \times R^n$. If f is C^1 in $x \in R^n$ for all $t \in [a, b]$ then f is globally Lipschitz in x if and only if $\partial f/\partial x$ is uniformly bounded on $[a, b] \times R^n$.*

Let us state the key existence result.

Proposition 2.6 (Local Existence and Uniqueness) *Let $f(x, t)$ be piecewise continuous in t and satisfy the Lipschitz condition*

$$\|f(x, t) - f(t, y)\| \le L \|x - y\|$$

for all $x, y \in B_r = \{x \in R^n \mid \|x - x_0\| < r\}$ and all $t \in [t_0, t_1]$. Then there exists a $\delta > 0$ such that the differential equation with initial condition

$$\dot{x} = f(x, t), \quad x(t_0) = x_0 \in B_r$$

has a unique solution over $[t_0, t_0 + \delta]$.

Proof: ([2], p. 74)

A continuation argument leads to the following global extension.

Proposition 2.7 (Global Existence and Uniqueness) *Suppose* $f(x,t)$ *is piecewise continuous in t and satisfies*

$$\|f(x,t) - f(t,y)\| \le L \|x - y\|$$

$$\|f(x,t_0)\| < h$$

for all $x, y \in R^n$ *and all* $t \in [t_0, t_1]$. *Then the equation*

$$\dot{x} = f(x,t), \quad x(t_0) = x_0$$

has a unique solution over $[t_0, t_1]$.

Continuous Dependence on Parameters and Initial Data

Let $\mu \in R^k$ and consider the parameter dependent differential equation

$$\dot{x} = f(x,t,\mu), \quad x(t_0) = x_0 \tag{2.3}$$

We will show that a solution $x(t; t_0, x_0, \mu)$ defined on a finite time interval $[t_0, t_1]$ is continuously dependent on the parameter μ and the initial data t_0, x_0.

Definition 2.8 *Let* $x(t; t_0, \xi_0, \mu_0)$ *denote a solution of (2.3) defined on the finite interval* $t \in [t_0, t_1]$ *with* $\mu = \mu_0$ *and* $x(t_0; t_0, \xi_0, \mu_0) = \xi_0$. *The solution is said to depend continuously on* μ *at* μ_0 *if for any* $\varepsilon > 0$ *there is a* $\delta > 0$ *such that such that for all* μ *in the neighborhood* $U = \{\mu \in R^k \,|\, \|\mu - \mu_0\| < \delta \}$, *(2.3) has a solution* $x(t; t_0, \xi_0, \mu)$ *such that*

$$\|x(t; t_0, \xi_0, \mu) - x(t; t_0, \xi_0, \mu_0)\| < \varepsilon$$

for all $t \in [t_0, t_1]$. *Similarly, the solution is said to depend continuously on* ξ *at* ξ_0 *if for any* $\varepsilon > 0$ *there is a* $\delta > 0$ *such that such that for all* ξ *in the neighborhood* $X = \{\xi \in R^k \,|\, \|\xi - \xi_0\| < \delta \}$, *(2.3) has a solution* $x(t; t_0, \xi, \mu_0)$ *such that*

$$\|x(t; t_0, \xi, \mu_0) - x(t; t_0, \xi_0, \mu_0)\| < \varepsilon$$

for all $t \in [t_0, t_1]$.

The following result establishes the basic continuity properties of (2.3) on finite time intervals.

Proposition 2.9 *Suppose* $f(x,t,\mu)$ *is continuous in* (x,t,μ) *and locally Lipschitz in x (uniformly in t and* μ) *on* $[t_0, t_1] \times D \times \{\|\mu - \mu_0\| < c\}$ *where* $D \subset R^n$

is an open connected set. Let $x(t; t_0, \xi_0, \mu_0)$ denote a solution of (2.3) that belongs to D for all $[t_0, t_1]$. Then given $\varepsilon > 0$ there is $\delta > 0$ such that

$$\|\xi - \xi_0\| < \delta, \ \|\mu - \mu_0\| < \delta$$

implies that there is a unique solution $x(t; t_0, \xi, \mu)$ of (2.3) defined on $t \in [t_0, t_1]$ and such that

$$\|x(t; t_0, \xi, \mu) - x(t; t_0, \xi_0, \mu_0)\| < \varepsilon, \ \forall t \in [t_0, t_1]$$

Proof: ([2], p. 86)

We emphasize that the results on existence and continuity of solutions hold on finite time intervals $[t_0, t_1]$. Stability, as we shall see below, requires us to consider solutions defined on infinite intervals. We will often tacitly assume that they are so defined. Continuity issues with respect to both initial conditions and parameters for solutions on infinite time intervals are quite subtle.

Invariant Sets

In the following paragraphs we shall restrict attention to autonomous systems

$$\dot{x} = f(x), \quad x(t_0) = x_0 \tag{2.4}$$

In many instances the results can be extended to nonautonomous systems by extending the nonautonomous differential equation with the addition of a new state $\dot{x}_{n+1} = 1$ to replace t in the right side of the differential equation.

Let us denote by $\Psi(x, t)$ the flow of the vector field f on R^n defined by (2.4) i.e. $\Psi(x, t)$ is the solution of (2.4) with $\Psi(0, x) = x$:

$$\frac{\partial \Psi(x, t)}{\partial t} = f(\Psi(x, t)), \quad \Psi(0, x) = x$$

Definition 2.10 *A set of points $S \subset R^n$ is* invariant *with respect to f if trajectories beginning in S remain in S both forward and backward in time, i.e., if $s \in S$, then $\Psi(t, s) \in S, \forall t \in R$.*

Obviously, any entire trajectory of (2.4) is an invariant set. Such an invariant set is minimal in the sense that it does not contain any proper subset which is itself an invariant set.

A set S is invariant if and only if $\Psi(t, S) \mapsto S$ for each $t \in R$.

Nonwandering Sets

Definition 2.11 *A point $p \in R^n$ is a* nonwandering point *with respect to the flow Ψ if for every neighborhood U of p and $T > 0$, there is a $t > T$ such that*

$\Psi(t, U) \cap U \neq \emptyset$. *The set of nonwandering points is called the* nonwandering set, *and denoted* Ω. *Points that are not nonwandering are called* wandering points.

The nonwandering set is a closed, invariant set. For proofs and other details see, for example, Guckenheimer and Holmes [3], Arrowsmith and Place [4] or Sibirsky [5]. The detailed structure of the nonwandering set is an important aspect of the analysis of strange attractors.

Obviously, fixed points and periodic trajectories belong to Ω.

Limit Sets

Definition 2.12 *A point* $q \in R^n$ *is said to be an* ω-limit *point of the trajectory* $\Psi(t, p)$ *if there exists a sequence of time values* $t_k \to +\infty$ *such that*

$$\lim_{t_k \to \infty} \Psi(t_k, p) = q$$

q *is said to be an* α-limit *point of* $\Psi(t, p)$ *if there exists a sequence of time values* $t_k \to -\infty$ *such that*

$$\lim_{t_k \to -\infty} \Psi(t_k, p) = q$$

The set of all ω-*limit points of the trajectory through* p *is the* ω-limit set, $\Lambda_\omega(p)$, *and the set of all* α-*limit points of the trajectory through* p *is the* α-limit set, $\Lambda_\alpha(p)$.

Hirsch and Smale [6] remind us that α, ω are the first and last letters of the Greek alphabet and, hence, the terminology.

Proposition 2.13 *The* α-, ω- *limit sets of any trajectory are closed invariant sets and they are subsets of the nonwandering set* Ω.

Proof: Hirsch and Smale [6] or Sibirsky [5] for closed, invariant sets. That they are subsets of Ω is obvious.

We can make some simple observations

1. if $r \in \Psi(t, p)$, then $\Lambda_\omega(r) = \Lambda_\omega(p)$ and $\Lambda_\alpha(r) = \Lambda_\alpha(p)$, i.e., any two points on a given trajectory have the same limit points.

2. if p is an equilibrium point, i.e., $f(p) = 0$ or $p = \Psi(t, p)$, then $\Lambda_\omega(p) = \Lambda_\alpha(p) = p$.

3. If $\Psi(t, p)$ is a periodic trajectory $\Lambda_\omega(p) = \Lambda_\alpha(p) = \Psi(R, p)$, i.e., the α and ω limit sets are the entire trajectory.

Finally, let us state the following important result.

Proposition 2.14 *A homeomorphism of a dynamical system maps ω-, α- limit sets into ω-, α- limit sets.*

Proof: [5].

2.4 Lyapunov Stability

2.4.1 Autonomous Systems

In the following paragraphs we consider autonomous differential equations and assume that the origin is an equilibrium point:

$$\dot{x} = f(x), \quad f(0) = 0 \tag{2.5}$$

with $f : D \to R^n$, locally Lipschitz in the domain D.

Definition 2.15 *The origin of (2.5) is*

1. *a stable equilibrium point if for each $\varepsilon > 0$, there is a $\delta(\varepsilon) > 0$ such that*

$$\|x(0)\| < \delta \Rightarrow \|x(t)\| < \varepsilon \ \forall t > 0$$

2. *unstable if it is not stable, and*

3. *asymptotically stable if δ can be chosen such that*

$$\|x(0)\| < \delta \Rightarrow \lim_{t \to \infty} x(t) = 0$$

The concept of Lyapunov stability is depicted in Figure (2.1).

The next seemingly trivial observation is nontheless useful. Among other things, it highlights the distinction between stability and asymptotic stability.

Lemma 2.16 (Necessary condition for asymptotic stability) *Consider the dynamical system $\dot{x} = f(x)$ and suppose $x = 0$ is an equilibrim point, i.e., $f(0) = 0$. Then $x = 0$ is asymptotically stable only if it is an isolated equilibrium point.*

Proof: If $x = 0$ is not an isolated equilibrium point, then in every neighborhood U of 0 there is at least one other equilibrium point. Thus, that not all trajectories beginning in U tend to 0 as $t \to \infty$. ∎

For linear systems the following result is easily obtained.

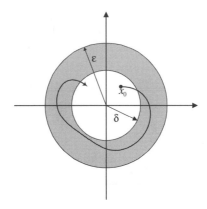

Figure 2.1: Definition of Lyapunov stability.

Proposition 2.17 *The origin of the linear system* $\dot{x} = Ax$ *is a stable equilibrium point if and only if*

$$\left\| e^{At} \right\| \leq N < \infty \ \forall t > 0$$

It is asymptotically stable if and only if, in addition, $\left\| e^{At} \right\| \to 0, \ t \to \infty$

Proof: Exercise (choose $\delta = \varepsilon/N$)

Positive Definite Functions

Definition 2.18 *A function* $V : R^n \to R^n$ *is said to be*

1. positive definite *if* $V(0) = 0$ *and* $V(x) > 0$, $x \neq 0$,

2. positive semidefinite *if* $V(0) = 0$ *and* $V(x) \geq 0$, $x \neq 0$,

3. negative definite *(negative semidefinite) if* $-V(x)$ *is positive definite (positive semidefinite)*

For a quadratic form $V(x) = x^T Q x$, $Q = Q^T$, the following statements are equivalent

1. $V(x)$ is positive definite

2. the eigenvalues of Q are positive real numbers

3. all of the principal minors of Q are positive

$$|q_{11}| > 0, \ \begin{vmatrix} q_{11} & q_{12} \\ q_{21} & q_{22} \end{vmatrix} > 0, \dots, |Q| > 0$$

Definition 2.19 *A C^1 function $V(x)$ defined on a neighborhood D of the origin is called a Lyapunov function relative to the flow defined by $\dot{x} = f(x)$ if it is positive definite and it is nonincreasing along trajectories of the flow, i.e.,*

$$V(0) = 0, \ V(x) > 0, \ x \in D - \{0\}$$

$$\dot{V} = \frac{\partial V(x)}{\partial x} f(x) \le 0$$

2.4.2 Basic Stability Theorems

Stability of a dynamical system may determined directly from an examination of the trajectories of the system or from a study of Lyapunov functions. The basic idea of the Lyapunov method derives from the idea of energy exchange in physical systems. A general physical conception is that stable systems dissipate energy so that the stored energy of a stable system decreases or at least does not increase as time evolves. The notion of a Lyapunov function is thereby an attempt to formulate a precise, energy-like theory of stability.

Proposition 2.20 (Lyapunov Stability Theorem) *If there exists a Lyapunov function $V(x)$ on some neighborhood D of the origin, then the origin is stable. Furthermore, if $\dot{V}$ is negative definite on D then the origin is asymptotically stable.*

Proof: Given $\varepsilon > 0$ choose $r \in (0, \varepsilon]$ such that

$$B_r = \{x \in R^n \, | \|x\| < r\} \subset D$$

Now, we can find a level set $C_\alpha = \{x \in R^n \, | V(x) = \alpha\}$ which lies entirely within B_r. Refer to Figure (2.2). The existence of such a set follows from the fact that since V is positive and continuous on B_r, it has a positive minimum, α, on ∂B_r. The level set C_α defined by $V(x) = \alpha$ must lie entire in B_r.

Now, since V is continuous and vanishes at the origin, there exists a $\delta > 0$ such that B_δ lies entirely within the set bounded by C_α, i.e.,

$$\Omega_\alpha = \{x \in R^n \, | V(x) \le \alpha\}$$

Since V is nonincreasing along trajectories, trajectories which begin in B_δ must remain in Ω_α, $\forall t > 0$. Hence they remain in B_ε. In the event that $\dot{V}$ is negative definite, V decreases steadily along trajectories. For any $0 < r_1 < r$ there is a $\beta < \alpha$ such that B_β lies entirely within B_{r_1}. Since $\dot{V}$ has a strictly negative maximum in the annular region $B_r - B_{r_1}$, any trajectory beginning in the annular region must eventually enter B_{r_1}. Thus, all trajectories must tend to the origin as $t \to \infty$. ∎

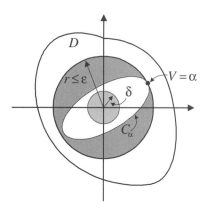

Figure 2.2: Sets used in proof of the Lyapunov stability theorem, Proposition (2.20).

Unlike linear systems, an asymptotically stable equilibrium point of a nonlinear system may not attract trajectories from all possible initial states. It is more likely that trajectories beginning at states in a restricted vicinity of the equilibrium point will actually tend to the equilibrium point as $t \to \infty$. The above theorem can be used to establish stability and also to provide estimates of the *domain of attraction* using level sets of the Lyapunove function $V(x)$.

The following theorem due to LaSalle allows us to more easily characterize the domain of attraction of a stable equilibrium point and is a more powerful result than the basic Lyapunov stability theorem because the conditions for asymptotic stability do not require $\dot{V}$ to be negative definite.

Proposition 2.21 (LaSalle Invariance Theorem) *Consider the system defined by equation (2.5). Suppose $V(x) : R^n \to R$ is C^1 and let Ω_c designate a component of the region $\{x \in R^n \,|V(x) < c\}$. Suppose Ω_c is bounded and that within Ω_c $\dot{V}(x) \leq 0$. Let E be the set of points within Ω_c where $\dot{V} = 0$, and let M be the largest invariant set of (2.5) contained in E. Then every solution $x(t)$ of (2.5) beginning in Ω_c tends to M as $t \to \infty$.*

Proof: (following [7]) $\dot{V}(x) \leq 0$ implies that $x(t)$ starting in Ω_c remains in Ω_c. $V(x(t))$ nonincreasing and bounded implies that $V(x(t))$ has a limit c_0 as $t \to \infty$ and $c_0 < c$. By continuity of $V(x)$, $V(x) = c_0$ on the positive limit set $\Lambda_\omega(x_0)$ of $x(t)$ beginning at $x_0 \in \Omega_c$. Thus, $\Lambda_\omega(x_0)$ is in Ω_c and $\dot{V}(x) = 0$ on $\Lambda_\omega(x_0)$. Consquently, $\Lambda_\omega(x_0)$ is in E, and since it is an invariant set, it is in M. ∎

Note that the theorem does not specify that $V(x)$ should be positive definite, only that it have continuous first derivatives and that there exist a bounded region on which $V(x) < c$ for some constant c. A number of useful results follow directly from this one.

Corollary 2.22 *Let $x = 0$ be an equilibrium point of (2.5). Suppose D is a neighborhood of $x = 0$ and $V : D \to R$ is C^1 and positive definite on D such that $\dot{V}(x) \leq 0$ on D. Let $E = \left\{ x \in D \,\middle|\, \dot{V}(x) = 0 \right\}$ and suppose that the only entire solution contained in E is the trivial solution. Then the origin is asymptotically stable.*

Corollary 2.23 *Let $x = 0$ be an equilibrium point of (2.5). Suppose*

1. *$V(x)$ is C^1*

2. *$V(x)$ is radially unbounded (Barbashin-Krasovskii condition), i.e.,*

$$\|x\| \to \infty \Rightarrow V(x) \to \infty$$

3. *$\dot{V}(x) \leq 0, \; \forall x \in R^n$*

4. *the only entire trajectory contained in the set $E = \left\{ x \in D \,\middle|\, \dot{V}(x) = 0 \right\}$ is the trivial solution.*

Then the origin is globally asymptotically stable.

The stability theorems provide only sufficient conditions for stability and construction of a suitable Lyapunov function may require a fair amount of ingenuity. In the event that attempts to establish stability do not bear fruit it may be useful to try to confirm instabilty.

Proposition 2.24 (Chetaev Instability Theorem) *Consider equation (2.5) and suppose $x = 0$ is an equilibrium point. Let D be a neighborhood of the origin. Suppose there is a function $V(x) : D \to R$ and a set $D_1 \subset D$ such that*

1. *$V(x)$ is C^1 on D,*

2. *the origin belongs to the boundary of D_1, ∂D_1,*

3. *$V(x) > 0$ and $\dot{V}(x) > 0$ on D_1,*

4. *On the boundary of D_1 inside D, i.e. on $\partial D_1 \cap D$, $V(x) = 0$*

Then the origin is unstable

Proof: Choose an r such that $B_r = \{x \in R^n \,|\|x\| \leq r\}$ is in D. Refer to Figure (2.3). For any trajectory beginning inside $U = D_1 \cap B_r$ at $x_0 \neq 0$, $V(x(t))$ increases indefinitely from $V(x_0) > 0$. But by continuity, $V(x)$ is bounded on U. Hence $x(t)$ must leave U. It cannot do so across its boundary interior to B_r so it must leave B_r. ∎

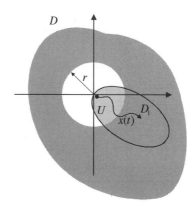

Figure 2.3: Sets used in proof of the Chetaev instability theorem, Proposition (2.24).

Stability of Linear Systems

Consider the linear system

$$\dot{x} = Ax \tag{2.6}$$

Proposition 2.25 *Consider the Lyapunov equation*

$$A^T P + PA = -Q \tag{2.7}$$

(a) *If there exists a positive definite pair of symmetric matrices P, Q satisfying the Lyapunov equation then the origin of the system (2.6) is asymptotically stable.*

(b) *If there exists a pair of symmetric matrices P, Q such that P has at least one negative eigenvalue and Q is positive definite, then the origin is unstable.*

Proof: Consider (a) first. Choose $V(x) = x^T P x$ and compute $\dot{V} = x^T (A^T P + PA)x = -x^T Q x$. The assumptions and the LaSalle stability theorem lead to the conclusion that all trajectories tend to the orgin as $t \to \infty$. Case (b) requires application of Chetaev's instability theorem. In this case consider $V(x) = -x^T P x$. Recall that for symmetric P the eigenvalues of P are real, they may be positive, negative or zero. On the positive eigenspace, $V < 0$, on the negative eigenspace, $V > 0$, and on the zero eigenspace, $V = 0$. Since P has at least one negative eigenvalue, the negative eigenspace is nontrivial and there is a set of points, D, for which $V > 0$. Let B_ε be an open sphere of small radius ε centered at the origin. Since V is continuous, the boundary of D in B_ε, $\partial D \cap B_r$, consists of points of points at which $V = 0$. It includes the origin and is never nonempty (even if all eigenvalues of P are negative). For $V(x) = -x^T P x$, $\dot{V} = x^T Q x$ and it is always positive since Q is assumed positive definite. Thus, the conditions of Proposition (2.24) are satisfied. ∎

Suppose $Q > 0$ and the P has a zero eigenvalue. If the matrix P has a zero eigenvalue then there are points $x \neq 0$ such that $V(x) = x^T P x = 0$. But at such points $\dot{V}(x) = -x^T Q x < 0$. Since V(x) is continuous this means that there must be points at which V assumes negative values. Thus, P must also have a negative eigenvalue. Thus, we have the following corollary to Proposition (2.25).

Corollary 2.26 *The linear system (2.6) is asymptotically stable if and only if for every positive definite symmetric Q there exists a positive definite symmetric P that satisfies the Lyapunov equation (2.7).*

Lagrangian Systems

The Lyapunov analysis of the stability of nonlinear dynamical systems evolved from a tradition of stability analysis via energy functions that goes back at least to Lagrange and Hamilton. We will consider a number of examples which are physically motivated and for which there are energy functions that serve as natural Lyapunov function candidates. Consider the class of Lagrangian systems characterized by the set of second order differential equations

$$\frac{d}{dt} \frac{\partial L(x, \dot{x})}{\partial \dot{x}} - \frac{\partial L(x, \dot{x})}{\partial x} = Q^T \tag{2.8}$$

where

1. $x \in R^n$ denotes a vector of *generalized coordinates* and $\dot{x} = dx/dt$ are the *generalized velocities*.

2. $L : R^{2n} \to R$ is the *Lagrangian*. It is constructed from the kinetic energy function $T(x, \dot{x})$ and the potential energy function $U(x)$, via $L(x, \dot{x}) = T(x, \dot{x}) - U(x)$.

3. The kinetic energy has the form

    $$T(x, \dot{x}) = \tfrac{1}{2} \dot{x}^T M(x) \dot{x}$$

    where for each fixed x, the matrix $M(x)$ is positive definite.

4. The potential energy is related to a force vector $f(x)$ via

    $$U(x) = \int f(x) dx$$

5. $Q(x, \dot{x}, t)$ is a vector of generalized forces.

Occasionally it is convenient to write the second order equations in first order form by defining new variable $v = \dot{x}$ to obtain

$$\begin{bmatrix} \dot{x} \\ \dot{v} \end{bmatrix} = \begin{bmatrix} v \\ -M^{-1}(x)f(x) - \frac{1}{2}M^{-1}(x)\left[\partial M(x)/\partial x\right] + M^{-1}(x)Q \end{bmatrix} \qquad (2.9)$$

Another useful first order form is Hamilton's equations obtained as follows. Define the *generalized momentum* as

$$p^T = \frac{\partial L}{\partial \dot{x}} = \dot{x}^T M(x) \Rightarrow \dot{x} = M^{-1}(x)p \qquad (2.10)$$

Define the *Hamiltonian* $H : R^{2n} \to R$

$$H(x,p) = \left[p^T\dot{x} - L(x,\dot{x})\right]_{\dot{x} \to M^{-1}p} = \frac{1}{2}p^T M^{-1}(x)p + U(x) \qquad (2.11)$$

The Hamiltonian is the total energy expressed in momentum rather than velocity coordinates. Notice that Lagrange's equation can be written

$$\dot{p}^T - \frac{\partial L}{\partial x} = Q^T \qquad (2.12)$$

Now, using the definition of H, (2.11), write

$$dH = \frac{\partial H}{\partial x}dx + \frac{\partial H}{\partial p}dp = dp^T\dot{x} + p^T d\dot{x} - \frac{\partial L}{\partial x}dx - \frac{\partial L}{\partial \dot{x}}d\dot{x} = dp^T\dot{x} - \frac{\partial L}{\partial x}dx$$

Using (2.12) we have

$$\frac{\partial H}{\partial x}dx + \frac{\partial H}{\partial p}dp = \dot{x}^T dp - (\dot{p} - Q)^T dx$$

Comparing coefficients of dp and dx, we have Hamilton's equations.

$$\dot{x} = \frac{\partial H(x,p)}{\partial p^T}, \quad \dot{p} = -\frac{\partial H(x,p)}{\partial x^T} + Q \qquad (2.13)$$

Example 2.27 (Soft Spring) *Consider a system of with kinetic and potential energy functions*

$$T = \frac{x_2^2}{2}, \quad U = \frac{x_1^2}{1 + x_1^2}$$

Lagrange's equations in first order form ($\dot{x}_1 = x_2$) with viscous damping are

$$\begin{bmatrix} \dot{x}_1 \\ \dot{x}_2 \end{bmatrix} = \begin{bmatrix} x_1 \\ -2\frac{x_1}{(1+x_1^2)^2} - cx_2 \end{bmatrix}$$

If we take the total energy as a candidate Lyapunov function,

$$V(x_1, x_2) = \frac{1}{2}x_2^2 + \frac{x_1^2}{1 + x_1^2}$$

an easy calculation shows that $\dot{V} = -cx_2 \leq 0$ for $c > 0$. Furthermore, the set $\dot{V} = 0$ consists of the x_1-axis and the only entire solution contained therein is the trivial solution. We conclude that the origin is asymptotically stable. We can not, however, conclude global aymptotic stability because the Lyapunov function is not radially unbounded. Let us look at the level sets of V:

In[7]:= ContourPlot[V, {x1, −5, 5}, {x2, −2, 2}, PlotPoints → 50,
   Contours → 15, ColorFunction → (GrayLevel[((# + 0.1)/1.1)^(1/4)]&),
   FrameLabel → {x1, x2}]

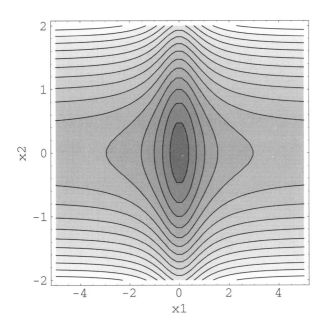

and, finally, at the trajectories:

In[8]:= Show[graphs2, AxesLabel → {x1, x2}, PlotRange → {{−5, 5}, {−2, 2}},
   DisplayFunction → $DisplayFunction]

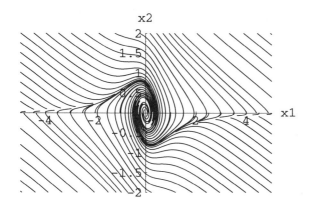

Example 2.28 (Variable Mass) *Consider a system with variable inertia, typical of a crankhaft. The kinetic and potential energy functions are*

$$T = (2 - \cos 2x_1)\, x_2^2, \quad U = x_1^2 + \tfrac{1}{4}x_1^4$$

Systems with variable mass are much easier to analyze using Hamilton's equations, so we define the generalized momentum $p = (2 - \cos 2x_1)\, x_2$ and the Hamiltonian

$$H(x_1, p) = \frac{p^2}{2(2 - \cos 2x_1)} + x_1^2 + \tfrac{1}{4}x_1^4$$

Again with viscous damping, Hamilton's equations are

$$\begin{bmatrix} \dot{x}_1 \\ \dot{p} \end{bmatrix} = \begin{bmatrix} \dfrac{p}{2 - \cos 2x_1} \\ -2x_1 - x_1^3 - \dfrac{p^2 \sin 2x_1}{(2 - \cos 2x_1)^2} + \dfrac{2cp}{(2 - \cos 2x_1)^2} \end{bmatrix}$$

It is not difficult to compute $\dot{H}$, indeed,

$In\,[9]:=$ $Simplify[Jacob[H, \{p, x1\}].\{-D[H, x1] - D[R, p], D[H, p]\}]$

$Out\,[9]=$ $\dfrac{2 \text{ c } p^2}{(-2 + Cos[2 \text{ x1}])^3}$

We conclude that $\dot{H} \leq 0$ for $c > 0$. Moreover, the only entire trajectory in the set $\dot{H} = 0$ is the trivial solution, and since H is radially unbounded, we can cnclude that the origin is globally asymptotically stable. Let is look at the level curves of H

$In\,[10]:=$ $ContourPlot[H, \{x1, -6, 6\}, \{p, -12, 12\}, PlotPoints \rightarrow 100, Contours \rightarrow 25,$
     $ColorFunction \rightarrow (GrayLevel[((\# + 0.1)/1.1)\,\hat{}\,(1/4)]\&), FrameLabel \rightarrow \{x1, p\}]$

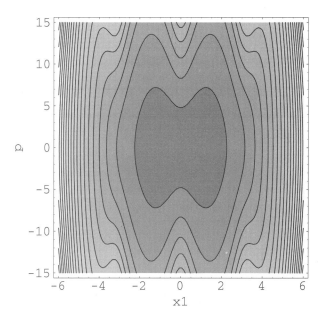

and at the state space trajectories

$In[11]:=$ $graphs3 = PhasePortrait[f, x, 15, \{\{-6, 6, 2\}, \{-5, 5, 5\}\}];$
$Show[graphs3, AxesLabel \to \{x1, p\}, PlotRange \to \{\{-4, 4\}, \{-4, 4\}\},$
$DisplayFunction \to \$DisplayFunction]$

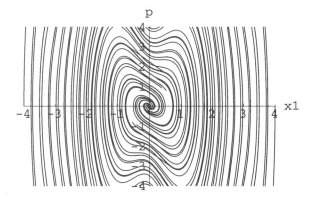

Example 2.29 (Multiple Equilibria) *Consider the system*

$$\ddot{x} + \left|x^2 - 1\right|\dot{x}^3 - x + \sin\left(\frac{\pi x}{2}\right) = 0$$

Notice that the system has three equilibria $(x,\dot{x}) = (0,0),(-1,0),(1,0)$. *We can determine their stability by examining the system phase portraits or using a Lyapunov analysis based on total energy as the candidate Lyapunov function. First let us examine the phase portraits.*

$In[12]:=$ $U = -\dfrac{2}{\pi} + \dfrac{x1^2}{2} + \dfrac{2\,\cos\left[\frac{\pi\,x1}{2}\right]}{\pi}$;

$In[13]:=$ $T = x2\hat{\,}2/2; V = T + U;$

$In[14]:=$ $F = Simplify[D[U, x1]];$
$\qquad\quad f = \{x2, -\ Abs[x1\hat{\,}2 - 1]\ x2\hat{\,}3 - F\}; x = \{x1, x2\};$

$In[15]:=$ $graphs3 = PhasePortrait[f, x, 2, \{\{-2, 2, 0.5\}, \{-1, 1, 1\}\}];$

$In[16]:=$ $Show[graphs3, AxesLabel \rightarrow \{x1, x2\}, PlotRange \rightarrow \{\{-1.5, 1.5\}, \{-1, 1\}\},$
$\qquad\quad DisplayFunction \rightarrow \$DisplayFunction]$

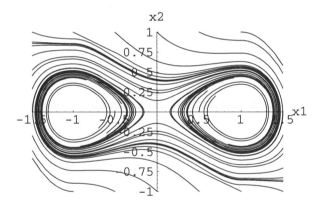

Thus, we see that the equilibrium point $(0,0)$ *is unstable and that the other two,* $(\pm 1, 0)$, *are asymptotically stable.*

Now, let us consider the Lyapunov viewpoint. The total energy is

$$V = \frac{\dot{x}^2}{2} + \frac{x^2}{2} - \frac{2}{\pi}\left(1 - \cos\left(\frac{\pi x}{2}\right)\right)$$

A straightforward calculation leads to

$$\dot{V} = -\left|x^2 - 1\right|\dot{x}^2$$

The LaSalle theorem (2.21) can now be applied. Let us view the level surfaces.

*In[17]:= ContourPlot[V, {x1, −2, 2}, {x2, −1, 1}, PlotPoints → 50,
Contours → 15, ColorFunction → (GrayLevel[((# + 0.1)/1.1)ˆ(1/4)]&),
FrameLabel → {x1, x2}]*

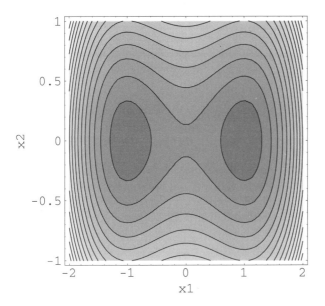

Notice that there are level surfaces that bound compact sets that include the equilibrium point $(1, 0)$. Pick one and designate it Ω_{c_1}. Moreover, $\dot{V} \leq 0$ everywhere, hence specifically in Ω_{c_1}, and the maximum invariant set contained in Ω_{c_1} is the equilibrium point. Consequently, all trajectories beginning in Ω_{c_1} tend to $(1, 0)$ so it is an asymptotically stable equilibrium point. A similar conclusion can be reached for the equilibrium point $(−1, 0)$.

2.4.3 Stable, Unstable, and Center Manifolds

Consider the autonomous system (2.5) and suppose $x = 0$ is an equilibrium point so that $f(0) = 0$. Let $A := \partial f(0)/\partial x$. Define three subspaces of R^n:

1. the *stable subspace, E^s*: the eigenspace of eigenvalues with negative real parts

2. the *unstable subspace, E^u*: the eigenspace of eigenvalues with positive real parts

3. the *center subspace, E^c*: the eigenspace of eigenvalues with zero real parts

An equilibrium point is called *hyperbolic* if A has no eigenvalues with zero real part, i.e., there is no center subspace, E^c. In the absence of center subspace the linearization is a reliable predictor of important qualitative features of the nonlinear system. The basic result is given by the following theorem. First, some definitions.

Let f, g be C^r vector fields on R^n with $f(0) = 0, g(0) = 0$. M is an open subset of the origin in R^n.

Definition 2.30 *Two vector fields f and g are said to be C^k-equivalent on M if there exists a C^k diffeomorphism h on M, which takes orbits of the flow generated by f on M, $\Phi(x,t)$, into orbits of the flow generated by g on M, $\Psi(x,t)$, preserving orientation but not necessarily parameterization by time. C^0-equivalence is referred to as* topological equivalence. *If there is such an h which does preserve parameterization by time then f, g are said to be C^k -conjugate. C^0-conjugacy is referred to as* topological-conjugacy.

Proposition 2.31 (Hartman-Grobman Theorem) *Let $f(x)$ be a C^k vector field on R^n with $f(0) = 0$ and $A := \partial f(0)/\partial x$. If A is hyperbolic then there is a neighborhood U of the origin in R^n on which the nonlinear flow of $\dot{x} = f(x)$ and the linear flow of $\dot{x} = Ax$ are topologically conjugate.*

Proof: (Chow & Hale [8], p108)

Definition 2.32 *Let U be a neighborhood of the origin. We define the* local stable manifold *and* local unstable manifold *of the equilibrium point $x = 0$ as, respectively,*

$$W_{loc}^s = \{x \in U \,|\, \Psi(x,t) \to 0 \text{ as } t \to \infty \wedge \Psi(x,t) \in U \,\forall t \geq 0\}$$

$$W_{loc}^u = \{x \in U \,|\, \Psi(x,t) \to 0 \text{ as } t \to -\infty \wedge \Psi(x,t) \in U \,\forall t \leq 0\}$$

Proposition 2.33 (Center Manifold Theorem) *Let $f(x)$ be a C^r vector field on R^n with $f(0) = 0$ and $A := \partial f(0)/\partial x$. Let the spectrum of A be divided into three sets $\sigma_s, \sigma_c, \sigma_u$ with*

$$\text{Re}\,\lambda = \begin{cases} < 0 & \lambda \in \sigma_s \\ = 0 & \lambda \in \sigma_c \\ > 0 & \lambda \in \sigma_u \end{cases}$$

Let the (generalized) eigenspaces of $\sigma_s, \sigma_c, \sigma_u$ be E^s, E^c, E^u, respectively. Then there exist C^r stable and unstable manifolds W^s and W^u tangent to E^s and E^u, respectively, at $x = 0$ and a C^{r-1} center manifold W^c tangent to E^c at $x = 0$. The manifolds W^s, W^c, W^u are all invariant with respect to the flow of $f(x)$. The stable and unstable manifolds are unique, but the center manifold need not be.

Proof: [9].

Example 2.34 (Center Manifold) *Consider the system*

$$\dot{x} = x^2, \quad \dot{y} = -y$$

from which it is a simple matter to compute

$$x(t) = x_0/(1 - tx_0), \quad y(t) = y_0 e^{-t} \Rightarrow y(x) = \left[y_0 e^{-1/x_0} \right] e^{1/x}$$

The phase portrait is shown below. Observe that $(0,0)$ is an equilibrium point with:

$$A = \begin{bmatrix} 0 & 0 \\ 0 & -1 \end{bmatrix} \Rightarrow E^s = span \left\{ \begin{matrix} 0 \\ 1 \end{matrix} \right\}, \quad E^c = span \left\{ \begin{matrix} 1 \\ 0 \end{matrix} \right\}$$

Notice that the center manifold can be defined using any trajectory beginning with $x < 0$ and joining with it the positive x-axis. Also, the center manifold can be chosen to be the entire x-axis. This is the only choice which yields an analytic center manifold.

In[18]:= $f = \{x1^2, -x2\}; x = \{x1, x2\};$

In[19]:= $graphs = PhasePortrait[f, x, 8, \{\{-0.2, 0.1, 0.05\}, \{-0.5, 0.5, 0.5\}\}];$

In[20]:= $Show[graphs, AxesLabel \rightarrow \{x, y\}, PlotRange \rightarrow \{\{-0.2, 0.2\}, \{-0.5, 0.5\}\},$
$DisplayFunction \rightarrow \$DisplayFunction]$

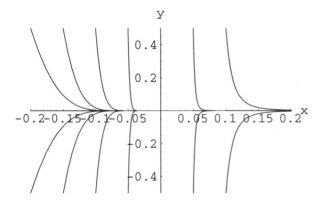

There are some important properties of these manifolds that will not be examined here. See, for example, [3] and [4]. Let us note, however, that existence and uniqueness of solutions insure that two stable (or unstable) manifolds cannot intersect or self-intersect. However, a stable and an unstable manifold can intersect. The global stable and unstable manifolds need not be simple submanifolds of R^n, since they may wind around in a complex manner, approaching themselves arbitrarily closely.

Motion on the Center Manifold

Consider the system of differential equations

$$\dot{x} = Bx + f(x, y)$$
$$\dot{y} = Cy + g(x, y)$$
(2.14)

where $(x, y) \in R^{n+m}$, f, g and their gradients vanish at the origin, and the eigenvalues of B have zero real parts, those of C negative real parts. The center manifold is tangent to E^c:

$$E^c = \text{span} \begin{bmatrix} I_n \\ 0_{m \times n} \end{bmatrix}$$

It has a local graph

$$W^c = \left\{ (x, y) \in R^{n+m} \mid y = h(x) \right\}, \quad h(0) = 0, \quad \frac{\partial h(0)}{\partial x} = 0$$

Once h is determined, the vector field on the center manifold W^c (i.e., the surface defined by $y = h(x)$) can be projected onto the Euclidean space E^c as

$$\dot{x} = Bx + f(x, h(x))$$
(2.15)

These calculations lead to the following result (see [3]).

Proposition 2.35 (Center Manifold Stability Theorem) *If the origin of (2.15) is asymptotically stable (resp. unstable) then the origin of (2.14) is asymptotically stable (resp. ustable).*

To compute h, we use the fact that on W^c it is required that $y = h(x)$ so that

$$\dot{y} = \frac{\partial h(x)}{\partial x} \dot{x} = \frac{\partial h(x)}{\partial x} [Bx + f(x, h(x))]$$

But $\dot{y}$ is also governed by (2.14) so we have the partial differential equation

$$\frac{\partial h(x)}{\partial x} [Bx + f(x, h(x))] = Ch(x) + g(x, h(x))$$
(2.16)

that needs to be solved along with the boundary conditions $h(0) = 0$, $\frac{\partial h(0)}{\partial x} = 0$.

Example 2.36 *Consider the following two dimensional system from Isidori [10].*

$$\dot{x} = cyx - x^3$$

$$\dot{y} = -y + ayx + bx^2$$

where a, b, c are real numbers. It is easy to see that the origin is an equilibrium point and that it is in the form of (2.14) with $B = 0$ and $C = -1$. To compute h we need to solve the partial differential equation

$$\frac{\partial h}{\partial x} \left[cxh(x) - x^3\right] + h(x) - ah(x)x - bx^2 = 0 \tag{2.17}$$

with boundary conditions $h(0) = 0$, $\frac{\partial h(0)}{\partial x} = 0$.

Assume a polynominal solution of the form

$$h(x) = a_0 + a_1 x + a_2 x^2 + a_3 x^3 + O(x^4)$$

In view of the boundary conditions we must have $a_0 = 0$ and $a_1 = 0$. Substitute h into (2.17) as follows, using Mathematica,

In[21]:= h = a2 x^2 + a3 x^3 + O[x]^4;
          F = D[h, x](c x h - x^3) + h - a h x - b x^2
Out[21]= (a2 - b) x² + (-a a2 + a3) x³ + O[x]⁴

Thus, we have

In[22]:= a2 = b; a3 = a a2; h
Out[22]= b x² + a b x³ + O[x]⁴

Now, obtain the motion on the center manifold

In[23]:= dx = c h x - x^3
Out[23]= (-1 + b c) x³ + a b c x⁴ + O[x]⁵

The last result can be rewritten as

$$\dot{x} = (-1 + bc)x^3 + abcx^4 + O(x^5)$$

Thus, we have the following results,

(a) if $bc < 1$, the motion on the center manifold is asymptotically stable,

(b) if $bc > 1$, it is unstable,

(c) if $bc = 1$, and $a \neq 0$, it is unstable

(d) if $bc = 1$, and $a = 0$, the above calculations are inconclusive. But in this special case it is easy to verify that $h(x) = x^2$ and the center manifold dynamics are $\dot{x} = 0$. So the motion is stable, but not asymptotically stable.

2.5 Problems

Problem 2.37 *Plot the level sets of the following norms on R^2:*

(a) $\|x\| = \sqrt{x_1^2 + x_2^2}$

(b) $\|x\| = |x_1| + |x_2|$

(c) $\|x\| = \sup(|x_1|, |x_2|)$

Problem 2.38 *Consider a system described by the differential equation:*

$$\ddot{x} + g(x) = 0$$

that describes a unit point mass with spring force $g(x)$. Show that $(x, \dot{x}) = (0, 0)$ is a stable equilibrium point if

(a) $xg(x) > 0$, $x \neq 0$

(b) $g(0) = 0$

Problem 2.39 *Consider the dissipative system*

$$\ddot{x} + a\dot{x} + 2bx + 3x^2 = 0, \quad a, b > 0$$

(a) *Show that there are two equilibrium points $(x, \dot{x}) = (0, 0)$ and $(x, \dot{x}) = (-2b/3, 0)$*

(b) *By linear approximation show that $(0, 0)$ is asymptotically stable and that $(-2b/3, 0)$ is unstable.*

(c) *Use the total energy of the undamped system $(a = 0)$ as a Lyapunov function and identify a largest region of attraction for $(0, 0)$. Show that the boundary of this region passes through the point $(-2b/3, 0)$.*

(d) *Pick values for $a, b > 0$ and plot state trajectories and level surfaces for the energy.*

Problem 2.40 *Consider the system*

$$\dot{x}_1 = x_2$$
$$\dot{x}_2 = -x_1 - x_2\text{sat}(x_2^2 - x_3^2)$$
$$\dot{x}_3 = x_3\text{sat}(x_2^2 - x_3^2)$$

(a) *Show that the origin is the unique equilibrium point.*

(b) *Taylor linearize at the origin and show that it is asymptotically stable.*

(c) *Using $V(x) = x^T x$, show that the origin is globally asymptotically stable.*

Problem 2.41 *Investigate the stability of the origin, including estimates of the domain of attraction, of the following systems:*

(a) $\ddot{x} = x - \text{sat}(2x + \dot{x})$, *Hint: Use $V(x) = x\dot{x}$,*

(b) $\ddot{x} + \dot{x}|\dot{x}| + x - x^3 = 0$, *Hint: Use total energy for $V(x)$.*

2.6 References

1. Hale, J.K., Ordinary Differential Equations. 1969, New York: John Wiley and Sons.

2. Khalil, H.K., Nonlinear Systems. 1992, New York: MacMillan.

3. Guckenheimer, J. and P. Holmes, Nonlinear Oscillations, Dynamical Systems, and Bifurcation of Vector Fields. 1983, New York: SpringerVerlag.

4. Arrowsmith, D.K. and C.M. Place, An Introduction to Dynamical Systems. 1990, Cambridge: Cambridge University Press.

5. Sibirsky, K.S., Introduction to Topological Dynamics. 1975, Leyden: Noordhoff International Publishing.

6. Hirsch, M.W. and S. Smale, Differential Equations, Dynamical Systems, and Linear Algebra. 1974, New York: Academic Press.

7. LaSalle, J. and S. Lefschetz, Stability by Lyapunovs Direct Method. 1961, New York: Academic Press.

8. Chow, S.N. and J.K. Hale, Methods of Bifurcation Theory. 1982, New York: SpringerVerlag.

9. Marsden, J.E. and M. McCracken, The Hopf Bifurcation and its Applications. 1976, New York: Springer-Verlag.

10. Isidori, A., Nonlinear Control Systems. 3 ed. 1995, London: Springer-Verlag.

Chapter 3

Introduction to Differential Geometry

3.1 Introduction

This chapter provides an overview of the differential geometry concepts necessary for a modern discussion of nonlinear control and analytical mechanics. We need to develop a basic understanding of manifolds, vector fields and flows, distributions and integral submanifolds along with tools that allow us to compute and manipulate these objects. The material described only very briefly here is deep and rich and a more thorough discussion can be found in many text books, e.g., [1-4]. It has its roots in analytical mechanics but it has become a cornerstone of nonlinear control.

In essence this chapter develops the tools required to address the evolutionary behavior of systems whose state spaces are curved rather than flat surfaces. Examples of the importance of this generalization abound. We have already considered the flow of a dynamical system on a curved surface when evaluating stability on a center manifold in the last chapter. Sometimes it is convenient to consider the state space of a pendulum to be a cylinder rather than the flat Euclidean space R^2. Electric power systems are typically modeled by systems of differential-algebraic equations in the form

$$\dot{x} = f(x, y)$$
$$0 = g(x, y)$$

where $x \in R^n$, $y \in R^m$, and $f : R^{n+m} \rightarrow R^n$, $g : R^{n+m} \rightarrow R^m$ are smooth functions. Clearly the motion is constrained to the set of points in R^{n+m} that satisfy the algebraic equation. Under the right circumstances, this set is an n-dimensional smooth surface. In other applications from robotics to spacecraft models often have state spaces that are not flat. Control design itself imposes the need to consider flows on general surfaces. Unique aspects of the navigation of ships and aircraft between points on earth arise because the motion takes place on a sphere. Control theoretic examples include the generalization of the concept of zero dynamics to nonlinear systems and the study of sliding modes in variable structure control systems.

While the subject matter might seem abstract on first acquaintance, its underlying concepts are quite intuitive and appealing. The formalism provides a precise basis for working with geometric ideas and the reader new to this material will no doubt find justification and clarification for familiar calculations.

We begin with a discussion of manifolds in Section 2 and then proceed to tangent spaces, vector fields and covector fields in Section 3. Section 4 introduces distributions, codistributions and the Frobenius theorem. Distributions play a role in nonlinear system theory much like that of linear subspaces in linear system theory. The Frobenius theorem answers classical questions of integrability central to problems of mechanics and partial differential equations. It turns out to be equally important to nonlinear control. Important tranformations of state equations are derived, based on distributions possessing certain properties and tools for computing such distribution are described. Section 5 provides a brief introduction to Lie Groups and Lie Algebras with the addition of some algebraic structure to the geometric objects of Sections 2, 3 and 4.

3.2 Manifolds

Roughly speaking a manifold is a smooth surface embedded in a Euclidean space of some dimension. We will need a more precise definition in order to work effectively with manifolds, but before proceeding formally let us examine how we ordinarily characterize such surfaces. Consider the Euclidean space R^2 and suppose x, y are its coordinates. The set of points that comprise a surface in R^2, e.g., the unit circle, is generally modeled in one of three ways:

- *explicitly*, by a mapping $y = g(x)$ (or, $x = g(y)$),

- *implicitly*, by a relation $f(x, y) = 0$,

- *parametrically*, by a mapping $x = h_1(s)$, $y = h_2(s)$, $s \in U \subset R$.

The explicit model is typically inadequate. For example, the unit circle, does not admit a global explicit representation. We would have to represent two pieces of the circle by separate expressions. Representations of the unit circle are

- explicit, top half: $y = \sqrt{1 - x^2}$ and bottom half: $y = -\sqrt{1 - x^2}$,

- implicit, $x^2 + y^2 = 1$,

- parametric, $y = \sin s$, $x = \cos s$, $s \in [0, 2\pi)$

In practice, interesting manifolds require either an implicit or a parametric representation even for a local characterization. Explicit representations, however, can also be useful as we have already seen in the computation of center manifolds in the previous chapter.

Now, let us turn to the formalities. Recall that a *homeomorphism* between any two topological spaces is a one-to-one continuous mapping with a continuous inverse. A homeomorphism not only maps points in a one-to-one manner, but it maps open sets in a one-to-one manner. Thus, if $\varphi : M \to N$ is a homeomorphism, M and N are topologically the same.

A differentiable manifold N of dimension n is a set of points that is locally topologically equivalent to the Euclidean space R^n. This concept can be made precise by introducing a set of local coordinate systems called charts. Each chart consists of an open set $U \subset N$ and a mapping φ that maps U homeomorphically onto $\varphi(U)$. A general discussion of differential manifolds and their application can be found in many texts including [1-3]. The key elements of the following definition are depicted in Figure (3.1).

Definition 3.42 *An m-dimensional manifold is a set M together with a countable collection of subsets $U_i \subset M$ and one-to-one mappings $\varphi_i : U_i \to V_i$ onto open subsets V_i of R^m, each pair (U_i, φ_i) called a* coordinate chart, *with the following properties:*

1. *the coordinate charts cover M,* $\bigcup_i U_i = M$

2. *on the overlap of any pair of charts the composite map*

$$f = \varphi_j \circ \varphi_i^{-1} : \; \varphi_i(U_i \cap U_j) \to \varphi_j(U_i \cap U_j)$$

   is a smooth function.

3. *if $p \in U_i$, $\bar{p} \in U_j$ are distinct points of M, then there are neighborhoods, W of $\varphi_i(p)$ in V_i and $\bar{W}$ of $\varphi_j(\bar{p})$ in V_j such that*

$$\varphi_i^{-1}(W) \cap \varphi_j^{-1}(\bar{W}) = \emptyset$$

The coordinate charts provide the set M with a topological structure so that the manifold is a topological space. Condition 3. of the definition is a form of the so-called Hausdorff separation axiom so that these manifolds are Hausdorff topological spaces. The coordinates in R^m of the image of a coordinate map $\varphi(p)$, $p \in M$ are called the coordinates of p. A chart (U, φ) is called a local coordinate system. If the overlap functions $f = \varphi_j \circ \varphi_i^{-1}$ are k-times continuously differentiable, then the manifold is called a C^k-manifold. If $k = \infty$, then the manifold is said to be smooth. It is analytic if the overlap functions are analytic. A local coordinate system is called a cubic coordinate system if $\varphi(U)$ is an open cube about the origin in R^m. If $p \in M$ and $\varphi(p) = 0$, the coordinate system is said to be centered at p.

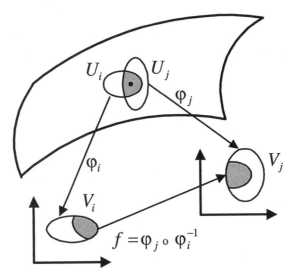

Figure 3.1: This figure illustrates coordinate charts, local coordinate maps and compatibility (overlap) functions on manifold M.

Example 3.43 (Differentiable Manifolds) *The following are simple examples of differentiable manifolds.*

1. *The Euclidean space R_m is an m-dimensional manifold. There is a single chart, $U = R_m$. The corresponding coordinate map is simply the identity map $I_d : R^m \to R^m$.*

2. *Any open subset $U \subset R^m$ is an m-dimensional manifold with single chart U and coordinate map again the identity.*

3. *The unit circle $S_1 = \{(x, y) | x^2 + y^2 = 1\}$ can be viewed as a one dimensional manifold with two coordinate charts. Define the charts $U_1 = S_1 - \{(-1, 0)\}$ and $U_2 = S_1 - \{(1, 0)\}$. Now we define the coordinate maps by isometric projection.*

$$\varphi_1 = \frac{2y}{1 - x} \; : \; S^1 - \{(-1, 0)\} \to R^1 \cong \{(1, y)\}$$

$$\varphi_2 = \frac{2y}{1 - x} \; : \; S^1 - \{(1, 0)\} \to R^1 \cong \{(-1, y)\}$$

The overlap functions are given by

$$f_1 = \varphi_2 \circ \varphi_1^{-1} = \frac{1+x}{1-x} : R^1 - \{0\} \to R^1 - \{0\}$$

$$f_2 = \varphi_1 \circ \varphi_2^{-1} = \frac{1-x}{1+x} : R^1 - \{0\} \to R^1 - \{0\}$$

Another description is obtained if we identify a point on S^1 by its angular coordinate θ, where $(x, y) = (cos\theta, sin\theta$, with two angles equivalent if they differ by an integral multiple of 2π. Thus, we have a single chart $U = \{\theta|\ 0 < \theta \leq 2\pi\}$.

4. *The unit sphere $S^2 = \{(x, y, z)|x^2 + y^2 + z^2 = 1\}$ is another basic example of a manifold with two coordinate charts. We may choose*

$$U_1 = S^2 - \{(0, 0, 1)\}$$

$$U_2 = S^2 - \{(0, 0, -1)\}$$

obtained by deleting the north and south poles, respectively from the sphere. Local coordinate functions can be defined as stereographic projections (from, respectively, the north or south poles) onto the horizontal plane that passes through the origin:

$$\varphi_1(x, y, z) = \left\{ \frac{x}{1-z}, \frac{y}{1-z} \right\}$$

$$\varphi_2(x, y, z) = \left\{ \frac{x}{1+z}, \frac{y}{1+z} \right\}$$

The compatibility function is

$$f_{2,1} = \varphi_1 \circ \varphi_2^{-1} : R^2 - \{0\} \to R^2 - \{0\}$$

$$f_{2,1}(x, y) = \left\{ \frac{x}{x^2 + y^2}, \frac{y}{x^2 + y^2} \right\}$$

5. *In general, if M and N are smooth manifolds of dimension m and n, then their Cartesian product $M \times N$ is a smooth manifold of dimension $m + n$. If their respective coordinate maps are $\varphi_i : U_i \to V_i \subset R^m$ and $\bar{\varphi}_j : \bar{U}_j \to \bar{V}_j \subset R^n$, then the induced coordinate charts on $M \times N$ are the Cartesian products*

$$\varphi_i \times \bar{\varphi}_j : U_i \times \bar{U}_j \to V_i \times \bar{V}_j \subset R^{n+m}$$

For our purposes a differentiable manifold can always be conceived as a smooth surface embedded in a Euclidean space. The unit sphere S^2 and the torus T^2 are manifolds embedded in R^3. As noted, manifolds are ordinarily specified as submanifolds of Euclidean space in one of two ways, parametrically or implicitly. Before describing these repesentations formally we define the notion of maximal rank of maps.

Definition 3.44 (Maximal Rank Condition) *Let $F : R^m \to R^n$ be a smooth map. The rank of F at $x_0 \in R^m$ is the rank of the Jacobian $D_x F(x_0)$. F is of maximal rank on $S \subset R^m$ if the rank of F is maximal (the minimum of m and n) for each $x_0 \in S$.*

A submanifold can be defined parametrically as follows.

Definition 3.45 *A submanifold embedded in R^n is a set $M \subset R^n$, together with a smooth one-to-one map $\phi : \Pi \subset R^m \to M$, $m \leq n$, which satisfies the maximal rank condition everywhere, where Π is called the parameter space and $M = \phi(\Pi)$ is the image of ϕ. If the maximal rank condition holds but the map is not one-to-one, then the set M (or the function ϕ) is called an immersion.*

Example 3.46 (One Dimensional Submanifolds in R^2 and R^3) *The following are some examples for parametrically defined one dimensional submanifolds. Notice that plots of parametrically defined submanifolds in R^2 and R^3 can easily be generated using the Mathematica functions* ParametricPlot *and* ParamtricPlot3D.

Consider a submanifold N embedded in R^3 defined by mapping $f : R \to R^3$

$$f(t) = (\cos t, \sin t, t)$$

N is a helix which spirals up the z axis. It is one-to-one and $D_t f = (-\sin t, \cos t, 1)$ so that the maximal rank condition is satisfied.

$In\,[24] :=$ *ParametricPlot3D*[{cos[t], sin[t], t}, {t, −3 π, 3 π},
    BoxRatios → {1, 1, 2}, *Boxed* → *False*, *Ticks* → *None*, *Axes* → *False*];

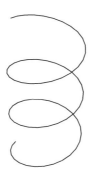

Now, consider a submanifold embedded in R^2 defined by the mapping

$$f(t) = ((1 + e^{-t/4}) \cos t, (1 + e^{-t/4}) \sin t)$$

Then as $t \to \infty$, N spirals in to the circle $x2 + y2 = 1$.

In[25]:= $D[\{(1 + \exp[-t/4]) \cos[t], (1 + \exp[-t/4]) \sin[t]\}, t]$

Out[25]= $\left\{ -\dfrac{1}{4} \ e^{-t/4} \ \mathrm{Cos}[t] - (1 + e^{-t/4}) \ \mathrm{Sin}[t], (1 + e^{-t/4}) \ \mathrm{Cos}[t] - \dfrac{1}{4} \ e^{-t/4} \ \mathrm{Sin}[t] \right\}$

In[26]:= $ParametricPlot[\{(1 + \exp[-t/4]) \cos[t], (1 + \exp[-t/4]) \sin[t]\},$
        $\{t, 0, 8 \ \pi\}, AspectRatio \to Automatic, Axes \to False];$

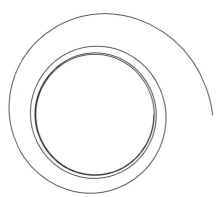

As another example, consider the mapping $f : R \to R^2$

$$f(t) = (\sin t, 2 \sin 2t)$$

$N = f(R)$ *is a figure eight which is self intersecting at the origin of* R^2*. Thus, f is not one-to-one although the maximum rank condition is satisfied since* $D_t f = (-\cos t, -4\cos 2t)$ *never vanishes.*

In[27]:= *ParametricPlot[{ sin[t], 2 sin[2 t]}, {t, 0, 2 π}, Axes $\to$ False];*

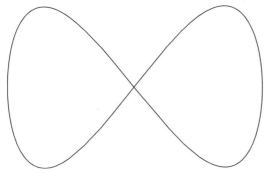

Let us modify the previous example by defining

$$f(t) = (\sin(2\arctan t), 2\sin(4\arctan t))$$

Now f is one-to-one and N passes through the origin only once. The maximal rank condition holds.

In[28]:= *ParametricPlot[{ sin[2 arctan[t]], 2 sin[4 arctan[t]]},*
            {t, 0, 100 π}, PlotPoints $\to$ 200, PlotRange $\to$ All, Axes $\to$ False];

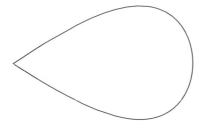

In[29]:= *D[{ sin[2 arctan[t]], 2 sin[4 arctan[t]]}, t]*

Out[29]= $\{\dfrac{2\ \text{Cos}[2\ \text{ArcTan}[t]]}{1+t^2}, \dfrac{8\ \text{Cos}[4\ \text{ArcTan}[t]]}{1+t^2}\}$

Example 3.47 (The sphere S^2**)** *Now consider the sphere* S^2*. First, we generate a graph using a parametric specification of the sphere:*

$$f(t, u) = \{\cos t \sin u, \sin t \sin u, \cos u\}$$

```
In[30]:= ParametricPlot3D[{Cos[t] Sin[u], Sin[t] Sin[u], Cos[u]},
           {t, 0, 2Pi}, {u, 0, Pi}]
```

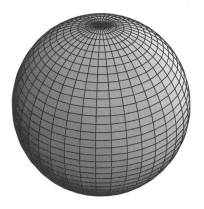

The maximal rank condition can be tested by computing the Jacobian using the ProPac function Jacob *and then testing for its rank by examining the span of its columns using* Span.

```
In[31]:=  AA = Jacob[{cos[t] sin[u], sin[t] sin[u], cos[u]}, {t, u}]
Out[31]= {{− Sin[t]  Sin[u], Cos[t]  Cos[u]}, {Cos[t]  Sin[u], Cos[u]  Sin[t]},
           {0, − Sin[u]}}
```

```
In[32]:=  Simplify[cos[u] Span[Transpose[AA]]]
Out[32]= {{Cos[u], 0, − Cos[t]  Sin[u]}, {0, Cos[u], − Sin[t]  Sin[u]}}
```

We see that the rank condition is satified everywhere except at the poles. Thus, the given mapping defines a submanifold of R^3 that is the sphere with the poles removed.

Example 3.48 (The torus T^2) *Now consider the torus T^2. First, we generate a graph of the torus using the prametric representation:*

$$f(t, u) = \{\cos t(3 + \cos u), sint(3 + \cos u), sinu\}$$

In[33]:=   *ParametricPlot3D[*
        {cos[t] (3 + cos[u]), sin[t] (3 + cos[u]), sin[u]},
        {t, 0, 2π}, {u, 0, 2π}, *PlotPoints* → 40, *Axes* → *False,*
        Boxed → *False*]

Now, let us exam the maximal rank condition.

In[34]:=   *AA = Jacob[*{cos[t] (3 + cos[u]), sin[t] (3 + cos[u]), sin[u]}, {t, u}]
Out[34]= {{−(3 + Cos[u]) Sin[t], − Cos[t] Sin[u]},
        {Cos[t] (3 + Cos[u]), − Sin[t] Sin[u]}, {0, Cos[u]}}

In[35]:=   *Simplify[Span[Transpose[AA]]]*
Out[35]= {{Sin[u], 0, − Cos[t] Cos[u]}, {0, Sin[u], − Cos[u] Sin[t]}}

The mapping fails to have have maximum rank when u = 0, i.e., on the outer edge (in the x − y plane) of the torus. The torus with these points removed is a properly parametrically defined submanifold of R^3 as specified by the given mapping.

It is illustrative to consider generating one dimensional submanifolds of R^3 by drawing curves on the surface of the torus (these would be submanifolds of the torus as well). We will consider mappings $f : R → R^3$ of the form:

$$f(t) = \{\cos t(3 + \cos \alpha t), \sin t(3 + \cos \alpha t), \sin t\}$$

where α is a parameter. These mappings produce curves on the torus surface. If α is a rational number they are closed curves. If α is irrational then, even though the mapping is one-to-one, its image is dense in the torus and its closure is the torus. The following computations illustrate these two cases.

In[36]:= *ParametricPlot3D*[
{cos[*t*] (3 + cos[10 *t*]), sin[*t*] (3 + cos[10*t*]), sin[10 *t*]},
{*t*, 0, 2π}, *PlotPoints* → 200, *Axes* → *False*, *Boxed* → *False*];

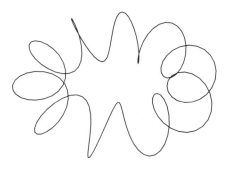

In[37]:= *ParametricPlot3D*[
{cos[*t*] (3 + cos[π 10/3 *t*]), sin[*t*] (3 + cos[π 10/3*t*]), sin[π 10/3 *t*]},
{*t*, 0, 20π}, *PlotPoints* → 2000, *Axes* → *False*, *Boxed* → *False*];

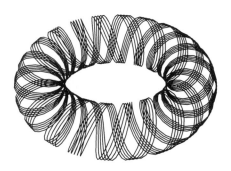

Among these examples two of them illustrate submanifolds that are somewhat pathological. Namely, the spiral in Example (3.46) and the irrational mapping onto the torus in Example (3.48). In these cases, although the map f is one-to-one and satisfies the maximal rank condition, it is not a homeomorphism (see Figure (3.2). This is the source of the complex topology of these submanifolds. We define a class of submanifolds with a more congenial topological structure.

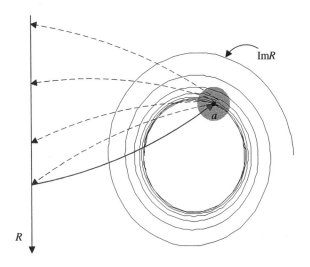

Figure 3.2: The mapping is not a homeomorphism because the inverse is not continuous. Every neighborhood of points on the spiral arbitrarily close to the limiting circle contain points that map back into distinct points in the parameter space.

Definition 3.49 *A* regular manifold *M embedded in R^n is a manifold parameterized by a smooth mapping $\phi : \Pi \subset R^m \to M \subset R^n$, such that for each $x \in M$ there exists a neighborhood U of x in R^n such that $\phi^{-1}(U \cap M)$ is a connected open subset of Π (equivalently, ϕ maps homeomorphically onto its image).*

In applications, manifolds are sometimes specified by a parameterization, but it is equally common to define manifolds implicitly. The following theorem is a consequence of the implicit function theorem.

Proposition 3.50 *Consider a smooth mapping $F : R^n \to R^m$, $m \leq n$. If F satisfies the maximal rank condition on the set $S = \{ x \in R^n |\ F(x) = 0 \}$, then S is a regular, $n - m$ dimensional manifold embedded in R^n.*

Let us make a few remarks and observations about Proposition (3.50).

1. Notice that it is only required that the maximal rank condition be satisfied on the set where F vanishes, i.e., on S itself. If the maximal rank condition

is satisfied everywhere, then each level set of F, $\{x|F(x) = c\}$ is a regular submanifold of R^n of dimension $m - n$.

2. A manifold S so defined is called an *implicit* submanifold .

3. Suppose (U, φ) is a local coordinate system on M and that k is an integer $0 \leq k < m$. Let $a \in \varphi(U)$ and define

$$S = \{p \in U \mid \varphi_i(p) = \varphi_i(a), \, i = k+1, \ldots, m\} \qquad (3.1)$$

The subset S of M together with the coordinate system $\{\varphi_j \backslash S, \, i = 1, \ldots, k\}$ forms a submanifold of M called a *slice* of the coordinate system (U, φ). The concept of a slice will be important in discusing controllability and observvbility.

4. Any smooth manifold can be implicitly defined in a Euclidean space of suitable dimension.

Example 3.51 *Consider the map $F : R^3 \to R$ defined by*

$$F(x, y, z) = x^2 + y^2 + z^2 - 2\sqrt{2(x^2 + y^2)}$$

F is of maximal rank everywhere except on the the circle $\{x^2 + y^2 = 2, z = 0\}$ where the Jacobian vanishes and on the z-axis where it does not exist. The level sets $\{(x, y, z) \in R^3 \mid F(x, y, z) = c\}$ are tori for $-2 < c < 0$, and like spheres with indented poles for $c > 0$. For c=-2, the level set is the circle $\{x^2 + y^2 = 2, z = 0\}$ on which the gradient of F vanishes.

Example 3.52 *Consider the set of orthogonal matrices*

$$O(2) = \{X \in R^{2\times 2} | X^T X = I\}$$

Such matrices form a subset of R^4. In fact, among the four coordinates x_1, x_2, x_3, x_4,

$$X = \begin{bmatrix} x_1 & x_2 \\ x_3 & x_4 \end{bmatrix}$$

there are three independent constraints

$$x_1^2 + x_3^2 = 1$$

$$x_1 x_2 + x_3 x_4 = 0$$

$$x_2^2 + x_4^2 = 1$$

It is easy to check that the Jacobian is of full rank on $O(2)$ so that $O(2)$ is an implicitly defined regular submanifold of dimension 1 in R^4. We can obtain a deeper insight into the structure of this manifold by seeking a (one-dimensional) parameterization. Let us attempt to identify

$$x_1 = \cos\theta, x_3 = \sin\theta$$

which clearly satisfies the first equation. The second equation is then satisfied by

$$x_2 = -\sin\theta, x_4 = \cos\theta$$

or

$$x_2 = \sin\theta, x_4 = -\cos\theta$$

either of which satisfies the fourth relation. It follows that the matrices can be either of the form

$$X_1 = \begin{bmatrix} \cos\theta & -\sin\theta \\ \sin\theta & \cos\theta \end{bmatrix}, \ 0 \le \theta < 2\pi$$

or

$$X_2 = \begin{bmatrix} \cos\theta & \sin\theta \\ \sin\theta & -\cos\theta \end{bmatrix} = \begin{bmatrix} \cos\theta & -\sin\theta \\ \sin\theta & \cos\theta \end{bmatrix} \begin{bmatrix} 1 & 0 \\ 0 & -1 \end{bmatrix}, \ 0 \le \theta < 2\pi$$

First, note that each family of matrices is in one-to-one correspondence with the points on a circle. They are disconnected because they have no common elements. Thus, we say that the manifold $O(2)$ has the structure of two disconnected copies of S^1. Second, note that the matrix X_1 (viewed as an operator on R^2) represents a rotation in the plane (through an angle θ) whereas, X_2 represents a relection in the x-axis followed by a rotation. Third, note that $\det\{X_1\} = 1$ and $\det\{X_2\} = -1$. The determinant distinguishes the two components of the manifold $O(2)$. Fourth, observe that the set of matrices $X_1(\theta)$ contains the identity element, in particular, $X_1(0) = I_2$. However, the set of matrices $X_2(\theta)$ does not.

3.3   Tangent Spaces and Vector Fields

3.3.1   The Tangent Space and Tangent Bundle

Consider a smooth two dimensional surface embedded in R^3. At each point on this surface it is easy to envision a tangent plane. Suppose a particle moves along a path in the surface. Then its velocity vector at a specified point on the path lies in the tangent plane to the surface at the prescribed particle location. The generalization of this concept to motion in more abstract manifolds is of central importance.

Definition 3.53 *Let $p : R \to M$ be a C^k, $k \ge 1$ map so that $p(t)$ is a curve in a manifold M. The tangent vector v to the curve $p(t)$ at the point $p_0 = p(t_0)$ is defined by*

$$v = \dot{p}(t_0) = \lim_{t \to t_0} \left\{ \frac{p(t) - p(t_0)}{t - t_0} \right\}$$

The set of tangent vectors to all curves in M passing through p_0 is the tangent space to M at p_0, denoted TM_{p_0}.

If M is an implicit submanifold of dimension m in R^{m+k}, i.e., $F : R^{m+k} \to R^k$, $M = \{x \in R^{m+k} | F(x) = 0\}$ and $D_x F$ satisfies the maximum rank condition on M, Then TM_p is the ker $D_x F(p)$ (translated, of course to the point p). That is TM_p is the tangent hyperplane to M at p. See Figure (3.3).

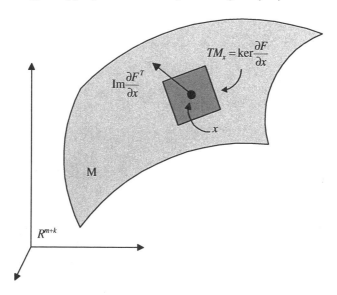

Figure 3.3: The tangent space to the manifold M at the point x, TM_x.

Let (U, φ) be a local coordinate chart that contains the point $p_0 \in M$ and suppose $p(t)$, with $p(t_0) = p_0$, denotes a curve in M and $x(t) = \varphi(p(t) \cap U)$ its image in $V \subset \varphi(U) \subset R^m$, as depicted in Figure (3.4). The tangent vector to the curve $p(t)$ at p_0 in M is v and the corresponding tangent vector to $x(t)$ at x_0 in R^m is $\bar{v}$. Thus, we have the following definition.

Definition 3.54 *The components of the tangent vector v to the curve $p(t)$ in M in the local coordinates (U, φ) are the m numbers $v_1, \ldots, v_m$ where $v_i = d\varphi_i/dt$.*

Another interpretation of a tangent vector is as an operator on scalar valued smooth functions. Consider the C^k map $F : M \to R$. Let $y = f(x)$, $x \in \varphi(U) \subset R^m$ denote the realization of F in the local coordinates (U, φ). Again, suppose $p(t)$ denotes a curve in M with $x(t)$ its image in R^m. Then the rate of change of F at a point p on this curve is

$$\frac{df}{dt} = v_1 \frac{\partial f}{\partial x_1} + \cdots + v_m \frac{\partial f}{\partial x_m} \tag{3.2}$$

Let us remark on some alternative views of the tangent vector. Notice that there are many curves in M that pass through p and have the same tangent vector.

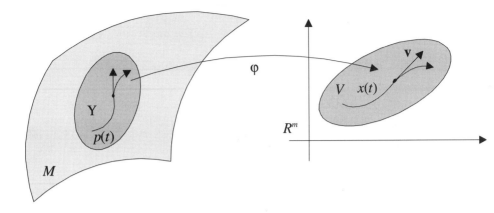

Figure 3.4: Motion along a curve in the manifold M is quantified in a local coordinate system.

We can define equivalence classes of curves by defining an equivalence relation among curves: two curves passing through p with the same tangent vector are equivalent. Each class is associated with a unique tangent vector. Thus, a tangent vector at p is sometimes defined as an equivalence class of curves through p. All curves belonging to the same class obviously produce the same value for dF/dt. Conversely, the tangent vector $(v_1, \ldots, v_m)$, is uniquely determined by the action of the directional derivative operator (called a *derivation*)

$$\mathbf{v} = v_1 \frac{\partial}{\partial x_1} + \cdots v_m \frac{\partial}{\partial x_m} \tag{3.3}$$

on such an equivalence class of curves. Thus, it is also possible to define the tangent vector as a map (the directional derivative) from the space of (equivalence classes of) differentiable functions passing through p to the real line. As a derivation, the tangent vector satisfies two important properties

1. linearity $\mathbf{v}(f + g) = \mathbf{v}(f) + \mathbf{v}(g)$

2. Liebniz' rule $\mathbf{v}(f \circ g) = \mathbf{v}(f) \circ g + f \circ \mathbf{v}(g)$

Both concepts of the tangent vector, 1) as a velocity vector - the curves approach, or 2) as a directional derivative - the derivation approach, are useful and may be used interchangeably. We can easily reconcile these viewpoints by noting that a tangent vector $(v_1, .., v_i, .., v_m) = (0, .., 1, .., 0)$ corresponds to the operator $\mathbf{v} = \partial/\partial x_i$.

Thus, we have the following definition.

Definition 3.55 *The set of partial derivative operators constitute a basis for* *the tangent space TM_p for all points $p \in U \subset M$ which is called the* natural basis.

The above definition makes sense when the tangent space is viewed as a space of differential operators. The formulation of the tangent vector as a directional derivative requires a local coordinate system, i.e., a chart (U, φ). We are free, however, to define a coordinate system for the tangent space TM_p. The 'natural' coordinate system on TM_p induced by (U, φ) has basis vectors which are tangent vectors to the coordinate lines on M passing through p. When taking the curves viewpoint, the tangent space TM_p is simply R^m and its elements may be thought of as column vectors and the symbols $\mathbf{v}_i = \partial/\partial x_i$ represent the (unit) basis vectors.

At each point $p \in M$, we have defined the tangent space. Taken together, these spaces form the tangent bundle.

Definition 3.56 *The union of all the tangent spaces to M is called the* tangent bundle *and is denoted TM,*

$$TM = \bigcup_{p \in M} TM_p$$

The tangent bundle is a manifold with $\dim \{TM\} = 2\dim \{M\}$. A point in TM is a pair (x, v) with $x \in M$, $v \in TM_x$. If $(x_1, .., x_m)$ are local coordinates on M and $(v_1, .., v_m)$ components of the tangent vector in the natural coordinate system on TM_x, then natural local coordinates on TM are $(x_1, \ldots, x_m, v_1, \ldots, v_m) = (x_1, \ldots, x_m, \dot{x}_1, \ldots, \dot{x}_m)$. Recall the natural 'unit vectors' on TM_x are $\mathbf{v}_1 = \partial/\partial x_1, ..., \mathbf{v}_m = \partial/\partial x_m$.

The mapping $\pi : TM \rightarrow M$ which takes the point (x, v) in TM to x in M is called the *natural projection* on TM. The inverse image of x under π is the tangent space at x, $\pi^{-1}(x) = TM_x$. TM_x is called the *fiber* of the tangent bundle over the point x. A *section* of TM is a mapping $\sigma : M \rightarrow TM$ such that the composite mapping $\pi \circ \sigma : M \rightarrow M$ is the identity. The mapping $\iota : M \rightarrow TM$ such that $\iota(x)$ is the zero vector of TM_p is called the *null section*.

One of the most important applications of the idea of tangent bundle occurs in analytical mechanics where the tangent bundle generalizes the concept of a state space. A *mechanical system* is a collection of mass particles which interact through physical constraints or forces, such as the pendulum of Figure (3.5). A *configuration* is a specification of the position for each of its constituent particles. The *configuration space* is a set M of elements such that any configuration of the system corresponds to a unique point in the set M and each point in M corresponds to a unique configuration of the system. The configuration space of a mechanical system is a differentiable manifold called the *configuration manifold*. Any system of local coordinates q on the configuration manifold are called *generalized coordinates*. The *generalized velocities* $\dot{q}$ are elements of the tangent spaces to M, TM_q represented in the natural basis. The *state space* is the tangent bundle TM which has local coordinates $(q, \dot{q})$.

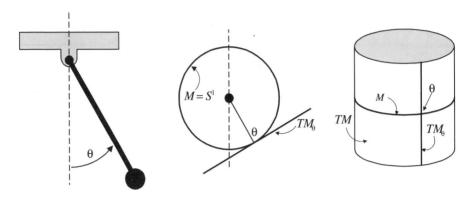

Figure 3.5: The pendulum illustrates the relationships between the configuration manifold M, the tangent space TM_θ, and the state space TM.

3.3.2   The Differential Map

Let C be a curve in M parameterized by the mapping $\phi : R \to M$. Let $F : M \to N$ be a smooth mapping. The image of C under the mapping F is the curve $\bar{C}$ in N which has a parameterization $\bar{\phi} = F \circ \phi$. Refer to Figure (3.6). At any point p on C there is a tangent vector $v \in TM_p$ which maps to a tangent vector $\bar{v} \in TN_{F(p)}$. We wish to determine the induced mapping $F_* : TM_p \to TN_{F(p)}$ which takes tangent vectors into tangent vectors.

In local coordinates, the chain rule provides

$$\frac{d\bar{\phi}}{dt} = \frac{\partial F}{\partial x}\frac{d\phi}{dt} \tag{3.4}$$

$$\bar{v} = \frac{\partial F}{\partial x}v \tag{3.5}$$

This relation defines the desired mapping, F_*, in local coordinates. We note that it is linear and that its matrix representation is simply the Jacobian of the mapping F. F_* is called the *differential map* of F and is sometimes denoted dF. Notice that

$$\bar{v}_i = v_1\frac{\partial F_i}{\partial x_1} + \cdots + v_m\frac{\partial F_i}{\partial x_m} = \mathbf{v}(F_i), \quad i = 1, \ldots, n \tag{3.6}$$

So that

$$\bar{v}(y) = (\mathbf{v}(F_1(p)), \ldots, \mathbf{v}(F_n(p))), \quad y = F(p) \in N \tag{3.7}$$

From the derivational point of view, the mapping dF is realized in local coordinates by

$$\bar{\mathbf{v}} = \mathbf{v}(F_1(p))\frac{\partial}{\partial y_1} + \cdots + \mathbf{v}(F_n(p))\frac{\partial}{\partial y_n} = \sum_{i=1}^{n} \mathbf{v}(F_i(p))\frac{\partial}{\partial y_i} \tag{3.8}$$

where y are the local coordinates on N. Because the differential map takes tangent vectors in TM_p to $TN_{F(p)}$ it is defined at the point p. Thus, it would be appropriate to write $F_*|_p$ or $dF|_p$. However, the point of evaluation is typically not indicated and is normally clear from the context.

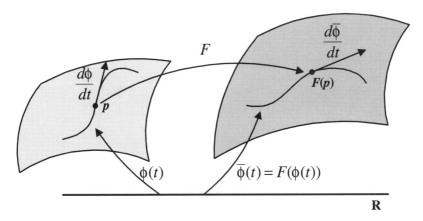

Figure 3.6: Motion along a curve in the manifold M maps into a manifold N. The mapping induces a map between the tangent spaces TM_p and $TN_{F(p)}$.

The following is a useful result which provides for computing the differential map for composite functions

Lemma 3.57 *Suppose $F : M \to N$ and $H : N \to P$ are smooth maps between manifolds, then*

$$d(H \circ F) = dH \circ dF$$

where $dF : TM_x \to TN_{y=F(x)}$, $dH : TN_y \to TP_{z=H(y)}$ and $d(H \circ F) : TM_x \to TP_{z=H(F(x))}$.

Proof: Direct computation in local coordinates provides the matrix representation of $d(H \circ F)$. For any $v \in TM_x$

$$d(H \circ F)(v) = \frac{\partial H(F(x))}{\partial x}(v) = \frac{\partial H(y)}{\partial y}\frac{\partial F(x)}{\partial x}(v) = dH(dF(v)) = dH \circ dF(v)$$

∎

3.3.3   Cotangent Spaces

The dual space to TM_p is denoted T^*M_p and it is called the *contagent space* to M at $p \in M$. The elements of T^*M_p are called *tangent covectors* . Recall that the dual space V^* of a linear vector space V is the space of linear functions from V to R. Thus, if we identify the elements of TM_p as column vectors of dimension m, the covectors may be thought of as row vectors of dimension m. There is a natural relationship between the differential mapping and the cotangent space. Let ϕ be a smooth mapping $\phi : M \rightarrow R$. Its differential at $p \in M$ is a linear map $\phi_* : TM_p \rightarrow TR_{\phi(p)} \cong R$. Since ϕ_* is a linear map from TM_p to the real line, it is an element of of the dual space to TM_p, that is, the cotangent space T^*M_p. In local coordinates ϕ_* is realized by

$$\tilde{v} = \left\lfloor \frac{\partial \phi}{\partial x_1}, \cdots, \frac{\partial \phi}{\partial x_m} \right\rfloor = \mathbf{v}(\phi)$$

Let $e^i, i = 1,, m$ denote the natural basis vectors of TM_p. The natural basis vectors e_i^* for T^*M_p are defined by the relations

$$e_i^* e_j = \delta_{ij}, i, j = 1, ..., m \tag{3.9}$$

Recall, that

$$e_1 = \begin{bmatrix} 1 \\ 0 \\ \vdots \\ \vdots \\ 0 \end{bmatrix}, \; e_2 = \begin{bmatrix} 0 \\ 1 \\ 0 \\ \vdots \\ 0 \end{bmatrix}, \; ..., \; e_m = \begin{bmatrix} 0 \\ \vdots \\ \vdots \\ 0 \\ 1 \end{bmatrix} \tag{3.10}$$

so that the basis vectors for T^*M_p are the row vectors

$$e_1^* = [1, 0, .., 0], \; e_2^* = [0, 1, .., 0], \; ..., \; e_m^* = [0, 0, .., 1] \tag{3.11}$$

Correspondingly, in the derivations viewpoint, TM_p is a linear vector space of differential operators which act on scalar valued functions on M and the basis elements for TM_p are $\mathbf{e}_i = \partial/\partial x_i$, $i = 1,, m$. The differential map associated with the smooth mapping $\phi : M \rightarrow R$ is

$$\phi_*(\mathbf{v}) = \mathbf{v}(\phi)\frac{d}{dy} \tag{3.12}$$

Of course, there is a one to one correspondence between $\mathbf{v}(\phi)d/dy \in TR_{\phi(p)}$ and $\mathbf{v}(\phi) \in R$. Thus, it is convenient to write $d\phi(\mathbf{v}) = \mathbf{v}(\phi)$ and refer to $d\phi$ as the differential mapping. In this way we again identify the differential mappings of scalar valued functions as elements of T^*M_p, i.e., the differential mappings $d\phi(\mathbf{v})$ are cotangent vectors. We seek basis elements in T^*M_p, $\mathbf{e}_i^*$, such that

$$\mathbf{e}_i^*(\mathbf{e}_j) = \mathbf{e}_i^* \left(\frac{\partial}{\partial x_j} \right) = \delta_{ij}, i, j = 1,, m \tag{3.13}$$

Let $\phi = x_i$, the coordinate map, and notice that

$$dx_i \frac{\partial}{\partial x_j} = \frac{\partial x_i}{\partial x_j} = \delta_{ij} \tag{3.14}$$

Thus, the set of cotangent vectors $\{dx_1, .., dx_m\}$ constitute a basis for the cotangent space T^*M_p. Now, any smooth function $\phi : M \to R$ gives rise to a cotangent vector $d\phi$ which can be expressed

$$d\phi = \sum_{i=1}^{m} v_i^* dx_i \tag{3.15}$$

$$v_i^* = \frac{\partial \phi}{\partial x_i} \tag{3.16}$$

so that

$$d\phi = \sum_{i=1}^{m} \frac{\partial \phi}{\partial x_i} dx_i \tag{3.17}$$

Suppose $F : M \to N$ is a smooth map. Recall that its differential, F_* or $dF : TM_p \to TN_{F(p)}$, is a linear map that takes tangent vectors in M to tangent vectors in N. This map is realized in local coordinates by the Jacobian of F. Consequently, there is an induced map between cotangent spaces called the *codifferential* and denoted F^* or $\delta F : T^*N_{F(p)} \to T^*M_p$. This map, realized by the transpose of the Jacobian of F, is sometimes called the *pull-back* as it takes covectors on N back to covectors on M. To see this, consider a covector $\omega \in TM_p^*$. Then ω is a linear mapping that takes each $v \in TM_p$ to R. Let $F : M \to N$ be a smooth mapping. Then the induced differential map dF takes vectors in TM_p into $TN_{F(p)}$, in local coordinates

$$w = \frac{\partial F}{\partial x} v \tag{3.18}$$

We seek a covector $\mu \in TN_{F(p)}^*$ such that $\omega v = \mu w$ for all $v \in TM_p$, i.e.,

$$\omega v = \mu \frac{\partial F}{\partial x} v, \ \forall v \in TM_p \tag{3.19}$$

Clearly, in local coordinates, ω and μ are related by

$$\omega = \mu \frac{\partial F}{\partial x} \tag{3.20}$$

This is the pull-back mapping.

3.3.4 Vector Fields and Flows

A vector field $f(x)$ on a manifold M is a mapping that assigns a vector $f \in TM_x$, where TM_x is the tangent space to M at x, to each point $x \in M$. Formally, we define:

Definition 3.58 *A vector field v on M is a map which assigns to each point $p \in M$, a tangent vector $v(p) \in TM_p$. It is a C^k-vector field if for each $p \in M$ there exist local coordinates (U, φ) such that each component $v_i(x), i = 1, ..., m$ is a C^k function for each $x \in \varphi(U)$.*

A vector field can be viewed as a mapping $v : M \to TM$ with the property that the composite mapping $\pi \circ v : M \to M$ [1] is the identity mapping, i.e. $\pi \circ v(x) = x$. Thus, the vector field v is a section of the tangent bundle.

Definition 3.59 *An integral curve of a vector field v on M is a parameterized curve $p = \phi(t), t \in (t_1, t_2) \subset R$ whose tangent vector at any point coincides with v at that point.*

Consider local coordinates (U, φ) on M, and the induced natural coordinates on TM_p. Then if $\phi(t)$ is an integral curve, its image $x(t) = \varphi \circ \phi(t) \subset R^m$ must satisfy the differential equation $dx/dt = v(x)$.

If $v(x)$ is sufficiently smooth, standard existence and uniqueness theorems for systems of ordinary differential equations imply corresponding properties for integral curves on manifolds. First we define a maximal integral curve.

Definition 3.60 *Let I_p denote an open interval of R with $0 \in I_p$. Suppose $\phi : I_p \to M$ is an integral curve of the vector field v such that $\phi(0) = p$. The integral curve ϕ is maximal if for any other integral curve $\widehat{\phi} : \widehat{I}_p \to M$ with $\widehat{\phi}(0) = p$, then $\widehat{I}_p \subset I_p$ and $\widehat{\phi}(t) = \phi(t)$ for $t \in \widehat{I}_p$.*

Now, the existence and uniqueness theorem can be stated.

Proposition 3.61 *Suppose v is a smooth (C^k) vector field on M. Then there exists a unique maximal integral curve $\phi : I_p \to M$ passing through the point $p \in M$.*

Proof: The result follows from the standard results on differential equations. ∎

Definition 3.62 *Let v be a smooth vector field on M and denote the parameterized maximal integral curve through $p \in M$ by $\Psi(t, p)$ so that $\Psi : I_p \times M \to M$ where I_p is a subinterval of R containing the origin and $\Psi(0, p) = p$. $\Psi(t, p)$ is called the* flow generated by v.

The flow Ψ has the following basic property.

---

[1] recall that π is the natural projection.

Proposition 3.63 *The flow Ψ of a smooth vector field v satisfies the differential equation on M*

$$\frac{d}{dt}\Psi(t,p) \;=\; v(\Psi(t,p)), \quad \Psi(0,p) = p$$

and has the semigroup property

$$\Psi(t_2, \Psi(t_1,p)) \;=\; \Psi(t_1 + t_2, p)$$

for all values of $t_1, t_2 \in R$ and $p \in M$ such that both sides of the relation are defined.

Proof: The differential equation merely states that v is tangent to the curve $\Psi(t,p)$ for all fixed p. The initial condition is part of the definition of Ψ. The semigroup property follows from the uniqueness property of differential equations. Both sides of the equation satisfy the differential equation and have the same initial condition at $t_2 = 0$. ∎

We will adopt the following notation

$$\exp(t\mathbf{v})p := \Psi(t,p) \tag{3.21}$$

The motivation for this is simply that the flow satisfies the three basic properties ordinarily associated with exponentiation. In particular, we have (from the properties of the flow function):

$$\exp(0 \cdot \mathbf{v})p = p \tag{3.22}$$

$$\frac{d}{dt}[\exp(t\mathbf{v})p] \;=\; \mathbf{v}(\exp(t\mathbf{v})p) \tag{3.23}$$

$$\exp[(t_1 + t_2)\mathbf{v}]p \;=\; \exp(t_1\mathbf{v})\exp(t_2\mathbf{v})p \tag{3.24}$$

whenever defined. Note the distinction between the vector field used as a column vector (v) and as a derivation ($\mathbf{v}$).

The vector field is said to be *complete* if I_p coincides with R. Thus the flow is defined on all of $R \times M$.

Further justification for the exponential notation comes from the action of a vector field on functions. Let v be a vector field on M and $f : M \to R$ a smooth function. The value of f along the flow (along an integral curve of v passing through p) is given in local coordinates by $f(\exp(t\mathbf{v})x)$. Then the rate of change of f can be computed

$$\begin{aligned}
\frac{d}{dt}f(\exp(t\mathbf{v})x) &= \frac{\partial f(\exp(t\mathbf{v})x)}{\partial x}\frac{d\exp(t\mathbf{v})x}{dt} = \frac{\partial f(\exp(t\mathbf{v})x)}{\partial x}\mathbf{v}(\exp(t\mathbf{v})x) \\
&= \sum_{i=1}^{m} v_i(\exp(t\mathbf{v})x)\frac{\partial}{\partial x_i}f(\exp(t\mathbf{v})x) = \mathbf{v}(f)(\exp(t\mathbf{v})x)
\end{aligned} \tag{3.25}$$

where $v = \{v_1(x), ..., v_m(x)\}$ is also in local coordinates. Similarly,

$$\frac{d^2}{dt^2}f(\exp(t\mathbf{v})x) \;=\; \mathbf{v}^2(f)(\exp(t\mathbf{v})x) \tag{3.26}$$

where $\mathbf{v}^2 = \mathbf{v}(\mathbf{v}(f))$, and so on. Thus, the Taylor series about $t = 0$ is

$$f(\exp(t\mathbf{v})x) \;=\; f(x) + t\,\mathbf{v}(f)(x) + \frac{t^2}{2}\mathbf{v}^2(f)(x) + \cdots = \sum_{k=0}^{\infty} \frac{t^k}{k!}\mathbf{v}^k(f)(x) \quad (3.27)$$

A similar formula is valid for vector valued functions $F : M \to R^n$. Let us interpret the action of $\mathbf{v}$ on F component-wise, that is $\mathbf{v}(F) = (\mathbf{v}(F_1), \ldots, \mathbf{v}(Fn))^T$. Then we have

$$F(\exp(t\mathbf{v})x) \;=\; F(x) + t\,\mathbf{v}(F)(x) + \frac{t^2}{2}\mathbf{v}^2(F)(x) + \cdots = \sum_{k=0}^{\infty} \frac{t^k}{k!}\mathbf{v}^k(F)(x) \quad (3.28)$$

An important operation is the or *Lie derivative* of a map $F : M \to R^n$ with respect to a vector field v on M.

Definition 3.64 *Let $v(x)$ denote a vector field on M and $F(x)$ a mapping $F : M \to R^n$, both in local coordinates. Then the Lie derivative of F with respect to v of order $0, \ldots, k$ is*

$$L_v^0(F) = F, \quad L_v^k(F) = \frac{\partial L_v^{k-1}(F)}{\partial x}v \quad (3.29)$$

Using this notation we can write

$$\mathbf{v}^k(F)(x) \;=\; \mathcal{L}_v^k(F)(x) \quad (3.30)$$

so that

$$F(\exp(t\mathbf{v})x) \;=\; \sum_{k=0}^{\infty} \frac{t^k}{k!}L_v^k(F)(x) \quad (3.31)$$

In particular, suppose F is the coordinate map from M to R^m, $F(x) = x$, so that $F(\exp(t\mathbf{v})x) = \exp(t\mathbf{v})x$ and we have

$$\exp(t\mathbf{v})x \;=\; x + t\,\mathbf{v}(x)(x) + \frac{t^2}{2}\mathbf{v}^2(x)(x) + \cdots = \sum_{k=0}^{\infty} \frac{t^k}{k!}\mathbf{v}^k(x)(x) \;=\; \sum_{k=0}^{\infty} \frac{t^k}{k!}L_v^k(x)(x)$$

$$(3.32)$$

The Taylor expansion is identical to that of the classical exponential.

3.3.5   Lie Bracket

The Lie bracket is a binary operation on vector fields that is essential in the subsequent discussion.

Definition 3.65 *If v, w are differentiable vector fields on M, then their Lie bracket $[v, w]$ is the unique vector field defined in local coordinates by the formula*

$$[v, w] = \frac{\partial w}{\partial x}v - \frac{\partial v}{\partial x}w$$

In terms of the derivation viewpoint, the Lie bracket is the unique vector field satisfying

$$[\mathbf{v}, \mathbf{w}](f) = \mathbf{v}(\mathbf{w}(f)) - \mathbf{w}(\mathbf{v}(f)) \tag{3.33}$$

for all smooth functions $f : M \to R$. In local coordinates we can easily derive the formula

$$[\mathbf{v}, \mathbf{w}] = \sum_{i=1}^{m} \{\mathbf{v}(w_i) - \mathbf{w}(v_i)\} \frac{\partial}{\partial x_i} = \sum_{i=1}^{m} \sum_{j=1}^{m} \left\{ v_j \frac{\partial w_i}{\partial x_j} - w_i \frac{\partial v_i}{\partial x_j} \right\} \frac{\partial}{\partial x_i} \tag{3.34}$$

which is precisely the definition we have adopted.

The Lie bracket can be given useful geometric interpretations. First, let us consider the Lie bracket as a directional derivative. We will compute the rate of change of a vector field w as seen by an observer moving with the flow $\Psi(x, t)$ generated by a second vector field v.

Proposition 3.66 *Suppose v, w are smooth vector fields on M and $\Psi(x,t)$ is the flow generated by v. Then*

$$\frac{dw\left(\Psi(x,t)\right)}{dt}\bigg|_{t=0} = [v, w]|_x$$

Proof: We need to compare $w(\Psi(x, t))$ with $w(x)$ as $t \to 0$. Since these two vectors exist in different tangent spaces ($TM_{\Psi(x,t)}$ and TM_x, respectively) we need to 'pull back' $w(\Psi(x, t))$ to the tangent space TM_x. This is easily done using the differential map. Thus,

$$
\begin{aligned}
\frac{dw(\Psi(x,t))}{dt}\bigg|_{t=0} &= \lim_{t \to 0} \left[\frac{\Psi_x(x, -t)w(\Psi(x,t)) - w(x)}{t} \right] \\
&= \lim_{t \to 0} \left[\frac{(I - v_x(x)\, t)(w(x) + w_x(x)v(x)\, t) - w(x)}{t} \right] \\
&= w_x(x)v(x) - v_x(x)w(x) \\
&= [v, w]|_x
\end{aligned}
$$

∎

Now, let us consider the Lie bracket as a commutator of flows. Beginning at point x in M follow the flow generated by v for an infinitesimal time which we take as $\sqrt{\varepsilon}$ for convenience. This takes us to a point $y = \exp(\sqrt{\varepsilon}\, v)x$. Then follow $\mathbf{w}$ for the same length of time, then $-\mathbf{v}$, then $-\mathbf{w}$. This brings us to a point ψ given by (see Figure (3.7)):

$$\psi(\varepsilon, x) = e^{-\sqrt{\varepsilon}\mathbf{w}} e^{-\sqrt{\varepsilon}\mathbf{v}} e^{\sqrt{\varepsilon}\mathbf{w}} e^{\sqrt{\varepsilon}\mathbf{v}} x \tag{3.35}$$

Proposition 3.67 *Let $\mathbf{v}$ and $\mathbf{w}$ be smooth vector fields on M. Then $\psi(\varepsilon, x)$, as given by (3.35), with x fixed defines a continuous path in M. Moreover,*

$$\frac{d}{d\varepsilon}\psi(0^+, x) = [v, w]|_x$$

Proof: Our proof follows [4]. Let us write $y = e^{\sqrt{\varepsilon}\mathbf{v}}x$, $z = e^{\sqrt{\varepsilon}\mathbf{w}}y$, $u = e^{-\sqrt{\varepsilon}\mathbf{v}}$, $\psi = e^{-\sqrt{\varepsilon}\mathbf{w}}u$. Now, for any vector field $\mathbf{v}$ we can use the Taylor series representation for the flow function, i.e.

$$e^{t\mathbf{v}}x = x + v(x)t + \tfrac{1}{2}v_x(x)v(x)t^2 + O(t^3)$$

Applying this successively to ψ, we obtain

$$
\begin{aligned}
\psi &= u - w(u)\sqrt{\varepsilon} + \tfrac{1}{2}w_x(u)w(u)\varepsilon + O(\varepsilon^{3/2}) \\
&= z - \{w(z) + v(z)\}\sqrt{\varepsilon} + \{\tfrac{1}{2}w_x(z)w(z) + v_x(z)w(z) + \tfrac{1}{2}v_x(z)v(z)\}\varepsilon \\
&\quad + O(\varepsilon^{3/2}) \\
&= y - v(y)\sqrt{\varepsilon} + \{v_x(y)w(y) - w_x(y)v(y) + \tfrac{1}{2}v_x(y)v(y)\}\varepsilon + O(\varepsilon^{3/2}) \\
&= x + \{v_x(x)w(x) - w_x(x)v(x)\}\varepsilon + O(\varepsilon^{3/2})
\end{aligned}
$$

Differentiating with respect to ε we get the desired result:

$$\frac{d\psi}{d\varepsilon}\bigg|_{\varepsilon=0} = v_x(x)w(x) - w_x(x)v(x)$$

∎

As we will see in later chapters, this theorem has important implications for the control of certain types of nonlinear systems.

We say that the vector fields v, w (or their flows) *commute* if

$$\psi(\varepsilon, x) = x = e^{-\sqrt{\varepsilon}w}e^{-\sqrt{\varepsilon}v}e^{\sqrt{\varepsilon}w}e^{\sqrt{\varepsilon}v}x \tag{3.36}$$

for all $\varepsilon, v \in R$ and $x \in M$ such that both sides are defined. The two vector fields commute if and only if $[v, w] = 0$. See Figure (3.7).

We can define higher order Lie Bracket operations. For notational convenience we define the *ad* operator.

Definition 3.68 *If v, w are C^k vector fields on M we define the kth-order iterated Lie bracket or* ad *operation:*

$$ad_v^0(w) = w, \quad ad_v^k(w) = [v, ad_v^{k-1}(w)] \tag{3.37}$$

3.3.6   Covector Fields

Like vector fields, covector fields play an important role in our subsequent discussion.

Definition 3.69 *A covector field or one-form ω, on a smooth manifold M is a mapping that assigns to each point $p \in M$ a tangent covector $\omega(p)$ in $T_p^* M$.*

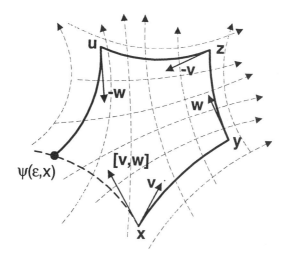

Figure 3.7: The commutation properties of the Lie bracket are illustrated in this diagram.

With any smooth function $\lambda : M \to R$ one can associate a covector field $d\lambda$. On the other hand it is not every covector field can be expressed as the differential of a scalar function.

Definition 3.70 *A covector field Ω on the manifold M is said to be* exact *if there exists a smooth, real vaued function $\lambda : M \to R$ such that $\omega = d\lambda$.*

Recall that the Lie bracket can be interpreted as a directional derivative ((3.66)). We can define a directional derivative for covector fields as well. Suppose ω is a smooth covector field and v a smooth vector field on a manifold M. As usual denote the flow generated by v, eminating from $x \in M$ at $t = 0$ by $\Psi(x,t)$. We wish to compute

$$\left. \frac{d\omega(\Psi(x,t))}{dt} \right|_{t=0} \tag{3.38}$$

Proposition 3.71 *Suppose v is a smooth vector field and ω a smooth covector field on M. For each $x \in M$ the derivative exists and in local coordinates is given by*

$$\left. \frac{d\omega(\Psi(x,t))}{dt} \right|_{t=0} = \left[\frac{\partial \omega^T}{\partial x} f \right]^T + \omega \frac{\partial f}{\partial x}$$

Proof: Again, we need to pull back $\omega(\Psi(x,t))$ from $T^*_{\Psi(x,t)}M$ to $T^*_x M$ and compute

$$\left. \frac{d\omega(\Psi(x,t))}{dt} \right|_{t=0} = \lim_{t \to 0} \left[\frac{\delta\Psi(x,t)\,\omega(\Psi(x,t)) - \omega(x)}{t} \right]$$

Now, for small t the pull back is approximated by

$$\delta\Psi(x,t)\,\omega(\Psi(x,t)) \approx \left(\omega(x) + \left[\frac{\partial\omega(x)^T}{\partial x}v(x)t\right]^T\right)(I + v_x(x)t)$$

so that

$$\frac{d\omega\,(\Psi(x,t))}{dt}\bigg|_{t=0} = \lim_{t\to 0}\frac{1}{t}\left\{\left[\frac{\partial\omega(x)^T}{\partial x}v(x)t\right]^T + \omega(x)v_x(x)t + O(t^2)\right\}$$

The result follows.   ∎

Thus, we give the following definition.

Definition 3.72 *Let f be a smooth vector field and ω a smooth covector field on M. The* Lie derivative *of ω along the f is the unique covector field denoted $L_f\omega$ and defined in local coordinates at the point at the point $x \in M$ by*

$$L_f\omega(x) = \left[\frac{\partial\omega^T}{\partial x}f\right]^T + \omega\frac{\partial f}{\partial x}$$

Remark 3.73 (Lie Derivative of a Differential Form) *Suppose $h(x)$ is a scalar function and let the covector field ω be defined by*

$$\omega = dh(x) = \left(\begin{array}{ccc}\frac{\partial h(x)}{\partial x_1} & \cdots & \frac{\partial h(x)}{\partial x_n}\end{array}\right)$$

Using the formula in Definition (3.72) it is straightforward to compute

$$L_f(dh) = f^T(x)\frac{\partial^2 h(x)}{\partial x^2} + \frac{\partial h(x)}{\partial x}\frac{\partial f(x)}{\partial x}$$

On the other hand, recall that the Lie derivative of the scalar function $h(x)$ with respect to a vector field f is

$$L_f h = \frac{\partial h(x)}{\partial x}f(x)$$

We can differentiate to compute

$$d(L_f h) = f^T(x)\frac{\partial^2 h(x)}{\partial x^2} + \frac{\partial h(x)}{\partial x}\frac{\partial f(x)}{\partial x}$$

Hence, we see that

$$d(L_f h) = L_f(dh)$$

3.4 Distributions and the Frobenius Theorem

3.4.1 Distributions

Let $v_1, \ldots, v_r$ denote a set of r vector fields on a manifold M of dimension m. $\Delta(p) = \operatorname{span}\{v_1(p), \ldots, v_r(p)\}$ is a subspace of $TM_p \sim R^m$.

Definition 3.74 *A smooth distribution Δ on M is a map which assigns to each point $p \in M$, a subspace of the tangent space to M at p, $\Delta(p) \subset TM_p$ such that Δ_p is the span of a set of smooth vector fields $v_1, \ldots, v_r$ evaluated at p. We write $\Delta = \operatorname{span}\{v_1, \ldots, v_r\}$. A distribution Δ has dimension $\dim \operatorname{span}\{v_1(p), \ldots, v_r(p)\}$ at p. It is nonsingular (or regular) at point $p \in M$ if there is a neighborhood of p in M on which the dimension of Δ is constant. Otherwise, the point p is a singular point of Δ.*

In general the *basis* vector fields $\{v_1, \ldots, v_r\}$ are not unique. If v is a vector field on M we say that v belongs to a given distribution Δ on M if $v(p) \in \Delta(p), \forall p \in M$. We write $v \in \Delta$. Suppose $U \subseteq M$ is an open set and Δ is of constant dimension r on U. For a smooth vector field $v \in \Delta = \operatorname{span}\{v_1, \ldots, v_r\}$ on U, it follows that there is a set of smooth coefficients, $c_1, \ldots, c_r$ such that

$$v(p) = \sum_{i=1}^{r} c_i(p) v_i(p), \ \forall p \in U \tag{3.39}$$

The notion of an integral curve of a single vector field can be generalized to that of an integral manifold of a set of vector fields or its corresponding distribution.

Definition 3.75 *An integral submanifold of a distribution $\Delta = \operatorname{span}\{v_1, \ldots, v_r\}$ is a submanifold $N \subset M$ such that $TN_p = \Delta(p)$ for each $p \in N$. The distribution Δ or the set of vector fields $\{v_1, \ldots, v_r\}$ is said to be (completely) integrable if through every point $p \in M$ there passes an integral manifold.*

Suppose N is an integral submanifold of $\Delta = \operatorname{span}\{v_1, \ldots, v_r\}$ on M. Then

$$TM_p \supset \Delta = TN_p \tag{3.40}$$

and $\dim(TN_p) = \dim(N)$ at each $p \in N$. But

$$\dim \Delta(p) = \dim \operatorname{span}\{v_1(p), .., v_r(p)\}$$

may vary as p varies throughout M. Thus, not all integral manifolds need be of the same dimension. Moreover, there may be a manifold N of smaller dimension than Δ that is tangent to it in the sense that Δ contains a subset of smooth vector fields that span TN_p at each point $p \in N$. This is the basis for a weaker notion of integral manifolds and integrability that is sometimes employed. For this reason the integral manifolds of the above definition are sometimes called *maximal* integral submanifolds and the terminology completely integrable requires the existence of maximal integral manifolds.

Definition 3.76 *A system of smooth vector fields* $\{v_1, \ldots, v_r\}$ *or the distribution* $\Delta = \text{span}\{v_1, \ldots, v_r\}$ *on M is said to be* in involution *or* involutive *if there exist smooth real valued functions* $c_k^{ij}(p)$, $p \in M$ *and* $i, j, k = 1, \ldots, r$ *such that for each* i, j

$$[v_i, v_j] = \sum_{k=1}^{m} c_{kij} v_k$$

The concept of an involutive distribution is key to many important results. The following Lemma provides a result that will prove extremely useful in applications.

Lemma 3.77 *Suppose* $\Delta = \{v_1(x), \ldots, v_k(x)\}$ *is a smooth, nonsingular and involutive distribution of dimension k on a neighborhood U of x_0 in R^n. Then there exists a smooth integral manifold of Δ, of dimension k, passing through the point x_0. Morover, the manifold is parametrically characterized, locally around x_0, by the mapping:*

$$\phi(s) = \phi_1^{s_1} \circ \phi_2^{s_2} \circ \cdots \circ \phi_k^{s_k}(x_0)$$

where $\phi_i^t(x) = \psi_i(x, t)$ *is the flow generated by the vector field $v_i(x)$ and 'o' denotes composition with respect to x.*

Proof: According to Definition (3.45) we need to show that there exists a neighborhood V of the origin in R^n such that

(i) $\phi(s)$ is a smooth one-to-one map on V

(ii) $\phi(s)$ satisfies the maximal rank condition for each $s \in V$

(iii) the manifold $\phi(V)$ is an integral manifold of Δ, i.e., each $\partial\phi(x)/\partial s_i \in \Delta(x)$, $i = 1, \ldots, k$.

Notice that the mapping is well defined on a neighborhood of the origin of R^k because each flow function $\phi_i^t(x0)$ is defined for sufficiently small t. Now, use the chain rule to compute

$$\begin{aligned}
\frac{\partial\phi}{\partial s_i} &= \frac{\partial\phi_1^{s_1}}{\partial x} \cdots \frac{\partial\phi_{i-1}^{s_{i-1}}}{\partial x} \frac{\partial}{\partial s_i} \left(\phi_i^{s_i} \circ \cdots \circ \phi_k^{s_k}(x_0)\right) \\
&= \frac{\partial\phi_1^{s_1}}{\partial x} \cdots \frac{\partial\phi_{i-1}^{s_{i-1}}}{\partial x} v_i \left(\phi_i^{s_i} \circ \cdots \circ \phi_k^{s_k}(x_0)\right)
\end{aligned}$$

In particular, at $s = 0$ we have $\phi_i^0 \circ \cdots \circ \phi_k^0(x_0) = x_0$, for each i, including $\phi(0) = x_0$, so that

$$\frac{\partial \phi(0)}{\partial s_i} = v_i(x_0)$$

Since the tangent vectors $v_i(x0)$, $i = 1, \ldots, k$ are independent, the mapping ϕ has rank k at $s = 0$. This establishes (i).

Notice that the point $x_i = \phi_i^{s_i} \circ \cdots \phi_n^{s_n}(x_0)$ reached by propogating forward from x_0 can also be reached by propogating backward from x, i.e., $x_i = \phi_{i-1}^{-s_{i-1}} \circ \cdots \phi_1^{-s_1}(x)$ so that

$$\frac{\partial \phi}{\partial s_i}(x) = (\phi_1^{s_1})_* \cdots (\phi_{i-1}^{s_{i-1}})_* v_i \left(\phi_{i-1}^{-s_{i-1}} \circ \cdots \phi_1^{-s_1}(x) \right)$$

where $x = \phi(s)$. In view of the fact that vectors $\partial \phi(0)/\partial s_i$ are linearly independent, for sufficiently small s, so are the vectors $\partial \phi(x)/\partial s_i$. What remains to be shown is that each $\partial \phi(x)/\partial s_i \in \Delta(x)$, $i = 1, \ldots, k$. This calculation is given in [7, p27]. ∎

Example 3.78 (Parametric from Implicit Manifold) *One useful application of Lemma (3.77) is the development of a parametic representation of a manifold from an implicit representation. Consider the mapping $F : R^k \to R^n$, $k < n$ and suppose* rank $\partial F/\partial x = k$ *on the set*

$$M = \{x \in R^n \,|\, F(x) = 0\}$$

Then M is a regular manifold of dimension $n - k$. It follows that

$$\Delta(x) = \ker \frac{\partial F(x)}{\partial x}$$

is a nonsingular, involutive distribution of dimension $n - k$. M is an integral manifold of Δ. Let $v_1(x), \ldots, v_{n-k}(x)$ be a set of basis vector fields for Δ and suppose $x_0 \in M$. Now, let $\phi_i^{s_i}(x)$ denote the flow corresponding to the vector field $v_i(x)$. The a parametric representation of the M is the map $\phi : R^{n-k} \to R^n$ given by $\phi(s) = \phi_1^{s_1} \circ \cdots \phi_{n-k}^{s_{n-k}}(x_0)$.

To compute the parametric $\phi(s)$ given $F(x)$ requires the following procedure. The calculations involve four steps:

(i) *compute the Jacobian DF,*

(ii) *generate a smooth basis set for $\ker DF$,*

(iii) *compute the flow functions (a local parameterization can be based on the exponential map),*

(iv) *form the composition.*

These steps have been implemented in the ProPac function:

```
ParametricManifold[f,x,x0,n].
```

The following illustration shows that the calculations – even though local, because of the use of the exponential map – capture interesting characteristics of the surface.

In[38]:= $F = \{x3^3 + x2 * x3 + x1\};$

In[39]:= $Surf = ParametricManifold[F, \{x1, x2, x3\}, \{0, 0, 0\}, 3]$

2 vector fields computed.

2 flow functions computed.

Out[39]= $\{\{-k1^3 - k1\ k2, k2, k1\}, \{k1, k2\}\}$

In[40]:= $ParametricPlot3D[Surf[[1]][[\{1, 2, 3\}]], \{k1, -1.5, 1.5\}, \{k2, -2, 2\},$
$PlotPoints- > \{25, 25\},$
$BoxRatios- > \{1, 1, 0.3\},$
$AxesEdge- > \{None, None, None\},$
$ViewPoint- > \{0.25, -1, 0.5\},$
$Boxed- > False]$

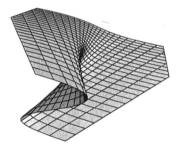

We are now in a position to state the key result of this section. One formulation of the original Frobenius theorem is as follows.

Proposition 3.79 *Let* $\{v_1, \ldots, v_r\}$ *be a nonsingular set of vector fields with*

$$\dim \text{span}\{v_1, \ldots, v_r\} = k$$

on M. *Then the set of vector fields or the distribution* $\Delta = \text{span}\{v_1, \ldots, v_r\}$ *is integrable with all integral manifolds of dimension* k *if and only if it is involutive.*

Proof: Sufficiency is established by Lemma (3.77). Necessity is proved as follows. Suppose x_0 is a point on an integral manifold M of Δ. Then there exists a neigborhood U of x_0 and a mapping $F: R^n \to R^{n-k}$ such that, around x_0, M is defined by

$$M = \{x \in U \,|\, F(x) = 0\}$$

Since M is an integral manifold, by definition we have

$$\frac{\partial F}{\partial x} v_i = 0, \quad i = 1, \ldots, k$$

or equivalently,

$$L_{v_i} F_j(x) = 0, \quad i = 1, \ldots, k, \ j = 1, \ldots, n-k, \ x \in U$$

Now, compute the Lie derivative of F_j along the vector field $[v_j, v_i]$.

$$L_{[v_j, v_i]} F_j(x) = L_{v_j} L_{v_i} F_j(x) - L_{v_i} L_{v_j} F_j(x) = 0$$

Thus, we have

$$\frac{\partial F}{\partial x} [v_i, v_j] (x) = 0$$

But the ker $\partial F / \partial x$ is spanned by the vectors $v_1(x), \ldots, v_k(x)$. Thus, we conclude that $[v_j, v_i] \in \Delta$ for $i, j = 1, \ldots, k$ so that Δ is involutive. ∎

These integral manifolds allow a partition of M into submanifolds of dimension k. The set of all integral manifolds is called a *foliation* of the manifold M, and the integral manifolds are the *leaves* of the foliation.

A stronger version of this theorem is due to Hermann.

Proposition 3.80 *Let* $\Delta = \mathrm{span}\{v_1, \ldots, v_r\}$ *be smooth distribution on* M. *Then the system is integrable if and only if it is in involution.*

Proof: See [5] ∎

This theorem provides necessary and sufficient conditions for integrability. Once again, the manifold M is filled with integral submanifolds. However, the integral submanifolds need not be of the same dimension. A more complete discussion of the Frobenius theorem and its implications can be found in [6] or [3].

Example 3.81 (Foliation of a Singular Distribution) *An example of an integrable, but singular, distribution is the following. Let* $M = R^3$ *and consider the distribution* $\Delta = \mathrm{span}\{v, w\}$ *with*

$$v = \begin{bmatrix} -y \\ x \\ 0 \end{bmatrix}, w = \begin{bmatrix} 2zx \\ 2yz \\ z^2 + 1 - x^2 - y^2 \end{bmatrix}$$

A simple calculation shows that $[v, w] \equiv 0$ *so that the distribution* Δ *is completely integrable. However, the distribution is singular because* $\dim \Delta = 2$ *everywhere except*

1. *on the z-axis* $(x = 0, y = 0)$

2. *on the circle* $x^2 + y^2 = 1, z = 0$

where $\dim \Delta = 1$. *The z-axis and the circle are one-dimensional integral manifolds. All others are the tori:*

$$T_c = \left\{ (x, y, z) \in R^3 \,\middle|\, (x^2 + y^2)^{-1/2}(x^2 + y^2 + z^2 + 1) = c > 2 \right\}$$

Set $r^2 = x^2 + y^2$ *to obtain a representation of the torus in polar coordinates:*

$$\left(r - \frac{c}{2} \right)^2 + z^2 = \left(\frac{c}{2} \right)^2 - 1$$

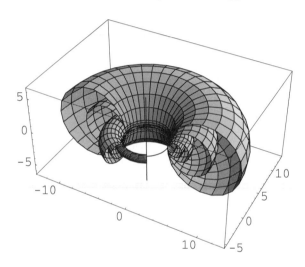

3.4.2   Codistributions

We may work with dual objects to vector fields and distributions. A *covector field* w on M assigns to each point $p \in M$ an element $w(p) \in T^*M_p$. A *codistribution* Ω on M is a mapping which assigns a subspace $\Omega(p)$ of T^*M_p to each point $p \in M$. As with distributions we write $\Omega = \text{span}\{w_1, \ldots, w_r\}$. Distributions are sometimes associated with special codistributions. As an example, for each $p \in M$, the *annihilator* of the distribution $\Delta(p)$ is the set of all covectors which annihilate vectors in $\Delta(p)$

$$\Delta^{\perp}(p) := \{w \in T^*M_p \,|\, w(v) = 0, \ \forall v \in \Delta(p)\} \tag{3.41}$$

It is sometimes more descriptive to denote the annhilator of Δ by annΔ as an alternative to $\Delta^\perp$. Conversely, given a codistribution Ω, we define its *kernel*, a distribution $\Omega^\perp$

$$\Omega^\perp(p) := \{v \in TM_p \,|\, v(w) = 0, \ \forall w \in \Omega(p)\,\} \qquad (3.42)$$

Sometimes we write kerΩ as an alternative to $\Omega^\perp$. It is not difficult to verify that the if p is a regular point of a smooth distribution Δ, then it is a regular point of the codistribution $\Delta^\perp$. Moreover, there is a neighborhood U of p such that $\Delta^\perp$ restricted to U is a smooth codistribution.

Remark 3.82 (Computing with Distributions & Codistributions) *The comments above are consistent with the association of a distribution with a matrix whose columns are its basis vector fields and the association of a codistribution with a matrix whose rows are its covector fields. Then it is possible to do formal pointwise geometric calculations (like projections) with distributions and codistributions using standard constructions from linear algebra. Some elemetary relationships between distributions and codistributions will prove particuarly useful in later calculations. For example, $[\Delta_1 \cap \Delta_2]^\perp = \Delta_1^\perp + \Delta_2^\perp$ provides a convenient way to compute the intersection of two distributions.*

Now, suppose $\Delta = \mathrm{span}\{v_1, \ldots, v_r\}$ is a smooth, involutive, and nonsingular distribution of dimension k on a neighborhood U of $p \in M$. The k-dimensional integral surfaces of Δ can be characterized on U by $m - k$ functions $\lambda_1(x) = c_1, \ldots, \lambda_{m-k}(x) = c_{m-k}$. Moreover, for each $x \in U$, the differentials $d\lambda_1, \ldots, d\lambda_{m-k}$ must must be orthogonal to Δ or equivalently, the codistribution $\Omega = \Delta^\perp$ is spanned by the exact covectors $d\lambda_1, \ldots, d\lambda_{m-k}$. These observations lead to another version of the classical Frobenius theorem.

Proposition 3.83 *Suppose $\Delta = \mathrm{span}\{v_1, \ldots, v_r\}$ is a smooth, involutive, and nonsingular distribution of dimension k on a neighborhood U of $p \in M$. Then there exist a set of functions $\lambda_1(x), \ldots, \lambda_{m-k}(x)$ on U that satify the first order partial differential equations*

$$\begin{bmatrix} \frac{\partial \lambda_1}{\partial x} \\ \vdots \\ \frac{\partial \lambda_{m-k}}{\partial x} \end{bmatrix} \begin{bmatrix} v_1 & \cdots & v_k \end{bmatrix} = 0$$

The ideas embodied in the Frobenius theorem will prove to be fundamental to the study of nonlinear control systems. The integral surfaces implied by the theorem set up a natural coordinate system that will be used below to study controllability and observability. For the moment, however, consider the following problem of finding a coordinate system 'matched' to a given distribution. Suppose $\Delta = \mathrm{span}\{v_1, \ldots, v_r\}$ is a smooth, nonsingular, involutive distribution of dimension k on some neighborhood U of a point x_0 in an m-dimensional manifold M. Assume that the distribution is characterized by a set of local coordinates, x. Then there are k-dimensional integral surfaces that form a foliation of U. Now, we wish to choose local coordinates in U, k of which locate points within these surfaces and the remaining $m - k$ coordinates identify the surface.

We can characterize the integral surfaces in two ways. First, adjoin to the given k vector fields an additional $m - k$ vector fields $v_{k+1}, \ldots, v_m$ such that

$$\text{span}\{v_1, \ldots, v_m\} = R^m \tag{3.43}$$

Let $\psi_i(t, x) = \psi_i^t(x)$ denote the flow generated by the vector field v_i. Then the composition

$$\Psi(z_1, \ldots, z_m) = \psi_1^{z_1} \circ \psi_2^{z_2} \circ \cdots \circ \psi_m^{z_m}(x_0) \tag{3.44}$$

defines a coordinate transformation $x = \Psi(z)$. The new coordinates are the flow lines associated with the m vector fields. In fact, $\Psi(z)|_{z_{k+1}=0,\ldots,z_m=0}$ is a parametric repesentation of the integral surface passing through the point x_0. Other leaves of the foliation are obtained by setting $z_{k+1} = c_{k+1}, \ldots, z_m = c_m$ where $c_{k+1}, \ldots, c_m$ are constants. Thus, the integral surfaces are naturally identified in these new coordinates.

Another characterization of the integral surfaces can be obtained by identifying the functions $\lambda_1, \ldots, \lambda_{n-k}$ of Theorem (3.83). Let $\Phi(x)$ denote the inverse coordinate transformation, i.e., $z = \Psi^{-1}(x) = \Phi(x)$. Then

$$\lambda_1(x) = \Phi_{k+1}(x), \ldots, \lambda_{m-k}(x) = \Phi_m(x) \tag{3.45}$$

Note that

$$\lambda_1(x) = c_1, \ldots, \lambda_{m-k}(x) = c_{m-k} \tag{3.46}$$

provide implicit representations of the integral surfaces. Choosing the constants $c_1 = 0, \ldots, c_{m-k} = 0$ produces the surface passing through x_0.

Example 3.84 *Consider the following example from Isidori ([7], Example 1.4.3). The given distribution*

$$\Delta = \text{span}\{v_1, v_2\}$$

is involutive. We add the vector field v_3 :

$$v_1 = \begin{bmatrix} 2x_3 \\ -1 \\ 0 \end{bmatrix}, \quad v_2 = \begin{bmatrix} -x_1 \\ -2x_2 \\ x_3 \end{bmatrix}, \quad v_3 = \begin{bmatrix} 1 \\ 0 \\ 0 \end{bmatrix}$$

and compute a new coordinate system as described in the previous paragraphs.

In[41]:= v1 = {2 x3, −1, 0}; v2 = {−x1, −2x2, x3}; v3 = {1, 0, 0};

First, let us check involutivity.

In[42]:= Involutive[{v1, v2}, {x1, x2, x3}]

Out[42]= True

Now, check to insure that v_3 as specified does indeed complete the set.

In[43]:= $Span[\{v1, v2, v3\}]$
Out[43]= $\{\{1, 0, 0\}, \{0, 1, 0\}, \{0, 0, 1\}\}$

The Mathematica function DSolve *to compute the flows. To do so, we need to convert the vector fields to ordinary differential equations in the form that Mathematica requires. The ProPac function* MakeODEs *does this.*

In[44]:= $Eqnf1 = MakeODEs[\{x1, x2, x3\}, v1, t]$

$Eqnf2 = MakeODEs[\{x1, x2, x3\}, v2, t]$

$Eqnf3 = MakeODEs[\{x1, x2, x3\}, v3, t]$
Out[44]= $BoxData(\{-2\ x3[t] + x1'[t] == 0, 1 + x2'[t] == 0, x3'[t] == 0\})$
Out[44]= $BoxData(\{x1[t] + x1'[t] == 0, 2\ x2[t] + x2'[t] == 0, -x3[t] + x3'[t] == 0\})$
Out[44]= $BoxData(\{-1 + x1'[t] == 0, x2'[t] == 0, x3'[t] == 0\})$

In[45]:= $sols1 = DSolve[Join[Eqnf1, \{x1[0] == y1, x2[0] == y2, x3[0] == y3\}],$
$\quad \{x1[t], x2[t], x3[t]\}, t];$
$sols2 = DSolve[Join[Eqnf2, \{x1[0] == y1, x2[0] == y2, x3[0] == y3\}],$
$\quad \{x1[t], x2[t], x3[t]\}, t];$
$sols3 = DSolve[Join[Eqnf3, \{x1[0] == y1, x2[0] == y2, x3[0] == y3\}],$
$\quad \{x1[t], x2[t], x3[t]\}, t];$

In[46]:= $psi1 = \{x1[t], x2[t], x3[t]\}/.sols1[[1]];$
$psi2 = \{x1[t], x2[t], x3[t]\}/.sols2[[1]];$
$psi3 = \{x1[t], x2[t], x3[t]\}/.sols3[[1]];$

The transformation is obtained via Equation (3.44) using the ProPac function FlowCompositon.

In[47]:= $Psi = FlowComposition[\{psi3, psi2, psi1\}, t, \{y1, y2, y3\},$
$\quad \{0, 0, 1\}, \{z3, z2, z1\}, \infty]$
Out[47]= $\left\{2\ e^{z2}\ z1 + e^{-z2}\ z3, -z1, e^{z2}\right\}$

The Mathematica functon Solve *is used to obtain the inverse transformaton.*

In[48]:= $Trans = Inner[Equal, \{x1, x2, x3\}, Psi, List]$
Out[48]= $\left\{x1 == 2\ e^{z2}\ z1 + e^{-z2}\ z3, x2 == -z1, x3 == e^{z2}\right\}$

In[49]:= $InvTrans = Solve[Trans, \{z1, z2, z3\}]$
Out[49]= $\{\{z3 \rightarrow x3\ (x1 + 2\ x2\ x3), z1 \rightarrow -x2, z2 \rightarrow Log[x3]\}\}$

The λ functions of Theorem (3.83) are obtained using Equation (3.45).

In[50]:= $\lambda = z3/.InvTrans$
Out[50]= $\{x3\ (x1 + 2\ x2\ x3)\}$

We easily confirm the conclusion of Theorem (3.83).

$In[51]:=$ $Jacob[\lambda, \{x1, x2, x3\}]$

$Out[51]=$ $\left\{\left\{x3, 2\ x3^2, x1 + 4\ x2\ x3\right\}\right\}$

$In[52]:=$ $Simplify[Jacob[\lambda, \{x1, x2, x3\}].Transpose[\{v1, v2\}]]$

$Out[52]=$ $\{\{0, 0\}\}$

3.4.3 Invariant Distributions

The importance of invariant subspaces in linear control theory is well known. For example, controllability, observability and modal subspaces all have a distinctive place in linear systems analysis. A corresponding role in nonlinear control theory is played by invariant distributions.

Definition 3.85 *A distribution* $\Delta = \mathrm{span}\{v_1, \dots, v_r\}$ *on* M *is invariant with respect to a vector field* f *on* M *if the Lie bracket* $[f, v_i]$, *for each* $i = 1, \dots, r$ *is a vector field of* Δ.

We will use the notation $[f, \Delta] = \mathrm{span}\{[f, v_i], i = 1, \dots, r\}$ so that Δ is invariant with respect to f may be stated $[f, \Delta] \subset \Delta$. Observe that in general

$$\Delta + [f, \Delta] \;=\; \Delta + \mathrm{span}\{[f, v_i],\ i = 1, .., r\ \} \;=\; \mathrm{span}\{v_1, .., v_r, [f, v_1], .., [f, v_r]\}$$

Example 3.86 (Invariant Linear Subspaces) *It is easily demonstrated that the notion of an invariant distribution is a natural generalization of the concept of an invariant linear subspace. Consider a subspace* $V = \mathrm{span}\{v_1, \dots, v_r\}$ *of* R^n, *where* $v_i \in R^n$, $i = 1, \dots, r$, *that is invariant under the linear mapping* A, *i.e.,* $AV \subset V$. *Define a distribution on* R^n

$$\Delta_V(x) = \mathrm{span}\{v_1, \dots, v_r\}$$

and a vector field

$$f_A(x) = Ax$$

at each $x \in R^n$. *We will prove that* Δ_V *is invariant under the vector field* f_A. *To do so, we need only show that* $[f_A, v_i] \in \Delta_V$ *for* $i = 1, \dots, r$. *Compute*

$$[f_A, v_i] = \frac{\partial v_i}{\partial x} f_A - \frac{\partial f_A}{\partial x} v_i = -Av_i$$

By assumption Av_i *is a vector of* $V = \mathrm{span}\{v_1, \dots, v_r\}$.

3.4.4 Transformation of Vector Fields

When a distribution Δ is integrable, invariance with respect to a vector field f takes on special significance. If N is an integral manifold of Δ, then the integral curve of f emanating from $p \in N$ remains in N. Using this fact, it is possible to construct a coordinate transformation that puts the vector field in a useful (block) triangular form. Such transformations will be employed to investigate controllability and observability of nonlinear systems as well as to establish feedback linearizaton methods for control system design. The main idea is established in the following lemma.

Lemma 3.87 *Let Δ be a be an involutive distribution of constant dimension d on an open subset U of R^n and suppose that Δ is invariant under a vector field f. Then at each point $x_0 \in U$ there exists a neighborhood U_0 of x_0 in U and a coordinate transformation $z = \Phi(x)$, defined on U_0, in which the vector field f is of the form*

$$\bar{f}(z) = \begin{bmatrix} f_1(z_1, \ldots z_d, z_{d+1}, \ldots, z_n) \\ \cdots \\ f_d(z_1, \ldots z_d, z_{d+1}, \ldots, z_n) \\ f_{d+1}(z_{d+1}, \ldots, z_n) \\ \cdots \\ f_n(z_{d+1}, \ldots, z_n) \end{bmatrix}$$

Proof: Δ is integrable because it is of constant dimension d and involutive. Thus, at each point x_0 there is a neighborhood U_0 of x_0 such that an integral manifold of Δ passes through each point $x \in U_0$. This implies that there exists a transformation of coordinates $z = \Phi(x)$, defined on U_0, with the property that

$$\text{span}\{d\Phi_{d+1}, \ldots, d\Phi_n\} = \Delta^\perp$$

i.e., the first d coordinates, are in the integral manifolds and the remaining $n - d$ are orthogonal to the integral manifolds.

Let $\bar{f}(z)$ denote the representation of f in the new coordinates. Define a family of vector fields that define bases for the tangent spaces TM_z

$$\tau^i(z) = \begin{bmatrix} \tau_1^i \\ \vdots \\ \tau_n^i \end{bmatrix}, \quad \tau_k^i = \begin{cases} 0 & k \neq i \\ 1 & k = i \end{cases}$$

Then

$$[\bar{f}, \tau^i] = -\frac{\partial \bar{f}}{\partial z}\tau^i = \frac{\partial \bar{f}}{\partial z_i}$$

Moreover, for $1 \leq i \leq d$, the vector field $\tau^i \in \Delta$. In fact, these vector fields form a basis for Δ. By construction, in the new coordinates every vector field of Δ has the property that the last $n - d$ coordinates vanish. Since Δ is invariant with respect to f, we have $[\bar{f}, \tau^i] \in \Delta$, so that its last $n - d$ components vanish in the new coordinates. Thus,

$$\frac{\partial \bar{f}_k}{\partial z_i} = 0$$

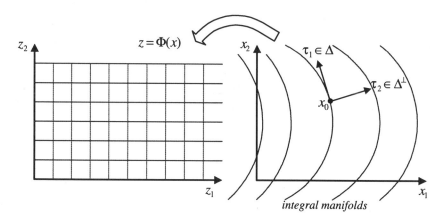

Figure 3.8: The transformation establishes new coordinates, d coordinates in the d-dimensional integral manifolds and $n - d$ orthogonal to them.

for all $d + 1 \leq k \leq n$ and $1 \leq i \leq d$.   ∎

The transformation of Lemma (3.87) is depicted in Figure (3.8) This result implies that under the stated conditions on $f(x)$, the dynamical system

$$\dot{x} = f(x)$$

can be locally represented by the triangular decomposition

$$\dot{z}_1 = f_1(z_1, z_2)$$

$$\dot{z}_2 = f_2(z_2)$$

Let $c \in U_0$ and notice that the set $S_c := \{x \in U_0 |\ z_2(x) = z_2(c)\}$ is a 'slice' (submanifold) of the neighborhood U_0 of dimension d passing through the point $c \in U_0$. Because of the triangular decomposition it is clear that the flow $f(x)$ carries slices into slices. This follows from the observation that all trajectories starting in S_c terminate in $S_{e^{\varepsilon f}c}$ after ε time units. If $z_2(c)$ satisfies $f_2(z_2(c)) = 0$, i.e., $z_2(c)$ is an equilibrium point of the second equation, then S_c is invariant with respect to the flow in the sense that any trajectory beginning in S_c remains therein at least until it leaves the neighborhood U_0.

Example 3.88 *Here is another example taken from Isidori ([7], Example 1.6.4). The distribution $\Delta = \mathrm{span}\{v_1, v_2\}$ on R^4 with*

$$v_1 = \begin{bmatrix} 1 \\ 0 \\ 0 \\ x_2 \end{bmatrix}, \quad v_2 = \begin{bmatrix} 0 \\ 1 \\ 0 \\ x_1 \end{bmatrix}$$

is easily shown to be involutive and invariant with repect to the vector field

$$f = \begin{bmatrix} x_2 \\ x_3 \\ x_3 x_4 - x_1 x_2 x_3 \\ \sin x_3 + x_2^2 + x_1 x_3 \end{bmatrix}$$

In[53]:= $v1 = \{1, 0, 0, x2\}; v2 = \{0, 1, 0, x1\};$
$\quad f = \{x2, x3, x3\ x4 - x1\ x2\ x3, \sin[x3] + x2\hat{}2 + x1\ x3\};$

In[54]:= $LieBracket[f, v1, \{x1, x2, x3, x4\}]$
Out[54]= $\{0, 0, 0, 0\}$

In[55]:= $LieBracket[f, v2, \{x1, x2, x3, x4\}]$
Out[55]= $\{-1, 0, 0, -x2\}$

Now, to obtain the new coordinate system we augment the basis fields of Δ with v_3, v_4:

In[56]:= $v3 = \{0, 0, 1, 0\}; v4 = \{0, 0, 0, 1\};$

and confirm that the expanded set does span R^4.

In[57]:= $Span[\{v1, v2, v3, v4\}]$
Out[57]= $\{\{1, 0, 0, 0\}, \{0, 1, 0, 0\}, \{0, 0, 1, 0\}, \{0, 0, 0, 1\}\}$

The ProPac function `TriangularDecomposition` *implements the procedure illustrated in Example (3.84) to obtain the required transformation and its inverse and applies it to the vector field f.*

In[58]:= $TriangularDecomposition[f, \{v1, v2, v3, v4\}, \{x1, x2, x3, x4\},$
$\quad\quad \{0, 0, 0, 0\}, \infty]$
Out[58]= $\{\{z1, z2, z3, z1\ z2 + z4\}, \{x1, x2, x3, -x1\ x2 + x4\},$
$\quad\quad \{z2, z3, z3\ z4, \operatorname{Sin}[z3]\}\}$

Note that as required the last two elements of the transformed field only depend on z_3, z_4.

3.4.5 Involutive Closure

In this section we describe two algorithms for computing distributions of fundamental importance to the subsequent discussion. We provide a description of them and summarize their essential properties. A more complete discussion can be found in [7].

When working with distributions, a fundamental problem is to find the 'smallest' distribution with the following properties:

1. it is nonsingular

2. it contains a given distribution Δ,

3. it is involutive,

4. it is invariant with respect to a given set of vector fields, $\tau_1, \ldots, \tau_q$.

First, let us establish the concept of a smallest distribution.

Definition 3.89 *Suppose $\mathcal{D}$ is a set of distributions on U. Then the* smallest *or* minimal *element in $\mathcal{D}$, if it exists, is the member of $\mathcal{D}$ that is contained in every other member. The* largest *or* maximal *element, if it exists, is the member that contains every other member.*

The following Lemma is given by Isidori [7].

Lemma 3.90 *Let Δ be a given smooth distribution and $\tau_1, \ldots, \tau_q$ a given set of smooth vector fields. The family of all distributions that are invariant with respect to $\tau_1, \ldots, \tau_q$ and contains Δ contains a minimal element and it is smooth.*

This distribution is denoted $\langle \tau_1, \ldots, \tau_q \mid \Delta \rangle$. An algorithm for finding will now be described. It proceeds by defining a nondecreasing set of distributions:

Algorithm 3.91

$$\begin{aligned} \Delta_0 &= \Delta \\ \Delta_k &= \Delta_{k-1} + \textstyle\sum_{i=1}^{q}[\tau_i, \Delta_{k-1}] \end{aligned} \qquad (3.47)$$

The essential properties of the sequence of distribution so generated are given by the following Lemma.

Lemma 3.92 *The distributions Δ_k generated by Algorithm (3.91) are such that*

$$\Delta_k \subset \langle \tau_1, \ldots, \tau_q \mid \Delta \rangle$$

for all k. If there exists an integer k^* such that $\Delta_{k^*} = \Delta_{k^*+1}$, then

$$\Delta_{k^*} = \langle \tau_1, \ldots, \tau_q \mid \Delta \rangle$$

Thus, Algorithm (3.91) produces a distribution that is invariant with respect to the given vector fields. Now, we give conditions under which it is also involutive.

Lemma 3.93 *Suppose Δ is spanned by a subset of the vector fields $\tau_1, \ldots, \tau_q$ and that $\Delta_{k^*} = \langle \tau_1, \ldots, \tau_q \mid \Delta \rangle$ is nonsingular on U. Then $\langle \tau_1, \ldots, \tau_q \mid \Delta \rangle$ is involutive on U.*

Definition 3.94 *The involutive closure of a given distribution Δ is the smallest involutive distribution containing Δ.*

It is obvious how Algorithm (3.91) can be used to compute the involutive closure of a given distribution.

The dual computation of finding the 'largest' distribution with the following properties is also important:

1. it is nonsingular

2. it is contained within a given distribuion Δ,

3. it is involutive,

4. it is invariant with respect to a given set of vector fields, $\tau_1, \ldots, \tau_q$.

The existence of a distribution with these properties implies the existence of a codistribution (namely, its annihilator) with the following properties:

1. it is nonsingular

2. it contains the given codistribution $\Delta^{\perp}$,

3. it is spanned locally around each $x \in U$ by a set of exact covector fields,

4. it is invariant with respect to a given set of vector fields, $\tau_1, \ldots, \tau_q$.

Thus, we seek the 'smallest' codistribution with these properties.

Lemma 3.95 *Let Ω be a given smooth distribution and $\tau_1, \ldots, \tau_q$ a given set of smooth vector fields. The family of all codistributions that are invariant with respect to $\tau_1, \ldots, \tau_q$ and contains Ω contains a minimal element and it is smooth.*

This codistribution is denoted $\langle \tau_1, \ldots, \tau_q \mid \Omega \rangle$. An algorithm for finding it is:

Algorithm 3.96

$$
\begin{aligned}
\Omega_0 &= \Omega \\
\Omega_k &= \Omega_{k-1} + \sum_{i=1}^{q} L_{\tau_i} \Omega_{k-1}
\end{aligned}
\tag{3.48}
$$

Lemma 3.97 *The codistributions Ω_k generated by Algorithm (3.96) are such that*

$$
\Omega_k \subset \langle \tau_1, \ldots, \tau_q \mid \Omega \rangle
$$

for all k. If there exists an integer k^* such that $\Omega_{k^*} = \Omega_{k^*+1}$, then

$$
\Omega_{k^*} = \langle \tau_1, \ldots, \tau_q \mid \Omega \rangle .
$$

Lemma 3.98 *Suppose Ω is spanned by a set $d\lambda_1, \ldots, d\lambda_s$ of exact covector fields and that $\langle \tau_1, \ldots, \tau_q \mid \Omega \rangle$. is nonsingular. Then $\langle \tau_1, \ldots, \tau_q \mid \Omega \rangle^\perp$ is involutive.*

These two algorithms have been implemented in the *ProPac* functons

1. `SmallestInvariantDistribution`, and

2. `LargestInvariantDistribution`.

Examples of their use will be deferred until Chapter 6.

3.5   Lie Groups and Algebras

The concepts of a linear vector space and its linear subspaces is central to the study of linear systems. As we have suggested, the appropriate generalization of the geometric structure of these objects is achieved by introducing manifolds, tangent spaces and distributions (and their integral submanifolds). However, linear vector spaces also have an important algebraic structure. By introducing an algebraic structure to manifolds we will make the transition: Manifolds $\rightarrow$ Lie groups and distributions $\rightarrow$ Lie (sub)algebras.

Lie groups and Lie algebras play an important role in mechanics and nonlinear control. In the following paragraphs we give a brief summary of the relevant concepts. Our goal here is simply to introduce essential terminology and notation and to provide some elementary examples. The interested reader should consult the many excellent references for more details.

Definition 3.99 *A group is a set G with a group operation (called multiplication) $m: G \times G \rightarrow G$, $m = g \cdot h$ for $g, h \in G$, having the following properties:*

1. if $g, h \in G$, then $m = g \cdot h$

2. associativity: if $g, h, k \in G$

$$g \cdot (h \cdot k) = (g \cdot h) \cdot k$$

3. identity element. There is an element $e \in G$ such that

$$e \cdot g = g = g \cdot e \forall g \in G$$

4. inverse. For each $g \in G$ there is an inverse denoted g^{-1} with the property

$$g \cdot g^{-1} = e = g^{-1} \cdot g$$

Example 3.100 (Groups) *1. $G = Z$, the set of integers with scalar addition the group operation:*

$$e = 0, g^{-1} = -g, \forall g \in Z$$

2. $G = R$, the real numbers with scalar addition the group operation.

3. $G = R^+$, the positive real numbers with ordinary scalar multiplication as the group operation.

4. $G = GL(n, Q)$, the set of invertible $n \times n$ matrices with rational numbers for elements and matrix multiplication the group operation.

5. $G = GL(n, R)$, as above but the elements are real numbers.

Definition 3.101 *An r-parameter Lie group is a group G which is also an r-dimensional smooth manifold such that both the group operation, $m : G \times G \to G, m(g, h) = g \cdot h$ for $g, h \in G$, and the inversion, $i : G \to G, i(g) = g^{-1}, g \in G$, are smooth mappings between manifolds.*

Example 3.102 (Lie Groups) *1. $G = R$ with scalar addition as the group operation is a 1-parameter Lie group.*

2. $GL(n, R)$ of invertible matrices with matrix multiplication the group operation is an n^2-parameter Lie group.

3. Let $G = R^r$ with vector addition the group operation. This is an r-parameter Lie group.

4. The set of nonzero complex numbers C^* form a two parameter Lie group under (complex) multiplication.

5. The unit circle $S^1 \subset C^*$ with multiplication induced from C^* is a one parameter Lie group. This is another characterization of $SO(2)$, the group of rotations in the plane.

6. The product $G \times H$ of two Lie groups is a Lie group with the product manifold structure and the direct product group structure, i.e., $(g_1, h_1) \cdot (g_2, h_2) = (g_1 \cdot g_2, h_1 \cdot h_2), g_i \in G, h_i \in H$.

7. Let K be the product manifold $Gl(nR) \times R^n$ and impose a group structure on K by defining group multiplication via $(A, v) \cdot (B, w) = (AB, v + w), A, B \in Gl(n, R)$ and $v, w \in R^n$. Then K is an $n^2 + n$ parameter Lie group. In fact K is the group of affine motions of R^n. If we identify the element (A, v) of K with the transformation $x \to Ax + v$ on R^n, then multiplication in K is the composition of affine motions.

Mappings between groups that preserve the algebraic structure of groups are of central importance:

Definition 3.103 *A map between Lie groups G and H, $\phi : G \to H$, is a (Lie group) homomorphism if ϕ is smooth and $\phi(a \cdot b) = \phi(a) \cdot \phi(b)$ for all $a, b \in G$. If, in addition, ϕ is a diffeomorphism, it is called an isomorphism.*

Lie groups which are isomorphic (connected by an isomorphism) are considered to be equivalent. Thus, for example, the multiplicative Lie group R^+ and the additive Lie group R are isomorphic and therefore equivalent. The isomorphism is $\phi(t) = e^t, t \in R$. Up to an isomorphism there are only two connected one-parameter Lie groups, R and $SO(2)$. Recall that N is a submanifold of M if there exists a parameter space $\bar{N}$ and a smooth one to one map $\phi : \bar{N} \to M$ such that $N = \phi(\bar{N}) \subset M$. Similarly, we can define Lie subgroups by requiring that the map ϕ respect the group operation.

Definition 3.104 *A Lie subgroup H of a Lie group G is a submanifold of G defined by $H = \phi(\bar{H}) \subset G$ in which the parameter space $\bar{H}$ is itself a Lie group and ϕ is a Lie group homomorphism.*

Example 3.105 *If ω is any real number, the submanifold*

$$H = \{ (t, \omega t) \mod 2\pi | \ t \in R \} \subset T^2$$

is a one parameter subgroup of the toroidal group $SO(2) \times SO(2)$. If ω is rational then H is isomorphic to the circle group $SO(2)$ and is closed, regular subgroup. If ω is irrational, then H is isomorphic to the Lie group R and is dense in the torus T^4. This Lie subgroup is not a regular submanifold of T^4.

The example illustrates that a Lie subgroup H of a Lie group G need not be a regular submanifold of G and hence a Lie subgroup need not be a Lie group in and of itself. However, the following is true.

Proposition 3.106 *If G is a Lie group, then the Lie subgroup $H = \phi(\bar{H})$ is a regular submanifold of G, and hence it is itself a Lie group, if and only if H is closed as a subset of G.*

Proof: (Warner [3])   ∎

Thus, rather than prove that H is a regular submanifold of G, it is sufficient to show that H is a closed subset of G in order to assure that H is a regular Lie subgroup, i.e., a Lie group in its own right (Olver [4]). If G is a Lie group there is a set of special vector fields on G which form a finite dimensional vector space called the Lie algebra of G.

Definition 3.107 *Let G be a Lie group. For any $g \in G$, left and right translation (or multiplication) by g are, respectively, the diffeomorphisms $R_g : G \to G$ and $L_g : G \to G$ defined by*

$$R_g(h) = h \cdot g$$

$$L_g(h) = g \cdot h$$

R_g is a diffeomorphism with inverse $R_{g^{-1}} = (R_g)^{-1}$. Note that

$$R_{g^{-1}}(R_g(h)) = R_g(h) \cdot g^{-1} = h \cdot g \cdot g^{-1} = h$$

Similarly, $L_{g^{-1}} = (L_g)^{-1}$.

Definition 3.108 *A vector field v on G is called right-invariant if*

$$dR_g(v(h)) = v(R_g(h)) = v(h \cdot g)$$

for all $g, h \in G$. It is left invariant if

$$dL_g(v(h)) = v(L_g(h)) = v(g \cdot h)$$

If v, w are right (left) invariant vector fields then so is $av + bw$ where a, b are real numbers. Thus, the set of right (left) invariant vector fields forms a vector space. If v, w are right (left) invariant vector fields on G, then so is their Lie bracket $[v, w]$.

$$dR_g([v, w]) = [v, w] \cdot g = [v \cdot g, w \cdot g] = [dR_g(v), dR_g(w)] = [v, w]$$

Example 3.109 (Right and left invariant vector fields) *Here are some examples of right and left invariant vector fields.*

1. *$G = R$. There is one right (or left) invariant vector field (up to a constant multiplier), $v = 1$ ($\mathbf{v} = \partial/\partial x$). To see this note that $R_y(x) = x + y$, for $x, y \in R$. Thus, the differential map is*

$$dR_y(v) = [\partial R_y(x)/\partial x]\, v = v, \ v \in TR_x$$

so that right invariance requires $v(x) = v(x + y)$ for all $x, y \in R$, which implies $v(x) =$ constant. Similarly, $L_y(x) = y + x$ implies $dL_y(v) = v, \ v \in TR_x$ so we arrive at the same conclusion.

2. *$G = R^+$ (the positive real numbers with ordinary scalar multiplication as the group operation). In this case right and left translation are $R_y(x) = xy$ and $L_y(x) = yx$, for $x, y \in R^+$. The corrsponding differential maps are $dR_y(v) = yv = dL_y(v)$ with $y \in R^+$ and $v \in TR_x^+$. Thus, right or left invariance requires that $yv(x) = v(yx)$ for all $x, y \in R^+$. The general solution to this relation is $v(x) = ax, \ a \in R$. Thus, the unique (up to scalar multiplication) right or left invariant vector field on R^+ is the linear vector field $v = x$.*

3. *$G = SO(2)$. The unique right or left invariant vector field is easily verified to be $v(\theta) = 1$ ($\mathbf{v} = \partial/\partial\theta$).*

Lemma 3.110 *The set of right (left) invariant vector fields of a group G is isomorphic to the tangent space to G at its identity element e, TG_e.*

Proof: First we show that any right invariant vector field on G is determined by its value at the identity element e and then that any tangent vector to G at e determines a right invariant vector field. Any right invariant vector field $v(g)$ on G satisfies $dR_g(v(h)) = v(R_g(h))$ for all $g, h \in G$. Since $R_g(e) = g$ for each $g \in G$, we set $h = e$ and obtain

$$v(g) = dR_g(v(e))$$

Conversely, any tangent vector to G at e determines a right invariant vector field by this same formula as we now show. First note that

$$dR_g(v(h)) = dR_g(dR_h(v(e))) = d(R_g \circ R_h)v(e)$$

Since $R_g \cdot R_h(k) = k \cdot h \cdot g$ for any $k \in G$, this leads to

$$dR_g(v(h)) = dR_{h \cdot g}(v(e))$$

By assumption $v(h \cdot g) = dR_{h \cdot g}(v(e))$, so that we reach the conclusion that

$$dR_g(v(h)) = v(R_g(h))$$

Consequently, $v(g) = dR_g(v(e))$ is a right invariant vector field. A similar computation establishes the result for left invariant vector fields. ∎

Definition 3.111 *The Lie algebra of a Lie group G, denoted* **g** *is the vector space of all left (or right) invariant vector fields on G.*

Since each left or right invariant vector field on G is uniquely associated with a vector tangent to G at e, we can identify the Lie algebra **g** of G with the tangent space to G at e, $\mathbf{g} \cong TG_e$. This implies that **g** is a vector space of the same dimension as the underlying Lie group. Moreover, as is convenient, we will view the Lie algebra of a Lie group either as the space of left or right invariant vector fields or as the tangent space to the group at the identity element.

As was done in the above proof, we will find it useful, from time to time, to contruct left and right invariant vector fields on a group G from an element of its Lie algebra **g** – viewed as the tangent space to G at the identity. This is accomplished using the formulas $v(g) = dL_g(\beta)$ or $v(g) = dR_g(\beta)$, $\beta \in \mathbf{g}$, respectively. We emphasize that in this application the differential maps take elements in $\mathbf{g} \cong TG_e \to TG_g$. In local coordinates the Jacobian is evaluated at the identity e.

Example 3.112 (Euclidean Space) *Suppose $G = R^n$ with vector addition the group operation. Then $L_g : R^n \to R^n$ is given by $R_g(x) = g + x$ with*

$x, g \in R^n$. The differential map $dL_g : TR_x^n \to TR_{x+g}^n$ is the identity (for all x and in particular for the identity element $x = 0$), i.e.,

$$\frac{\partial L_g}{\partial x} = I \Rightarrow \tilde{v} = v, \ v \in TR_x^n, \ \tilde{v} \in TR_{g+x}^n$$

Any left invariant vector field $v(x)$ must satisfy $dL_g(v(x)) = v(g + x)$ for all x and g, which in this case reduces to $v(x) = v(g + x)$ for all $x, g \in R^n$. Thus, every left invariant vector field is constant in the direction of g for arbitrary g. Thus, the set of left invariant vector fields, and, hence, the Lie algebra of R^n, is the set of constant vector fields.

Example 3.113 (Rotation Group and its Lie Algebra) Let $SO(3)$ denote the group of rotations of three-dimensional euclidean space. $SO(3)$ represents the configuraton space of a rigid body free to rotate about a fixed point. An element, which we denote by L is a rotation matrix (a real 3×3 matrix with $L^T L = I$). A motion of the rigid body corresponds to a path $L(t)$ in the group. The velocity $\dot{L}(t)$ is a tangent vector to the group at the point $L(t) \in SO(3)$. Recall that left and right translation on $SO(3)$ are the functions $L_A(L) = AL$ and $R_A(L) = LA$, $\forall A, L \in SO(3)$. We can easily compute the differential maps associated with these functions. Consider left translation and suppose $g(t)$ is a path in $SO(3)$ passing through the point L at $t = 0$. Its image under left translation by A is $\bar{g}(t) = Ag(t)$, so that $\dot{\bar{g}}(t) = A\dot{g}(t)$. $dL_A : TSO(3)_L \to TSO(3)_{AL}$ is $dL_A(L) = AL$. Similarly, $dR_A(L) = LA$.

We can translate the velocity vector to the group identity element by left or right translation, thereby identifying two different elements in the Lie algebra **so**(3):

$$\omega_b = dL_{L^T} \dot{L} = L^T \dot{L}, \ \omega_s = dR_{L^T} \dot{L} = \dot{L} L^T$$

Since, $L^T L = I$ implies $\dot{L}^T L + L^T \dot{L} = 0$, ω_b and ω_s are skew-symmetric matrices. Moreover, as we will see in the next chapter, ω_b is simply the angular velocity (as observed) in the body and ω_s is the angular velocity (as observed) in space.

The definition of a Lie algebra need not be based on the a priori reference to an underlying Lie group. In general

Definition 3.114 A Lie algebra is a vector space **g** together with a bilinear operation

$$[\cdot, \cdot] : \mathbf{g} \times \mathbf{g} \to \mathbf{g}$$

called the Lie bracket for **g**, satisfying the following axioms

1. bilinearity

$$[av_1 + bv_2, w] = a[v_1, w] + b[v_2, w]$$
$$[v, aw_1 + bw_2] = a[v, w_1] + b[v, w_2]$$

for $a, b \in R$ and $v, v_1, v_2, w, w_1, w_2 \in \mathbf{g}$.

2. *skew-symmetry*

$$[v, w] = -[w, v]$$

3. *Jacobi identity*

$$[u, [v, w]] + [w, [u, v]] + [v, [w, u]] = 0$$

for all $u, v, w \in \mathbf{g}$.

Notice that in our definition of a Lie algebra $\mathbf{g}$ of a Lie group G, the required bilinear operation occurs naturally and is, in fact, the ordinary Lie bracket of vector fields. The Lie algebra $\mathbf{g}$ of the Lie group G consists of the left invariant vector fields on G. But it has been shown that the Lie bracket of two left invariant vector fields is a left invariant vector field so that the ordinary Lie bracket of vector fields provides a mapping $[\cdot, \cdot] : \mathbf{g} \times \mathbf{g} \to \mathbf{g}$. Moreover, it satisfies the required properties 1), 2) and 3) of the above definition.

Example 3.115 (Lie algebras)   *1. The vector space of smooth vector fields on a manifold M forms a Lie algebra under the Lie bracket operation on vector fields.*

2. *The vector space $\mathbf{gl}(n, R)$ of all $n \times n$ real matrices forms a Lie algebra if we define $[A, B] = AB - BA$.*

3. *R^3 with the vector cross product as the Lie bracket is a Lie algebra.*

Definition 3.116 *A Lie subalgebra $\mathbf{h}$ of a Lie algebra $\mathbf{g}$ is a (vector) subspace of $\mathbf{g}$ which is closed under the Lie bracket, i.e., $[v, w] \in \mathbf{h}$ whenever $v, w \in \mathbf{h}$.*

If H is a Lie subgroup of a Lie group G, any left invariant vector field on H can be extended to a left invariant vector field on G (set $v(g) = dL_g(v(e))$, $g \in G$ and where $v(e) \in TH_e \subset TG_e$ defines the left invariant vector field on H). In this way the Lie algebra $\mathbf{h}$ of H is realized as a subalgebra of $\mathbf{g}$.

Proposition 3.117 *Let G be a Lie group with Lie algebra $\mathbf{g}$. If $H \subset G$ is a Lie subgroup, its Lie algebra $\mathbf{h}$ is a subalgebra of $\mathbf{g}$. Conversely, if $\mathbf{h}$ is any s-dimensional subalgebra of $\mathbf{g}$, there is a unique , connected subgroup H of G with Lie subalgebra $\mathbf{h}$.*

Proof: We outline the basic idea of the proof. If H is a Lie subgroup with of the Lie group G, then there is a common identity element e and TH_e is a subspace of TG_e. Consequently, $\mathbf{h}$ is a subalgebra of $\mathbf{g}$. To prove the converse, note that any basis of $\mathbf{h}$, say $\{v_1, \ldots, v_s\}$, defines a distribution on G. Since $\mathbf{h}$ is a subalgebra, each

$$[v_i, v_j] \in \mathbf{h} \Rightarrow [v_i, v_j] \in \text{span}\{v_1, \ldots, v_s\}$$

and therefore **h** defines an involutive distribution on G. Moreover, at each point $g \in G$, $\{v_1, \ldots, v_s\}$ is a linearly independent set of tangent vectors. Thus, the Frobenius theorem implies that there is an s-dimensional integral submanifold of this distribution passing through every point $g \in G$ and through the identity element e in particular. This is the Lie subgroup H corresponding to **h**. It remains only to verify that the manifold so defined is indeed a group. ∎

Example 3.118 (Lie subalgebras)   *1. Recall that $Gl(n, R) = Gl(n)$ is the set of invertible $n \times n$ matrices with real elements and that it is a group under matrix multiplication. In fact it is a Lie group of dimension n^2 The Lie algebra of $Gl(n)$ is denoted $\mathbf{gl}(n)$. Let H be a subgroup of $Gl(n)$. We wish to characterize its Lie algebra **h** which a subalgebra of $\mathbf{gl}(n)$. We can find $h \cong TH_e$ by looking at the one dimensional subgroups which are contained in H. That is, suppose $a \in \mathbf{gl}(n) \cong TG_e$ so that a is a (right invariant) vector field on G and the maximal integral manifold of a passing through e is $\{e^{\varepsilon a}, \ \varepsilon \in R\}$. Thus,*

$$\mathbf{h} = \{ a \in \mathbf{gl}(n) |\ e^{\varepsilon a} \in H \ \forall \varepsilon \in R \}$$

2. Recall the group of orthogonal matrices $O(n) = \{ X \in Gl(n) |\ X^T X = I \}$. This group is a subgroup of $Gl(n)$ with $\dim O(n) = n(n-1)/2$. Since $\mathbf{gl}(n) \cong TG_e$ we may view the elements of $\mathbf{gl}(n)$ as $n \times n$ matrices and its Lie bracket is then matrix commutation. Let the matrix $A \in \mathbf{gl}(n)$, then $A \in \mathbf{h}$ if and only if

$$\left(e^{\varepsilon A} \right)^T \left(e^{\varepsilon A} \right) = I$$

and this is satisfied if and only if $A^T + A = 0$, i.e., A is antisymmetric.

3. Another subgroup of $Gl(n)$ is the special orthogonal group

$$SO(n) = \{ X \in Gl(n) |\ \det X = 1 \}$$

This Lie group is also of dimension $n(n-1)/2$. It is one of the components of $O(n)$. In fact, it is the connected component of the identity. The Lie algebra of $SO(n)$ is the same as the Lie algebra of $O(n)$ (they have the same tangent space at e): $\mathbf{so}(n) = $ real skew symmetric $n \times n$ matrices

Remark 3.119 (Properties of Lie algebras) *In the following paragraphs, we briefly summarize some useful terminology and elementary properties associated with Lie algebras.*

1. An algebra is the direct sum *of two algebras $\mathbf{a} + \mathbf{b}$ if $\mathbf{g} = \mathbf{a} + \mathbf{b}$ is a vector space and $[\mathbf{a}, \mathbf{b}] = 0$. We then write $\mathbf{g} = \mathbf{a} \oplus \mathbf{b}$. It is the* semi-direct sum *if $[\mathbf{a}, \mathbf{b}] \subset \mathbf{a}$, i.e., if $[w, v] \in \mathbf{a}$ whenever $w \in \mathbf{a}$, $v \in \mathbf{b}$. We then write $\mathbf{g} = \mathbf{a} \oplus_s \mathbf{b}$.*

*2. A subalgebra **h** is an* ideal *of $\mathbf{g}$ if $[\mathbf{h}, \mathbf{g}] \subset \mathbf{h}$. If $\mathbf{g} = \mathbf{a} \oplus_s \mathbf{b}$, then $\mathbf{a}$ is an ideal of $\mathbf{g}$. If $\mathbf{g} = \mathbf{a}_1 \oplus \mathbf{a}_2 \oplus \cdots \oplus \mathbf{a}_n$, then each $\mathbf{a}_i$ is an ideal of $\mathbf{g}$.*

3. *If H is a subgroup of G, we define the following equivalence relation. For $a, b \in G$,*

$$a \equiv b(\mathrm{mod}\, H), \quad if\ a^{-1}b \in H$$

The equivalence classes under this relation are called the left cosets *of H and are denoted aH. Similarly, if we define the relation*

$$a \equiv b(\mathrm{mod}\, H), \quad if\ ab^{-1} \in H$$

The equivalence classes under this relation are called the right cosets *of H and are denoted Ha. H is normal if $aH = Ha$ for all $a \in G$.*

4. *If H is normal, the cosets of H form a group with group operation*

$$(aH) \cdot (bH) = a \cdot bH$$

This group is called the quotient group *and is denoted G/H. Consider the example in Figure (3.9).*

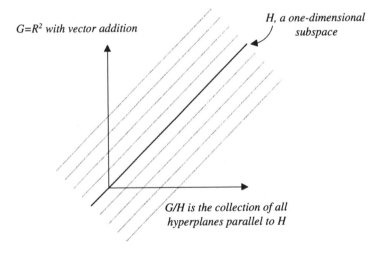

G=R² with vector addition

H, a one-dimensional subspace

G/H is the collection of all hyperplanes parallel to H

Figure 3.9: The quotient group associated with the group $G = R^2$ and its subgroup H, a linear subspace of R^2, is the collection of all translations of H.

5. *Suppose $\mathbf{h}$ is a subalgebra of $\mathbf{g}$. For any $w \in \mathbf{g}$, define the equivalence class of w in $\mathbf{g}$ by the relation*

$$w \equiv v(\mathrm{mod}\, h), \quad w - v \in \mathbf{h}$$

The equivalence class of w so defined is denoted $w + \mathbf{h}$. These equivalence classes form a Lie algebra if $\mathbf{h}$ is an ideal of $\mathbf{g}$. We can define a Lie bracket on the classes

$$[w + \mathbf{h}, v + \mathbf{h}] := [w, v] + \mathbf{h}$$

The set of equivalence classes now forms a new algebra called the quotient algebra , denoted **g/h**.

3.6 Problems

Problem 3.120 *Consider the set of affine vector fields $\mathcal{A}$ of the form $f(x) = Ax + b$, $A \in R^{n \times n}$, $b \in R^n$. Show that $\mathcal{A}$ is closed under the Lie bracket operation, i.e., $[f, g] \in \mathcal{A}$ for all $f, g \in \mathcal{A}$.*

Problem 3.121 *Determine the smallest distribution that is invariant with respect to the vector fields*

$$\tau_1(x) = \begin{bmatrix} 1 \\ x_1 \end{bmatrix}, \; \tau_2(x) = \begin{bmatrix} x_2 \\ x_1 \end{bmatrix}$$

and contains the distribution $\Delta(x) = \text{span} \{\tau_1(x)\}$, i.e., $\langle \tau_1, \tau_2 \, | \Delta \rangle$.

3.7 References

1. Boothby, W.M., An Introduction to Differentiable Manifolds and Riemannian Geometry. 1986, San Diego: Academic Press.

2. Hirsch, M.W., Differential topology. 1976, New York: SpringerVerlag.

3. Warner, F.W., Foundations of Differentiable Manifolds and Lie Groups. 1983, New York: SpringerVerlag.

4. Olver, P.J., Applications of Lie Groups to Differential Equations. 1986, New York: SpringerVerlag.

5. Hermann, R., Cartan Connections and the Equivalence Problem for Geometric Structures. Contributions to Differential Equations, 1964. 3: p. 199-248.

6. Abraham, R., J.E. Marsden, and T. Ratiu, Manifolds, Tensor Analysis, and Applications. 1988, New York: Springer-Verlag.

7. Isidori, A., Nonlinear Control Systems. 3 ed. 1995, London: Springer-Verlag.

Chapter 4

Kinematics of Tree Structures

4.1 Introduction

Multibody mechanical systems often assume the structure of a chain or a tree. Even when they do not (i. e., a system containing a closed loop), it is typically convenient to build a model for an underlying tree (by breaking the loop) and then to add the necessary constraints (to re-establish the loop). In this chapter we focus on the kinematics of tree structures. The next chapter will supplement the present discussion to accommodate constraints.

The systems we consider are composed of rigid bodies[1] connected together by joints. Each joint has a set of velocity variables and configuration parameters[2] equal to the number of degrees of freedom of the joint. The set of all joint velocities defines the (quasi-) velocity vector, p, for the system and the set of all joint parameters comprise the system generalized coordinate vector, q. Our main goal is to assemble the key kinematic equation that relates the quasi-velocities to the coordinate velocities:

$$\dot{q} = V(q)p \qquad (4.1)$$

In addition, we wish to establish formulas that allow the computation of the position, orientation and/or velocity of reference frames at various locations in the system.

We begin in the next section with an analysis of individual joints. The goal

---

[1]We do not discus flexible bodies in this book. However, the methods described here apply with some additional constructs that are implemented in *ProPac* . See *ProPac* help for more information.

[2]At least locally.

is to characterize the motion of an outboard reference frame with respect to an inboard reference frame. Joints are normally defined in terms of constraints on the relative velocity across the joint. Formulas will be derived that provide a natural parameterization of the joint configuraton and all other kinematic quantities. In Section 3 we turn to the kinematics of chain and tree structures. Various formulas are derived that allow the complete characterizaton of chain and tree configuration and velocities in terms of individual joint quantities. Computer implementation of the required calculations are also described.

4.2   Kinematics of Joints

A joint constrains the relative motion between two bodies. In this section we develop a mathematical desription of joints that is convenient for assembling multibody dynamical models.

4.2.1   The Geometry of Joints

We designate two rigid bodies and reference frames fixed within them s (space) and b (body). The configuration space M of relative motion between two unconstrained rigid bodies is the Special Euclidean group $SE(3)$ consisting of all rotations and translations of R^3. $SE(3)$ is the semi–direct product of the rotation group $SO(3)$ with the vector group R^3, [1]. An element in $SE(3)$ may be represented by a matrix

$$X = \begin{bmatrix} L^T & R \\ 0 & 1 \end{bmatrix}, \ L \in SO(3), \ R \in R^3 \tag{4.2}$$

Consider a space reference frame XYZ and a body reference frame xyz. The configuration of the body frame relative to the space frame is X as defined in (4.2). Recall that location of a point at position r in the body has location $\mathbf{R}$ in the space frame with

$$\mathbf{R} = R + L^T r$$

as illustrated in Figure 4.1. The inverse of X is

$$X^{-1} = \begin{bmatrix} L & -R \\ 0 & 1 \end{bmatrix} \tag{4.3}$$

Two successive relative motions X_1 and X_2 combine to yield

$$X = X_2 X_1 = \begin{bmatrix} L_2^T & R_2 \\ 0 & 1 \end{bmatrix} \begin{bmatrix} L_1^T & R_1 \\ 0 & 1 \end{bmatrix} = \begin{bmatrix} L_2^T L_1^T & L_2^T R_1 + R_2 \\ 0 & 1 \end{bmatrix}$$

as illustrated in Figure (4.2).

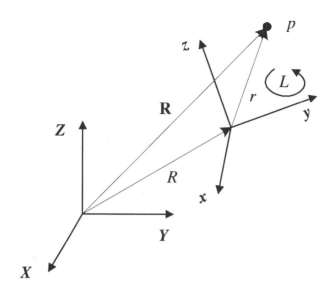

Figure 4.1: Point p can be represented in either the body frame or space frame.

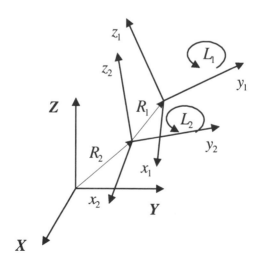

Figure 4.2: Two successive rigid body motions characterized by configuration matrices X_1, X_2 leave the body in configuration $X = X_2 X_1$ with respect to the space frame.

In general geometric terms, a joint is characterized by a relation on the tangent bundle $TSE(3)$. Such a relation is usually expressed in local coordinates by an equation of the type (see, for example [2,3])

$$f(q, \dot{q}) = 0 \qquad (4.4)$$

where $f : TSE(3) \to R^k$. Natural constraints almost always occur on one of two forms:

$$f(q) = 0 \qquad (4.5)$$

in which only the coordinates appear, or

$$F(q)\dot{q} = 0 \qquad (4.6)$$

in which the coordinate velocities appear linearly. Equation (4.5) defines a submanifold of $SE(3)$ which identifies admissible configurations. Constraints of this from are called geometric constraints because they restrict the relative geometry of the two bodies. Constraints of the form (4.6) are called kinematic because they restrict the relative velocity of two bodies. The geometric meaning of (4.6) is highlighted by restating it as

$$\dot{q} \in \Delta(q) \qquad (4.7)$$

where $\Delta(q)$ is a distribution on $SE(3)$ defined as $\Delta(q) = \ker [F(q)]$. If the constraint is of the form of (4.6), then it is holonomic [2, 3] if the distribution $\Delta(q)$ is integrable. General conditions for integrability of a distribution are well known and given by the Frobenius theorem. Recall, from Chapter 3, that local coordinates on TM constitute the pair (q, v) with q local coordinates on M and v local coordinates on TM_x. Thus, in general, TM is isomorphic to $M \times \mathbf{g}$, where $\mathbf{g}$ denotes the Lie algebra associated with M, it is possible to characterize joint constraints which involve velocities (i.e., (4.6)) by a smooth map $f : SE(3) \times \mathbf{se}(3) \to R^k$ so that the joint is defined by equations of the form:

$$A(q)p = 0, \qquad (4.8)$$

where $p \in \mathbf{se}(3)$ and $A(q)$ is a linear operator on $\mathbf{se}(3)$. The geometric meaning of (4.8) is

$$p \in \ker A(q) \qquad (4.9)$$

Equation (4.8) is a more general and will prove to be a more convenient characterization of kinematic joints than (4.6).

Let us take $M = SE(3)$ and consider the formal representation of objects belonging to its Lie algebra $\mathbf{g} = se(3)$. We can use either right or left translations on M to define $\mathbf{g}$. We choose left, so that

$$p := X^{-1}\dot{X} = \begin{bmatrix} L & -R \\ 0 & 1 \end{bmatrix} \begin{bmatrix} \dot{L}^T & \dot{R} \\ 0 & 0 \end{bmatrix} = \begin{bmatrix} L\dot{L}^T & L\dot{R} \\ 0 & 0 \end{bmatrix} = \begin{bmatrix} \widetilde{\omega}_b & v_b \\ 0 & 0 \end{bmatrix} \qquad (4.10)$$

Notice that in (6.47) we use the conventional notation, by which any vector $a \in R^3$ is converted into a skew-symmetric matrix $\tilde{a}(a)$:

$$\tilde{a}(a) = \begin{bmatrix} 0 & -a_3 & a_2 \\ a_3 & 0 & -a_1 \\ -a_2 & a_1 & 0 \end{bmatrix}$$

Thus, we see that $se(3)$ is isomorphic to R^6 and we can consider an element p of $se(3)$ to be a pair of objects– body angular velocity and linear velocity–(ω_b, v_b) or, equivalently, $(\tilde{\omega}_b, v_b)$. When doing formal group calculations however, we use the matrix form shown in (6.47).

4.2.2 Simple Kinematic Joints

Kinematic joints are joints that are described by velocity constraints such as (4.6) or (4.8). They are simple if the motion axes are fixed in (at least) one of the bodies–in which case the constraint can be formulated so that A is a constant (independent of the configuration). For lack of a general terminology we call such joints *simple kinematic joints*. We now focus on simple kinematic joints. It is convenient to define a matrix H whose columns form a basis for $\ker A$ so that

$$\ker A = \text{Im } H, \quad H \text{ is of full rank } r = \dim \ker A. \tag{4.11}$$

Solutions of (4.8) are of the form

$$p = H\beta, \ \beta \in R^r \tag{4.12}$$

β represents the joint quasi–velocity and r is the number of velocity degrees of freedom. H is called the joint map matrix .

Examples of Joint Map Matrices of Simple Joints - H

$$\begin{bmatrix} 0 \\ 0 \\ 1 \\ 0 \\ 0 \\ 0 \end{bmatrix} \quad \begin{bmatrix} 0 & 0 \\ 1 & 0 \\ 0 & 1 \\ 0 & 0 \\ 0 & 0 \\ 0 & 0 \end{bmatrix} \quad \begin{bmatrix} 0 \\ 0 \\ 0 \\ 0 \\ 0 \\ 1 \end{bmatrix} \quad \begin{bmatrix} 0 \\ 0 \\ 1 \\ 0 \\ 0 \\ s \end{bmatrix} \quad \begin{bmatrix} 0 & 0 \\ 0 & 0 \\ 1 & 0 \\ 0 & 0 \\ 0 & 0 \\ 0 & 1 \end{bmatrix}$$

1 *dof*	2 *dof*	1 *dof*	1 *dof*	2 *dof*
revolute	*universal*	*prismatic*	*screw*	*cylindrical*
body z − axis	*body y, z − axis*	*body z − axis*	*body z − axis*	*body z − axis*

The joint configuration is defined, in general, by the differential equations

$$\dot{X} = Xp \tag{4.13}$$

or, equivalently

$$\dot{L} = -\tilde{\omega}_b L, \quad \dot{R} = L^T v_b \qquad (4.14)$$

It is easy enough to replace $\tilde{\omega}_b$ and v_b by β using (4.12). Let H be partitioned so that H_1 contains the first 3 rows and H_2 the second three rows of H, then

$$\dot{X} = X \begin{bmatrix} \tilde{H_1}\beta & H_2\beta \\ 0 & 0 \end{bmatrix} \qquad (4.15)$$

or

$$\dot{L} = -(\tilde{H_1}\beta)\, L, \quad \dot{R} = L^T H_2 \beta. \qquad (4.16)$$

The joint kinematics are defined by (4.15) or (4.16). Given the quasi–velocities β, (4.15) and (4.16) can be integrated to provide the relative translational position and rotation matrix of the two bodies. However, this representation may not be the most informative and it certainly provides more information than necessary since it locates the relative position in the six dimensional group $SE(3)$ instead of the relevant subgroup. If the constraint is holonomic, precisely r dimensions would suffice. First, we provide a result for single degree of freedom joints.

Proposition 4.122 *Consider a simple single degree of freedom joint with joint map matrix $H = h \in R^6$. Then the joint configuration matrix can be parameterized by a parameter $\epsilon \in R$ in the form:*

$$X(\varepsilon) = \begin{bmatrix} L^T(\varepsilon) & R(\varepsilon) \\ 0 & 1 \end{bmatrix} \qquad (4.17)$$

with

$$L(\varepsilon) = e^{-\tilde{h}_1 \varepsilon}, \quad R(\varepsilon) = \int\limits_0^\varepsilon e^{-\tilde{h}_1 \sigma} h_2 d\sigma \qquad (4.18)$$

Proof: [4]. Consider a general one degree of freedom joint in which H is composed of the single column h. Then the distribution $\Delta(X)$ on $SE(3)$ consists of the single vector field

$$\begin{bmatrix} L^T \tilde{h}_1 & L^T h_2 \\ 0 & 0 \end{bmatrix}$$

This is an integrable distribution and we seek the integral manifold which passes through the point

$$X_0 = \begin{bmatrix} I & 0 \\ 0 & 1 \end{bmatrix}$$

The one dimensional manifold we seek can be characterized (at least locally) by a map $\xi : R \to SE(3)$. Let $\epsilon \in R$ be the parameter. Then we seek a solution to the differential equation

$$\frac{d\xi}{d\varepsilon} = \begin{bmatrix} L^T \tilde{h}_1 & L^T h_2 \\ 0 & 0 \end{bmatrix}, \quad \xi(0) = X_0 \qquad (4.19)$$

or equivalently

$$\frac{dL}{d\varepsilon} = -\tilde{h}_1 L, \ L(0) = I \tag{4.20}$$

and

$$\frac{dR}{d\varepsilon} = L^T h_2, \ R(0) = 0 \tag{4.21}$$

so that the conclusion follows. ∎

Note that if H is composed of several columns, say r columns, then we can consider this joint as a sequence of r single column joints and compute $X_i(\varepsilon_i)$ for each joint. Thus, we have

Corollary 4.123 *Consider a simple joint with r degrees of freedom and joint map matrix $H = [h_1 \ldots h_r] \in R^{6 \times r}$, then there is a parameter vector $\varepsilon \in R^r$ and the joint configuration matrix can be expressed in the form*

$$X(\varepsilon) = X_r(\varepsilon_r) \ldots X_2(\varepsilon_2) X_1(\varepsilon_1) \tag{4.22}$$

where each $X_i(\varepsilon_i)$ is of the form of Proposition (4.122) with $h = h_i$.

We conclude that any simple kinematic joint is holonomic and, in fact, we have explicitly computed a local representation of its configuration manifold. Now, any motion results in a velocity $\dot{X} = Xp$. We wish to characterize this relation (locally) in terms of the rate of change of the joint parameters. In other words, we seek to relate $\dot{\varepsilon}$ and β. The following proposition does that.

Proposition 4.124 *Consider a simple joint with joint map matrix H, and suppose the joint is parameterized according to Proposition (4.122) and Corollary (4.123). Then the joint kinematic equation is*

$$\dot{\varepsilon} = V(\varepsilon)\beta \tag{4.23}$$

where $V(\varepsilon)$ is defined by the following algorithm:

1. For $j = 1, ..r$ define and

$$U_j^T(\varepsilon_j, .., \varepsilon_1) = L_j^T(\varepsilon_j) U_{j-1}^T(\varepsilon_{j-1}, .., \varepsilon_1), \ U_0^T = I \tag{4.24}$$

$$\Gamma_j(\varepsilon_j, .., \varepsilon_1) = L_j^T(\varepsilon_j) \Gamma_{j-1}(\varepsilon_{j-1}, .., \varepsilon_1) + R_j, \ \Gamma_0 = 0 \tag{4.25}$$

2. Define $B(\varepsilon)$

$$B(\varepsilon) := \begin{bmatrix} b_{11} & \cdots & b_{1r} \\ b_{21} & \cdots & b_{2r} \end{bmatrix} \tag{4.26}$$

$$\tilde{b}_{1i} := U_{i-1}\tilde{h}_{i1}U_{i-1}^T \tag{4.27}$$

$$b_{2i} := U_{i-1}\tilde{h}_{i1}\Gamma_{i-1} + U_{i-1}h_{i2} \tag{4.28}$$

3. Define $V(\varepsilon)$

$$V(\varepsilon) := B^*(\varepsilon)H, \ B^*(\varepsilon) \ \text{denotes a left inverse of } B(\varepsilon) \qquad (4.29)$$

Proof: [4]. Any motion results in a velocity $\dot{X} = Xp$ which implies

$$\dot{X} = \sum \frac{\partial X}{\partial \varepsilon_i} \dot{\varepsilon}_i = X(\varepsilon)p$$

Now, we directly compute

$$\sum_{i=1}^{r} \frac{\partial X}{\partial \varepsilon_i} \dot{\varepsilon}_i = \sum_{i=1}^{r} \left\{ X_r(\varepsilon_r) \cdots X_{i+1}(\varepsilon_{i+1}) \frac{dX_i}{d\varepsilon_i} X_{i-1}(\varepsilon_{i-1}) \cdots X_1(\varepsilon_1) \dot{\varepsilon}_i \right\}$$

and premultiplying by X^{-1} we obtain

$$\sum_{i=1}^{r} \left\{ [X_{i-1}(\varepsilon_{i-1}) \cdots X_1(\varepsilon_1)]^{-1} X_i^{-1}(\varepsilon_i) \frac{dX_i}{d\varepsilon_i} X_{i-1}(\varepsilon_{i-1}) \cdots X_1(\varepsilon_1) \dot{\varepsilon}_i \right\} = p \tag{4.30}$$

Notice that

$$X_i^{-1}(\varepsilon_i) \frac{dX_i}{d\varepsilon_i} = \begin{bmatrix} \widetilde{h}_{i1} & h_{i2} \\ 0 & 0 \end{bmatrix}$$

Also, define $W_j(\varepsilon_j, \ldots, \varepsilon_1)$, $j = 1, .., r$ by the recursion

$$W_j(\varepsilon_j, \ldots, \varepsilon_1) := X_j(\varepsilon_j) W_{j-1}(\varepsilon_{j-1}, \ldots, \varepsilon_1) \tag{4.31}$$

$$W_1(\varepsilon_1) = X_1(\varepsilon_1) \tag{4.32}$$

so that (4.30) can be written

$$\sum_{i=1}^{r} W_{i-1}^{-1} \begin{bmatrix} \widetilde{h}_{i1} & h_{i2} \\ 0 & 0 \end{bmatrix} W_{i-1} \dot{\varepsilon}_i = p \tag{4.33}$$

We can easily determine, from (4.31), (4.32), that W_j is of the form

$$W_j = \begin{bmatrix} U_j^T(\varepsilon_j, \ldots, \varepsilon_1) & \Gamma_j(\varepsilon_j, \ldots, \varepsilon_1) \\ 0 & 1 \end{bmatrix}$$

with

$$U_j^T(\varepsilon_j, \ldots, \varepsilon_1) = L_j^T(\varepsilon_j) U_{j-1}^T(\varepsilon_{j-1}, \ldots, \varepsilon_1), \ U_0^T = I$$

$$\Gamma_j(\varepsilon_j, \ldots, \varepsilon_1) = L_j^T(\varepsilon_j) \Gamma_{j-1}(\varepsilon_{j-1}, \ldots, \varepsilon_1), \ \Gamma_0 = 0$$

Thus, (4.33) reduces to

$$\sum_{i=1}^{r} \begin{bmatrix} U_{i-1} \widetilde{h}_{i1} U_{i-1}^T & U_{i-1} \widetilde{h}_{i1} \Gamma_{i-1} + U_{i-1} h_{i2} \\ 0 & 0 \end{bmatrix} \dot{\varepsilon}_i = p = \begin{bmatrix} \widetilde{H_1 \beta} & H_2 \beta \\ 0 & 0 \end{bmatrix} \tag{4.34}$$

Each expression of the form $U_{i-1}\widetilde{h}_{i1}U_{i-1}^T$ is an antisymmetric matrix so we can define $b_{1i} \in R^3$ such that

$$\widetilde{b}_{1i} = U_{i-1}\widetilde{h}_{i1}U_{i-1}^T$$

We also define

$$b_{2I} = U_{i-1}\widetilde{h}_{i1}\Gamma_{i-1} + U_{i-1}h_{i2}$$

Then (4.34) can be written

$$B(\varepsilon)\dot{\varepsilon} = H\beta, \ B(\varepsilon) = \begin{bmatrix} b_{11} & \cdots & b_{1r} \\ b_{21} & \cdots & b_{2r} \end{bmatrix}$$

Let B^* denote the left inverse of B-which exists on a neighborhood of $\epsilon = 0$ because $B(0) = H$ is of full rank. Then

$$\dot{\varepsilon} = B^*(\varepsilon)H\beta = V(\varepsilon)\beta, \ V(\varepsilon) := B^*(\varepsilon)H$$

∎

4.2.3 Compound Kinematic Joints

Not all joints are simple kinematic joints. But in many cases it is possible to define the action of a joint in terms of a sequence of simple kinematic joints. We call such joints *compound kinematic joints*. In general, a compound joint is defined as a joint which can be characterized as the relative motion of a sequence of p reference frames such that relative motion between two successive frames is defined by a simple kinematic joint. Then each of the p simple joints is characterized by a joint map matrix H_i with r_i columns, a quasi–velocity vector, β_i, of dimension r_i, a parameter vector, ε_i, of dimension r_i, and a kinematic matrix $\Gamma_i(\varepsilon_i)$. Thus if we define $\varepsilon := [\varepsilon_1 \ldots \varepsilon_p]$ and $\beta := [\beta_1 \ldots \beta_p]$ we have the joint kinematics defined by

$$\dot{\varepsilon} = diag\,[\Gamma_1(\varepsilon_1), \ldots, \Gamma_p(\varepsilon_p)] \tag{4.35}$$

and, assuming the frames are indexed from the outermost, the overall joint configuration matrix is

$$X(\varepsilon) = X_p(\varepsilon_p) \cdots X_2(\varepsilon_2)X_1(\varepsilon_1) \tag{4.36}$$

Equations (4.35) and (4.36) provide the kinematic equations for compound joints. Figure (4.2) may be thought of as depicting a 2-frame compound joint.

Remark 4.125 *In view of equation (4.36), a p-frame compound joint with joint map matrices H_i, $i = 1, \ldots, p$, yields the same configuration manifold parameterization as a simple joint with joint map matrix $H = [H_1 \cdots H_p]$.*

As we will see below, the overall joint map matrix is also required in order to assemble the dynamical equations for multibody systems. The required constructions are provided in the following proposition.

Proposition 4.126 *Consider a compound joint composed of p simple joints with joint map matrices $H_i = [h_1^1 \cdots h_i^{r_i}] \in R^{6 \times r_i}$, $i = 1, \ldots, p$. Suppose $\varepsilon := [\varepsilon_1 \ldots \varepsilon_p]$ and $\beta = [\beta_1 \cdots \beta_p]$ are the corresponding simple joint parameters and quasi–velocities. Then the composite joint map matrix $H(\varepsilon) \in R^{6 \times (r_1 + \cdots + r_p)}$ is given by the following construction:*

$$H(\varepsilon) := \begin{bmatrix} h_{11} & \cdots & h_{1r} \\ h_{21} & \cdots & h_{2r} \end{bmatrix} \tag{4.37}$$

where

$$\tilde{h}_{j1} := U_{i-1} \tilde{h}_{i1}^j U_{i-1}^T$$

$$h_{j2} := U_{i-1} \tilde{h}_{i1}^j \Gamma_{i-1} + U_{i-1} h_{i2}^j \text{ for } i = 1, .., p, \quad j = 1, .., r_i \tag{4.38}$$

$$U_i^T(\varepsilon_i, .., \varepsilon_1) = L_i^T(\varepsilon_i) U_{i-1}^T(\varepsilon_{i-1}, .., \varepsilon_1), \quad U_0^T = I \tag{4.39}$$

$$\Gamma_i(\varepsilon_i, .., \varepsilon_1) = L_i^T(\varepsilon_i) \Gamma_{i-1}(\varepsilon_{i-1}, .., \varepsilon_1) + R_i, \quad \Gamma_0 = 0 \tag{4.40}$$

Proof: [4]. The overall joint velocity is

$$\dot{X} = \sum_{i=1}^{p} \sum_{j=1}^{r_i} \frac{\partial X}{\partial \varepsilon_i^j} \dot{\varepsilon}_i^j = X(\varepsilon) p \tag{4.41}$$

Notice that for each fixed $i > 2$,

$$\sum_{j=1}^{r_i} \frac{\partial X}{\partial \varepsilon_i^j} \dot{\varepsilon}_i^j = X_p(\varepsilon_p) \cdots X_{i+1}(\varepsilon_{i+1}) \left\{ \sum_{j=1}^{r_i} \frac{\partial X}{\partial \varepsilon_i^j} \dot{\varepsilon}_i^j \right\} X_{i-1}(\varepsilon_{i-1}) \cdots X_1(\varepsilon_1) \tag{4.42}$$

But, as computed above for simple joints,

$$\sum_{j=1}^{r_i} \frac{\partial X}{\partial \varepsilon_i^j} \dot{\varepsilon}_i^j = X_i(\varepsilon_i) \begin{bmatrix} \widetilde{h_{i1} \beta_i} & h_{i2} \beta \\ 0 & 0 \end{bmatrix} \tag{4.43}$$

Thus we have

$$
\begin{aligned}
\dot{X} &= X(\varepsilon) p \\
&= \begin{bmatrix} \widetilde{h_{11} \beta_1} & h_{12} \beta_1 \\ 0 & 0 \end{bmatrix} \\
&\quad + \sum_{i=2}^{p} X_p(\varepsilon_p) \cdots X_i(\varepsilon_i) \begin{bmatrix} \widetilde{h_{i1} \beta_i} & h_{i2} \beta_i \\ 0 & 0 \end{bmatrix} X_{i-1}(\varepsilon_{i-1}) \cdots X_1(\varepsilon_1)
\end{aligned} \tag{4.44}
$$

or, premultiplying through by $X(\varepsilon)$,

$$
\begin{aligned}
\begin{bmatrix} \widetilde{h_{11} \beta_1} & h_{12} \beta_1 \\ 0 & 0 \end{bmatrix} + & \\
\sum_{i=2}^{p} [X_{i-1}(\varepsilon_{i-1}) \cdots X_1(\varepsilon_1)]^{-1} & \begin{bmatrix} \widetilde{h_{i1} \beta_i} & h_{i2} \beta_i \\ 0 & 0 \end{bmatrix} X_{i-1}(\varepsilon_{i-1}) \cdots X_1(\varepsilon_1) \\
= p = & \begin{bmatrix} \tilde{\omega}_b & v_b \\ 0 & 0 \end{bmatrix}
\end{aligned} \tag{4.45}
$$

This important relationship gives the body rates across the compound joint in terms of the joint quasi-velocities. Now, we can also write

$$\begin{bmatrix} \widetilde{h_{i2}\beta_i} & h_{i1}\beta_i \\ 0 & 0 \end{bmatrix} = \sum_{j=1}^{r_i} \begin{bmatrix} \widetilde{h}_{i1}^j & h_{i2}^j \\ 0 & 0 \end{bmatrix} \tag{4.46}$$

So that (4.45) can be written in the form

$$p = H(\varepsilon)\beta$$

where $H(\varepsilon)$ is constructed as stated. ∎

Note that these equations differ from those of Proposition (4.124) only in that each ε_i is a vector of dimension r_i rather than a scalar.

4.2.4 Joint Computations

The computations described above have been implemented in *ProPac*. The function Joints computes all of the required joint quantities. Recall that simple joints are characterized by the number of degrees of freedom, r, an r-vector of joint quasi- velocities, p, and a $6 \times r$ joint map matrix, H. Across the joint, the relative velocity vector is Hp. Moreover, H is a constant (independent of the joint configuration) and the columns represent the joint action axes in the outboard frame (by convention). A compound joint is equivalent to a sequence of simple joints. Thus, it is necessary to define a set of numbers that represent the degrees of freedom associated with each intermediate frame and a corresponding set of (constant) joint map matrices. When defining a joint in *ProPac*, it is necessary to also assign names for both the joint quasi-velocities and joint configuration variables.

A k-frame compound joint with n degrees of freedom is defined by the data structure:

$$\{r, H, q, p\}$$

where

$\begin{aligned} r \quad &= \quad k - \text{vector whose elements define the number of degrees of freedom} \\ & \quad \text{for each simple joint, with } n = r_1 + \cdots + r_k. \\ H \quad &= \quad [H_1 \ldots H_k] \text{ a matrix composed of the k joint map matrices of the} \\ & \quad \text{simple joints.} \\ q \quad &= \quad n - \text{vector of joint coordinate names.} \\ p \quad &= \quad n - \text{vector of joint quasi-velocity names.} \end{aligned}$

Example 4.127 (2 dof simple and compound joints) *Here is a sample computation that illustrates the difference between simple and compound joints:*

In[59]:= (* spherical joint - a simple 2-dof revolute joint *)
    r1={2};H1={{1,0},{0,0},{0,1},{0,0},{0,0},{0,0}};
    q1={a1x,a1z};p1={w1x,w1z};
    (* universal joint - a compound 2-dof revolute joint *)
    r2={1,1};H2={{1,0},{0,0},{0,1},{0,0},{0,0},{0,0}};
    q2={a2x,a2z};p2={w2x,w2z};
    JointLst={{r1,H1,q1,p1},{r2,H2,q2,p2}};
    {V,X,H}=Joints[JointLst];

The results are given below.

Spherical Joint:

$$V = \begin{pmatrix} 1 & 0 \\ 0 & \cos a_{1x} \end{pmatrix}$$

$$X = \begin{pmatrix} \cos a_{1z} & -\cos a_{1x}\sin a_{1z} & \sin a_{1x}\sin a_{1z} & 0 \\ \sin a_{1z} & \cos a_{1x}\cos a_{1z} & -\cos a_{1z}\sin a_{1x} & 0 \\ 0 & \sin a_{1x} & \cos a_{1x} & 0 \\ 0 & 0 & 0 & 1 \end{pmatrix}$$

$$H = \begin{pmatrix} 1 & 0 \\ 0 & 0 \\ 0 & 1 \\ 0 & 0 \\ 0 & 0 \\ 0 & 0 \end{pmatrix}$$

Universal Joint:

$$V = \begin{pmatrix} 1 & 0 \\ 0 & 1 \end{pmatrix}$$

$$X = \begin{pmatrix} \cos a_{2z} & -\cos a_{2x}\sin a_{2z} & \sin a_{2x}\sin a_{2z} & 0 \\ \sin a_{2z} & \cos a_{2x}\cos a_{2z} & -\cos a_{2z}\sin a_{2x} & 0 \\ 0 & \sin a_{2x} & \cos a_{2x} & 0 \\ 0 & 0 & 0 & 1 \end{pmatrix}$$

$$H = \begin{pmatrix} 1 & 0 \\ 0 & \sin a_{2x} \\ 0 & \cos a_{2x} \\ 0 & 0 \\ 0 & 0 \\ 0 & 0 \end{pmatrix}$$

Example 4.128 (3 dof Universal Joint) *A widely used example of a compound joint is the 3 degree of freedom universal joint. Such a joint is illustrated in Figure (4.3). This joint is composed of three elements and requires three frames to describe the composite motion. The relative motion between each of them involves one degree of freedom. In our terminology*

$$H = \begin{bmatrix} 1 & 0 & 0 \\ 0 & 1 & 0 \\ 0 & 0 & 1 \\ 0 & 0 & 0 \\ 0 & 0 & 0 \\ 0 & 0 & 0 \end{bmatrix}$$

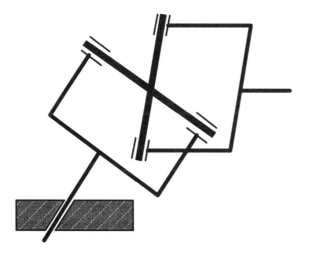

Figure 4.3: Diagram of a 3 dof universal joint. Note that the joint itself is composed of three bodies in addition to the fixed reference body.

$In[60]:=$ $H = Join[IdentityMatrix[3], DiagonalMatrix[\{0, 0, 0\}]];$
$r = \{1, 1, 1\};$
$q = \{t1, t2, t3\};$
$p = \{w1, w2, w3\};$
$JointLst = \{\{r, H, q, p\}\};$
$\{V, X, H\} = Joints[JointLst];$

The results of this calculation are:

$$H = \begin{bmatrix} 1 & 0 & -\sin t2 \\ 0 & \cos t1 & \cos t2 \sin t1 \\ 0 & -\sin t1 & \cos t1 \cos t2 \\ 0 & 0 & 0 \\ 0 & 0 & 0 \\ 0 & 0 & 0 \end{bmatrix}$$

$$X \; =$$

$$
\begin{bmatrix}
\cos t2 \cos t3 & \cos t3 \sin t1 \sin t2 - \cos t1 \sin t3 & \cos t1 \cos t3 \sin t2 + \sin t1 \sin t3 & 0 \\
\cos t2 \sin t3 & \cos t1 \cos t3 + \sin t1 \sin t2 \sin t3 & -\cos t3 \sin t1 + \cos t1 \sin t2 \sin t3 & 0 \\
-\sin t2 & \cos t2 \sin t1 & \cos t1 \cos t2 & 0 \\
0 & 0 & 0 & 1
\end{bmatrix}
$$

$$
V = \begin{bmatrix} 1 & 0 & 0 \\ 0 & 1 & 0 \\ 0 & 0 & 1 \end{bmatrix}
$$

4.2.5　Remarks on Configuration Coordinates

The joint quasi–velocities are naturally defined by the action of the joint. Joint configuration coordinates, however, are defined by the kinematic relation (4.23). While these equations formally define the coordinates (by defining $\dot{\varepsilon}$), they also provide a physical interpretation. Before examining some examples, note that $V(\varepsilon)$ itself follows directly from the joint definition. Therefore to the extent that there is some freedom in specifying the joint parameters (the vector r and the matrix H), the user sets up the physical meaning of the coordinates ε. To see how this works, consider a general six degree of freedom joint (unconstrained 6 dof relative motion) defined by:

```
In[61]:=  H = IdentityMatrix[6];
          r = {6};
          q = {ax, ay, az, x, y, z};
          p = {wx, wy, wz, ux, uy, uz};
```

Consider this joint as depicting the relative motion of a body with respect to a space frame. The velocity transformation matrix V is:

$$V \;=\; diag(V_1, V_2)$$

$$
V_1 \;=\; \begin{bmatrix}
1 & \sin ax \tan ay & \cos ax \tan ay \\
0 & \cos ax & -\sin ax \\
0 & \sec ay \sin ax & \cos ax \sec ay
\end{bmatrix}
$$

$$V_2 =$$

$$
\begin{bmatrix}
\cos ay \cos az & \cos az \sin ax \sin ay - \cos ax \sin az & \cos ax \cos az \sin ay + \sin ax \sin az \\
\cos ay \sin az & \cos ax \cos az + \sin ax \sin ay \sin az & -\cos az \sin ax + \cos ax \sin ay \sin az \\
-\sin ay & \cos ay \sin ax & \cos ax \cos ay
\end{bmatrix}
$$

Inspection and comparison with standard results (e.g., [5]) reveals that the coordinates ax, ay, az are Euler angles in the 3–2–1 convention, and the coordinates x, y, z define the position of the body frame relative to the space frame, as represented in the space frame. In other words, the quasi–velocity vector (ux, uy, uz) corresponds to the body linear velocity in the body frame whereas the coordinate velocity $(\dot{x}, \dot{y}, \dot{z})$ represent the same body linear velocity in the space frame. By interchanging the first three columns of H, the resultant angle parameters again turn out to be Euler parameters, but in different conventions. If the columns in H corresponding to angles and linear displacements are interchanged, then the representation of the linear velocity and displacement will switch from space to body frame (or vice–versa).

4.3   Chain and Tree Configurations

In general a multibody system can be viewed in terms of an underlying tree structure upon which is imposed additional algebraic and/or differential constraints. In this section we describe the data structures used to define multibody tree structures. In later sections we show how to compute velocities and configuration coordinates of reference frames at arbitrary locations in the tree. A tree can be defined in terms of a set of chains, each beginning at the root body.

Representation of Chain & Tree Structures

ProPac provides tools to build models for mechanical systems that have an underlying tree topology. Chain structures are a special case. Systems with closed loops are accommodated by adding constraints to the underlying tree. A tree consisting of n bodies also contains n joints. Every system contains a base reference frame that is designated body '0'. Otherwise, bodies and joints can be numbered arbitrarily. Joint data and body data are organized into lists by the analyst, i.e.:

```
JointList={JointData_1,...,BodyData_n}
BodyList={BodyData_1,...,BodyData_n}
```

The structure of the individual data objects will be described below. Joints and bodies are implicitly numbered by their position in the data lists. Each body contains a unique 'inboard' node, corresponding to (the outboard side of) a joint through which the body connects to an inner branch of the tree, or to the root (body 0). See Figures (4.4) and (4.5). Each body may also contain 'outboard' nodes. The outboard nodes are distinguished body locations that may be associated with a joint location (the inboard side of the joint), a sensor location, a point of application of an external force or any other feature of interest. Since one joint connects the tree to the root (the root node may be considered an outboard node of body 0), there must be at least $n - 1$ outboard nodes among the n bodies corresponding to the remaining $n - 1$ joints. These are the n 'outboard joint nodes'. The outboard joint nodes must be numbered 1 through n and must correspond to the associated joint number. The specific association of numbers to joints is not essential but by convention the root node is normally assigned the number 1. The remaining outboard notes can be numbered arbitrarily. The inboard nodes need not be assigned numbers.

In summary there are two important book-keeping principles:

- joints and bodies are numbered according to their position in the data lists,

- outboard joint nodes must be numbered consistently with their repective joints.

A tree is composed of a set of defining chains. For instance consider a tree composed of the following sequences of bodies:

$$0,1,2,4 \qquad 0,1,2,3,5 \qquad 0,1,2,3,6$$

All defining chains of any tree will start with body 0, so we need not list it. However, the body sequences alone do not adequately define a tree. For instance bodies 5 and 6 both connect to body 3, but they will do so through different joints. This information can be provided by defining each chain as an ordered list of pairs - each pair consisting of a body and its inboard joint: inboard joint, body. For example, consider the following three chains:

$$\{\{1,1\},\{2,2\},\{5,4\}\} \qquad \{\{1,1\},\{2,2\},\{3,3\},\{4,5\}\}$$
$$\{\{1,1\},\{2,2\},\{3,3\},\{6,6\}\}$$

Chain 1 consists of bodies 1, 2, and 4. Body 1 connects to the reference (Body 0) at Joint 1, Body 2 connects to Body 1 at Joint 2, and Body 4 connects to Body 2 at Joint 5. The data also indicates that body 5 connects to body 3 at joint 4 (in the second chain), and body 6 connects to body 3 at joint 6 (third chain). Recall that each joint is uniquely associated with an outboard node of a particular body. A tree is defined by the data structure:

```
Tree = {list of chains}
Chain = ordered list of pairs {inboard joint, body}
     = {{first inboard joint, first body},...,
                   {last inboard joint, last body}}
```

Reference Frames

It is assumed that there is a single inertially fixed reference frame whose origin is the inboard side of the root joint. Each body has a primary reference frame, fixed in the body with origin at the inboard node. Body data is defined in this frame. As appropriate, there may be other body fixed frames as well with origins at outboard node locations. Normally, the axes of these frames are parallel to the primary frame when the body is undeformed. For the system as a whole, there is a 'reference configuration' corresponding to the nominal joint configurations (associated with zero joint motion parameters) and undeformed bodies. In the reference configuration all reference frames (body and space) are alligned. The analyst sets up the reference configuration when choosing body frame orientations for joint and body data definitions. It is recommended that the analyst begin by defining a physically meaningful reference configuration from which data definitions will logically follow.

Rigid Body Data Structure

A rigid body is defined by its mass, inertia matrix and the location of distinguished points or nodes where joints or sensors may be located. We assume the following:

1. There is a distinguished point that corresponds to the inboard joint of the body. The body frame has its origin located there.

2. The center of mass and all other points of interest (nodes) including outboard joint locations are defined in the body frame.

3. The inertia matrix is defined in the body frame and it is the inertia matrix about the center of mass.

The data for a rigid body is organized in a list as follows. A rigid body with k outboard nodes is defined by the data structure:

$$\{\texttt{com},\{\texttt{out1},..,\texttt{outk}\},\texttt{m},\texttt{Inertia}\}$$

where

```
com is the center of mass location,
outi = {node number, location} for the ith outboard node,
m is the mass, and
Inertia is the inertia tensor (about the center of mass).
```

4.3.1   Configuration Relations

Consider a serial chain composed of $K + 1$ rigid bodies connected by joints as illustrated in Figure (4.4). The bodies are numbered 0 through K, with 0 denoting the base or reference body, which may represent any convenient inertial reference frame. The kth joint connects body $k - 1$ at the point C_{k-1} with body k at the point O_k.

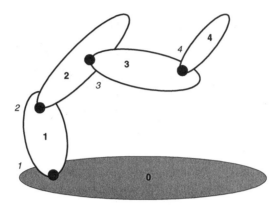

Figure 4.4: A serial chain composed of $K + 1$ rigid bodies numbered 0 through K and K joints numbered 1 through K.

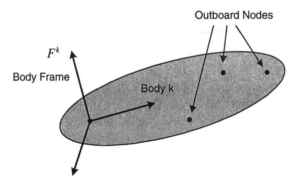

Figure 4.5: On an arbitrary kth link the inboard and outboard joint hinge points are designated O_k and C_k. The body fixed reference frame has its origin at O_k.

Let F^k denote a reference frame fixed in body k with origin at O_k. r_{co}^k denotes the vector from O_k to C_k in F^k and r^k denotes the vector from O_k to O_{k+1} in F^k. We will use a coordinate specific notation in which vectors represented in F^k (or its tangent space) will be identified with a superscript "k". Coordinate free relations carry no superscript. Sometimes it is convenient to employ a frame fixed in body k and aligned with F^k but with origin at some point P_k other than O_k. We use the designation $F_{P_k}^k$. Let r_{po}^k denote the vector from O_k to P_k

in F^k. Then the parallel translation of F^k to $F^k_{P_k}$ results in the configuraton matrix

$$X_{k,P_k} = \begin{bmatrix} I & r_{po} \\ 0 & 1 \end{bmatrix}$$

The kth joint has n_k, $1 \leq n_k \leq 6$ degrees of freedom which can be characterized by n_k coordinates $q(k)$ and, correspondingly, n_k quasi-velocities $\beta(k)$ and a configuration matrix $X_k(q(k))$. We wish to compute the Euclidean configuraton matrix for a reference frame fixed in the last body with origin at the terminal node of the chain, designated P_K. For example, this would be node 5 in Figure (4.4). We obtain the configuration relative to the space frame by successive motions: action of joint $1 \rightarrow$ translation to $C_1 \rightarrow \cdots \rightarrow$ action of joint $K \rightarrow$ translation to P_K:

$$X_{P_K} = X_{K,P}X_K \ldots X_{2,C_2}X_2 X_{1,C_1}X_1 \tag{4.47}$$

Equation (7.32) can be modified to compute the relative configuration between body fixed frames at any two nodes in a chain or tree. To accommodate trees in this calculation requires a simple procedure to find a chain connecting the two nodes.

4.3.2 Velocity Relations in Chains

Once again consider a chain composed of $K + 1$ bodies as illustrated in Figure (4.4). Rodriguez et al [6-8] define the spatial velocity at point C of any body-fixed reference frame with origin at point C as $V_c = [\omega, vc]$ where V_c is the velocity of point C and ω is the angular velocity of the body. Let O be another point in the same body and let r_{co} denote the location of C in the body frame with origin at O. Then the spatial velocity at point C is related to that at O by the relation

$$V_c = \phi(r_{co})V_o \tag{4.48}$$

where

$$\phi(r_{co}) = \begin{bmatrix} I & 0 \\ -\tilde{r}_{co} & I \end{bmatrix},$$

and its adjoint

$$\phi^*(r_{co}) = \begin{bmatrix} I & \tilde{r}_{co} \\ 0 & I \end{bmatrix} \tag{4.49}$$

Joint k has a joint map matrix $H(k) \in R^{6 \times n_k}$ so that

$$V_o(k) - V_c(k-1) = H(k)\beta(k) \tag{4.50}$$

Thus, sequential application of (4.48) and (5.30) leads to the following recursive velocity relation that we write in coordinate specific notation

$$V^i(k) = \phi(r^i_{co}(k-1))V^i(k-1) + H^i(k)\beta^i(k) \tag{4.51}$$

where the superscript i denotes the reference frame. Let us assume that $H(k)$ and $\beta(k)$ are specified in the frame F^k and $V(k-1)$ has been computed in the frame F^{k-1}. Then it is convenient to compute $V(k)$ in the kth frame

$$V^k(k) = diag(L_{k-1,k}, L_{k-1,k})\phi(r_{co}^{k-1}(k-1))V^{k-1}(k-1) + H^k(k)\beta^k(k) \quad (4.52)$$

If $V^0(0)$ is given, then equation (4.52) allows us to compute recursively, for $k = 1, .., K$, the linear velocity of the origin of F^k and the angular velocity of F^k, both represented in the coordinates of F^k. In what follows we take $V^0(0) = 0$. Abusing notation somewhat, it is convenient to define

$$\phi(k, k-1) = diag(L_{k-1,k}, L_{k-1,k})\phi(r_{co}^{k-1}(k-1)) \quad (4.53)$$

so that (4.52) can be written

$$V^k(k) = \phi(k, k-1)V^{k-1}(k-1) + H^k(k)\beta^k(k), \quad k = 1, \dots, K, \quad V^0(0) = 0 \quad (4.54)$$

Equations (4.48) and (4.54) allow a sequential computation of velocities at any point along a chain. Notice that ϕ and H depend on joint configuration parameters so that the joint configuration and velocity variables are needed to perform the computation.

4.3.3   Configuration and Velocity Computations

In addition to the function Joints described above, there are other kinematic computations implemented in *ProPac*. We will describe and illustrate three of them: EndEffector, RelativeConfiguration and NodeVelocity.

EndEffector[BodyList,X] returns the Euclidean configuration matrix of a frame in the last body of a chain, with origin at the outboard joint location. BodyList is list of body data in nonstandard chain form, X is a corresponding list of joint Euler configuration matrices. EndEffector can also be used in the form EndEffector[ChainList,TerminalNode,BodyList,X] where BodyList is the standard body data structure. ChainList identifies the system subchain that terminates with TerminalNode. EndEffector can also be used in the form EndEffector[TerminalNode,TreeList,BodyList,X] in the event that the appropriate chain has not been identified. This last form is probably the most useful.

RelativeConfiguration[Node1,Node2,TreeList,BodyList,X,q] returns the configuration matrix for a body fixed frame at node Node2 as seen by an observer in a body fixed frame at node Node1. Note that each node is defined in a specific body and the frame is fixed in the body in which the node is defined.

NodeVelocity[ChainList,TerminalNode,BodyList,X,H,p] returns the velocity at TerminalNode, where the body data, BodyList, the joint data X and H, and the quasivelocity names p corresponds to the chain defined by ChainList. The velocity is a six element vector defined in the body fixed frame. This function is used by GeneralizedForce that will be described later. The following syntax may also be used, similarly to the function EndEffector:

NodeVelocity[ChainList,TerminalNode,BodyList,X,H,q,p],

and

NodeVelocity[TerminalNode,TreeList,BodyList,X,H,q,p].

Example 4.129 (5 dof Robot Arm) *The functions described above will be illustrated with an example of a 5 dof robot arm as shown in Figure (4.6).*

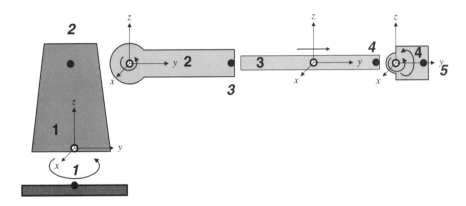

Figure 4.6: A 5 dof robot arm is illustrated. The body fixed reference frames located at the inboard joint nodes are illustrated. Note that in the reference configuration the reference frames are alligned.

First we define the joint data.

```
In[62]:= r1={1};H1={{0},{0},{1},{0},{0},{0}};
         q1={theta1};p1={w1};
         r2={1};H2=Transpose[{{1,0,0,0,0,0}}];
         q2={theta2};p2={w2};
         r3={1};H3=Transpose[{{0,0,0,0,1,0}}];
         q3={y};p3={v};
         r4={2};H4=Transpose[{{0,1,0,0,0,0},{1,0,0,0,0,0}}];
         q4={theta3,theta4};p4={w3,w4};
         JointList={{r1,H1,q1,p1},{r2,H2,q2,p2},{r3,H3,q3,p3},{r4,H4,q4,p4}};
```

Now, define the body data,

In[63]:= `com1={0,0,l1/2}; mass1=m1; out1={2,{0,0,l1}};`
          `Inertia1=DiagonalMatrix[{J1x,J1x,J1z}];`
          `com2={0,0,l2/2}; mass2=m2; out2={3,{0,0,l2}};`
          `Inertia2=DiagonalMatrix[{J2x,0,J2x}];`
          `com3={0,0,l3/2}; mass3=m3; out3={4,{0,0,l3}};`
          `Inertia3=DiagonalMatrix[{J3x,0,J3z}];`
          `com4={0,0,l3/2}; mass4=m4; out4={5,{0,0,l4}};`
          `Inertia4=DiagonalMatrix[{J4x,J4y,J4z}];`

          `BodyList={{com1,{out1},mass1,Inertia1},{com2,{out2},mass2,Inertia2},`
          `{com3,{out3},mass3,Inertia3},{com4,{out4},mass4,Inertia4}};`

and the interconnection structure

In[64]:= `TreeList={{{1,1},{2,2},{3,3},{4,4}}};`

The joint parameters are computed with the command:

In[65]:= `{V,X,H} = Joints[JointList];`

The joint velocity transformation matrices can be displayed as follows.

In[66]:= `V[[1]]//MatrixForm`
Out[66]= $(1 \quad)$

In[67]:= `V[[2]]//MatrixForm`
Out[67]= $(1 \quad)$

In[68]:= `V[[3]]//MatrixForm`
Out[68]= $(1 \quad)$

In[69]:= `V[[4]]//MatrixForm`
Out[69]= $\begin{pmatrix} 1 & 0 \\ 0 & \text{Cos[theta3]} \end{pmatrix}$

Using the function **EndEffector** *the configuration of a frame fixed at node 4 can be computed.*

In[70]:= `TerminalNode = 4;`
          `XE = EndEffector[TerminalNode, TreeList, BodyList, X];`

From the configuration matrix we can pull out the translation vector,

In[71]:= `XE[[{1,2,3},4]]//MatrixForm`
Out[71]=
$$\begin{pmatrix} -y\ \text{Cos[theta2]}\ \text{Sin[theta1]} + l2\ \text{Sin[theta1]}\ \text{Sin[theta2]} + l3\ \text{Sin[theta1]}\ \text{Sin[theta2]} \\ y\ \text{Cos[theta1]}\ \text{Cos[theta2]} - l2\ \text{Cos[theta1]}\ \text{Sin[theta2]} - l3\ \text{Cos[theta1]}\ \text{Sin[theta2]} \\ l1 + l2\ \text{Cos[theta2]} + l3\ \text{Cos[theta2]} + y\ \text{Sin[theta2]} \end{pmatrix}$$

and the rotation matrix:

In[72]:= `XE[[{1,2,3},{1,2,3}]]//MatrixForm`
Out[72]= $\begin{pmatrix} \text{Cos[theta1]} & -\text{Cos[theta2]}\ \text{Sin[theta1]} & \text{Sin[theta1]}\ \text{Sin[theta2]} \\ \text{Sin[theta1]} & \text{Cos[theta1]}\ \text{Cos[theta2]} & -\text{Cos[theta1]}\ \text{Sin[theta2]} \\ 0 & \text{Sin[theta2]} & \text{Cos[theta2]} \end{pmatrix}$

As another example, we can compute the relative configuration of a frame at node 4 as seen by an oberver in a frame at node 2. We use the function RelativeConfiguration.

In[73]:= $Node1 = 2; Node2 = 4; q = Flatten[\{q1, q2, q3, q4\}]$

Out[73]= $\{theta1, theta2, y, theta3, theta4\}$

In[74]:= *RelativeConfiguration[Node1, Node2, TreeList, BodyList, X, q]//*
            MatrixForm

Out[74]= $\begin{pmatrix} 1 & 0 & 0 & 0 \\ 0 & Cos[theta2] & -Sin[theta2] & y\ Cos[theta2] - (12+13)\ Sin[theta2] \\ 0 & Sin[theta2] & Cos[theta2] & (12+13)\ Cos[theta2] + y\ Sin[theta2] \\ 0 & 0 & 0 & 1 \end{pmatrix}$

The configuration of a frame at node 4 relative to node 3 is a pure translation:

In[75]:= *RelativeConfiguration[3, 4, TreeList, BodyList, X, q]//MatrixForm*

Out[75]= $\begin{pmatrix} 1 & 0 & 0 & 0 \\ 0 & 1 & 0 & y \\ 0 & 0 & 1 & 13 \\ 0 & 0 & 0 & 1 \end{pmatrix}$

Finally, we compute the spatial velocity of a frame fixed in body 3 at node 4.

In[76]:= $p = Flatten[\{p1, p2, p3, p4\}]$

Out[76]= $\{w2, w2, v, w3, w4\}$

In[77]:= $TerminalNode = 4;$

In[78]:= *NodeVelocity[TerminalNode, TreeList, BodyList, X, H, q, p]//MatrixForm*

Out[78]= $\begin{pmatrix} w2 \\ w1\ Sin[theta2] \\ w1\ Cos[theta2] \\ (12+13)\ w1\ Sin[theta2] \\ v + (-12-13)\ w2 \\ 0 \end{pmatrix}$

4.4 Problems

Problem 4.130 (Reconnaissance robot) *The reconnaissance robot shown in Figure (4.7) moves on a flat surface. The vehicle has three degrees of freedom, its linear coordinates x, y and its angular orientation θ. The radar system also has three degrees of freedom. It can move vertically, z, and rotate in both azimuth and elevation, ϕ, ψ, relative to the vehicle. Suppose the radar system is pointing at a target, the vehicle and radar configuration are known as well as the range to target. Compute the target coordinates in a space fixed frame.*

Problem 4.131 (Overhead crane) *The overhead crane shown in Figure (4.8) is used to move and position heavy loads in the $x - z$ plane. The cart moves in one (x) linear direction on rails, the arm connects to the cart via a revolute*

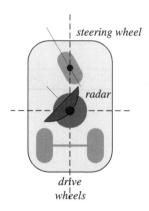

Figure 4.7: A reconnaissance vehicle carrying a range-finding radar system.

*joint (angle ϕ from downward z direction) and the cable length L is variable.
Assume that the cable is always in tension and treat the payload as a point mass.
The arm cable joint can be treated as a two degree of freedom compound joint
consisting of rotation and extension (to model cable playout). Determine the
spatial coordinates of the payload in terms of the four joint parameters.*

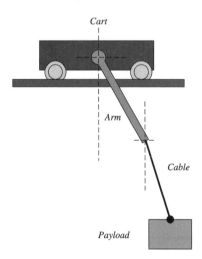

Figure 4.8: An overhead crane used for moving and positioning heavy loads.

4.5 References

1. Olver, P.J., Applications of Lie Groups to Differential Equations. 1986, New York: SpringerVerlag.

2. Meirovitch, L., Methods of Analytical Dynamics. 1970, New York: McGrawHill, Inc.

3. Rosenberg, R.M., Analytical Dynamics of Discrete Systems. 1977, New York: Plenum Press.

4. Kwatny, H.G. and G.L. Blankenship, Symbolic Construction of Models for Multibody Dynamics. IEEE Transactions on Robotics and Automation, 1995. 11(2): p. 271-281.

5. Goldstein, H., Classical Mechanics. 2nd ed. 1980, Reading: AddisonWesley.

6. Rodriguez, G., Kalman Filtering, Smoothing and Recursive Robot Arm Forward and Inverse Dynamics. IEEE Transactions on Robotics and Automation, 1987. RA3(6): p. 624639.

7. Rodriguez, G. and K. KreutzDelgado, Spatial Operator Factorization of the Manipulator Mass Matrix. IEEE Transactions on Robotics and Automation, 1992. 8(1): p. 6576.

8. Jain, A. and G. Rodriguez, Recursive Flexible Multibody System Dynamics Using Spatial Operators. AIAA Journal of Guidance, Control and Dynamics, 1992. 15(6): p. 14531466.

Chapter 5

Dynamics

5.1 Introduction

The purpose of this chapter is to describe symbolic computing tools for assembling and manipulating control system design models for constrained multibody mechanical systems. The methods introduced in [1] for chains and trees are summarized and extended to constrained systems. New computing tools that support the analysis of constrained systems are described and ilustrated.

The derivation of the explicit dynamical equations of motion for mechanical systems of even moderate complexity is difficult and time consuming.[1] Consequently, there has been a growing interest in automated derivation using computers [1-4]. Much of this work has focused on chain and tree structures that characterize important robotic and vehicular systems. Many systems, however, are not tree structures; they involve closed loops or other forms of algebraic and/or differential constraints imposed on top of an underlying tree. Typical examples would be a grasping robotic hand or a vehicle with rolling wheels. The additional complexity of such systems magnifies the utility of computer assembly of the governing equations.

Our discussion is based on *Poincaré's equations* [5-7] also referred to as *Lagrange's equations in quasi-coordinates*[8, 9] or *pseudo-coordinates* [10], and the *Euler-Poincaré's equations* in [11]. Poincaré's equations preserve the underlying theoretical structure and elegance of the Lagrange formulation, but they are often more natural and can be substantially simpler than Lagrange's equations. Furthermore, their assembly can be much easier to automate making them a practical choice for modeling 'industrial strength' systems. Perhaps most importantly, in the words of Neimark and Fufaev [9]: "The main advantage... is

---

[1]Note that modeling and simulation software such as ADAMS and DADS do not produce explicit nonlinear equations of motion required for control system analysis and design.

the unification of the form of the ordinary Lagrange equations, the equations of motion of nonholonomic systems, and also equations such as Euler's dynamical equations of motion of a rigid body with a fixed point."

The dynamical equations will be generated in the form

$$M(q)\dot{p} + C(p,q)p + F(p,q) = Q \tag{5.1}$$

where p is a vector of quasi-velocities, q is the generalized coordinate vector, Q is a vector of externally applied generalized forces and the functions $M(q)$, $C(p,q)$ and $F(p,q)$ are the parameters of the system. The assembly of Q, $M(q)$, $C(p,q)$ and $F(p,q)$ is the main topic of this chapter. In combination with the kinematic equations describe in the previous chapter

$$\dot{q} = V(q)p \tag{5.2}$$

these equations provide a consistent closed set of equations.

In Section 2 we develop Poincaré's formulation of Lagrange's equations and in Section 3 we apply it to general chain and tree structures. Then we consider constrained systems in Section 4 in which we treat both holonomic and nonholonomic differential constraints as well as configuraton constraints. We describe symbolic computing tools and give examples along the way. Finally, in Section 5, we describe numerical simulation.

5.2   Poincaré's Equations

It is well known that in some cases it is easier to formulate the equations of motion in terms of velocity variables that can not be expressed as the time derivatives of any corresponding configuration coordinates. Such velocities are called quasi-velocities and are often associated with so-called quasi-coordinates. Quasi-velocities are meaningful physical quantities. The angular velocity of a rigid body is a prime example. Quasi-coordinates are not meaningful physical quantities. They make sense only in terms of infinitesimal motions. The notion of quasi-velocities and quasi-coordinates leads to a generalization of Lagrange's equations which is applicable to systems with nonholonomic as well as holonomic constraints. Such generalizations were produced at the turn of the century and are associated with the names of Poincare, Appell, Maggi, Hamel, Gibbs and Boltzman (see, for example, Arnold et al [5] and Niemark and Fufaev [9], Gantmacher [10]).

5.2.1   Preliminaries

Consider a holonomically constrained system whose possible configurations correspond to the points of a smooth manifold M of dimension m called the *configuration manifold*. Local coordinates on M can be used to define the system configuration. They are called *generalized coordinates*. Any motion of the system over a time interval $[t_1, t_2]$ traces a path in M characterized in local coordinates by a map $q(t)$: $[t_1, t_2] \mapsto M$. At any point $q \in M$ the *generalized velocity* $\dot{q}$ belongs to the *tangent space* to M at q denoted by $T_q M$. The state space for the dynamical system is the $2m$ dimensional manifold $TM = \bigcup_{q \in M} T_q M$, called the *tangent bundle*.

A *virtual displacement* of the system at a configuration $q \in M$ is an infinitesimal displacement δq that takes the system to an admissible configuration $q' \in M, q' = q + \delta q$. Clearly, δq is a virtual displacement if and only if it is infinitesimal and satisfies $\delta q \in T_q M$. If a system in configuration q is acted upon by a *generalized force*, Q, then the *virtual work* performed by the force under a virtual displacement δq is $\delta W = Q^T \delta q$.

Let M be the m-dimensional configuration manifold for a Lagrangian system and suppose $v_1, \ldots, v_m$ constitute a system of m linearly independent vector fields on M. Then each commutator or Lie bracket can be expressed

$$[v_i, v_j] = \sum_{k=1}^{m} c_{ij}^k(q) v_k \qquad (5.3)$$

Indeed, the coefficients are easily computed in local coordinates.[2] Define

$$V := [v_1 \ v_2 \ \cdots v_m], \quad U = \begin{bmatrix} u_1 \\ u_2 \\ \vdots \\ u_m \end{bmatrix} := V^{-1}, \ \chi_{ij} = [c_{ij}^1 \ c_{ij}^2 \ \ldots c_{ij}^m]^t. \qquad (5.4)$$

Then (1) yields

$$\chi_{ij} = U[v_i, v_j] \quad \text{or} \quad c_{ij}^k = u_k[v_i, v_j]. \qquad (5.5)$$

Suppose $q(t)$: $[t_1, t_2] \mapsto M$ is a smooth path, then $\dot{q}(t)$ denotes the tangent vector to the path at the point $q(t) \in M$. Thus, we can always express $\dot{q}(t)$ as a linear combination of the tangent vectors $v_i, i = 1, \ldots, m$:

$$\dot{q} = \sum_{i=1}^{m} v_i(q) p$$

or:

$$\dot{q} = V(q) p \qquad (5.6)$$

where

$$p = U(q) \dot{q} \qquad (5.7)$$

The variables p are called *quasi-velocities*. Since these quantities are "velocities" we might try to associate them with a set of coordinates π, in the sense that

---

[2]In local coordinates, vector fields on a manifold of dimension m may be thought of as column vectors of length m and covector fields as row vectors of length m. We will use this device often to do calculatons.

$\dot{\pi} = p$. This is not always possible because in view of (5.7) we must have

$$\delta\pi = U(q)dq$$

but, in fact, the right hand side (of each $\delta\pi_i$) may not be an exact differential.

5.2.2   Poincaré's Form of Lagrange's Equations

First, let us review some elementary variational constructions that will be used in the drivation of Poincaré's eqauations. If $f : M \mapsto R$ is a smooth function, then $v_i(f)$ is the derivative of f in the direction of the vector field v_i. The rate of change of f along the path is given by

$$\dot{f} = \frac{\partial f}{\partial q}\dot{q} = \sum_{i=1}^{m} v_i(f)p_i \qquad (5.8)$$

where the variables p_i are the quasi–velocities defined above.

Once again, consider the smooth path and suppose the end points are q_1, and q_2, i.e., $q(t_1) = q_1, q(t_2) = q_2$. A variation of $q(t)$ is a smooth map $q(\varepsilon, t)$: $(-\varepsilon_0, \varepsilon_0) \times [t_1, t_2] \to M$ such that $q(0, t) = q(t)$. For every variation, we can define

$$w(t) = \frac{\partial q}{\partial \varepsilon}(0, t) \qquad (5.9)$$

$w(t)$ may be thought of as a vector field defined on M along $q(t)$. Conversely, let $w(t)$ be any smooth vector field defined on M along $q(t)$ with $w(t_1) = 0$ and $w(t_2) = 0$. For any such vector field, there is a variation such that (5.9) is satisfied. The implication of this is that we can define variations of $q(t)$ in the form

$$q(\varepsilon, t) = q(t) + \varepsilon w(t). \qquad (5.10)$$

It is always possible to write the Lagrangian in terms of q and p. Set $\widetilde{L}(p, q) = L(\dot{q}, q)$. In terms of $\widetilde{L}$ Lagrange's equations are attainable in the form given by the following lemma.

Proposition 5.132 *Hamilton's principle leads to the equations of motion in terms of the coordinates* q, p

$$\frac{d}{dt}\frac{\partial \widetilde{L}}{\partial p_k} - \sum_{i;j=1}^{m} c_{jk}^i \frac{\partial \widetilde{L}}{\partial p_i} p_j - v_k(\widetilde{L}) = Q^t v_k \qquad (5.11)$$

or, in local coordinates,

$$\frac{d}{dt}\frac{\partial \widetilde{L}}{\partial p} - \sum_{j=1}^{m} p_j \frac{\partial \widetilde{L}}{\partial p} U X_j - \frac{\partial \widetilde{L}}{\partial q} V = Q^t V \qquad (5.12)$$

where $V = [v_1 \; v_2 \cdots v_m]$ *and* $X_j = [[v_j, v_1][v_j, v_2] \cdots [v_j, v_m]]$.

Proof: (following Arnold et al [5]). Let $q(\varepsilon, t)$ be a variation of the path $q(t)$. Then we can set

$$\frac{\partial f(q(\varepsilon, t))}{\partial t} = \sum_i v_i(f)p_i \tag{5.13}$$

$$\frac{\partial f(q(\varepsilon, t))}{\partial \varepsilon} = \sum_j v_j(f)w_j \tag{5.14}$$

Among other things, this associates with each fixed t along the path $q(t)$, a virtual displacement $\delta q = \sum_k v_k w_k = Vw$. Differentiating (5.13) with respect to ε and using the fact that differentiation with respect to q and ε commute we obtain:

$$\frac{\partial^2 f(q(\varepsilon, t))}{\partial \varepsilon \partial t} = \sum_i v_i(\frac{\partial f}{\partial \varepsilon})p_i + v_i(f)\frac{\partial p_i}{\partial \varepsilon} = \sum_{i,j} v_i(v_j(f))w_j p_i + \sum_i v_i(f)\frac{\partial p_i}{\partial \varepsilon}$$

Similarly, differentiating (5.14) with respect to t results in

$$\frac{\partial^2 f(q(\varepsilon, t))}{\partial t \partial \varepsilon} = \sum_{j,i} v_j(v_i(f))p_i w_j + \sum_j v_j(f)\frac{\partial w_j}{\partial t}$$

Equating these expressions, we obtain

$$\sum_i v_i(f)\frac{\partial p_i}{\partial \varepsilon} = \sum_{j,i} (v_j(v_i(f)) - v_i(v_j(f)))p_i w_j + \sum_j v_j(f)\frac{\partial w_j}{\partial t}$$

$$\sum_i v_i(f)\frac{\partial p_i}{\partial \varepsilon} = \sum_{j,i} [v_i, v_j](f)p_i w_j + \sum_j v_j(f)\frac{\partial w_j}{\partial t}$$

$$\sum_i v_i(f)\frac{\partial p_i}{\partial \varepsilon} = \sum_{j,i}\sum_k c_{ij}^k v_k(f)p_i w_j + \sum_j v_j(f)\frac{\partial w_j}{\partial t}$$

Now, renaming some summation indices

$$\sum_k v_k(f)\frac{\partial p_k}{\partial \varepsilon} = \sum_k v_k(f)\sum_{j,i} c_{ij}^k p_i w_j + \sum_k v_k(f)\frac{\partial w_k}{\partial t}$$

In view of the fact that this relation must hold for any f and any variation $q(\varepsilon, t)$, we have:

$$\frac{\partial p_k}{\partial \varepsilon} = \sum_{j,i} c_{ij}^k p_i w_j + \frac{\partial w_k}{\partial t}$$

We can use this formula to calculate the variation of the action integral

$$\delta \int_{t_1}^{t_2} \widetilde{L}(p,q)dt = \lim_{\varepsilon \to 0} \frac{d}{d\varepsilon} \int_{t_1}^{t_2} L(p(\varepsilon,t),q(\varepsilon,t))dt$$

$$\begin{aligned}
\frac{d}{d\varepsilon} \int_{t_1}^{t_2} \widetilde{L}(p(\varepsilon,t),q(\varepsilon,t))dt &= \int_{t_1}^{t_2} \left\{ \widetilde{L}_p \frac{\partial p}{\partial \varepsilon} + \widetilde{L}_q \frac{\partial q}{\partial \varepsilon} \right\} dt \\
&= \int_{t_1}^{t_2} \left\{ \sum_k \widetilde{L}_{p_k} \left(\sum_{j,i} c_{ij}^k p_i w_j + \frac{\partial w_k}{\partial t} \right) + \widetilde{L}_q \frac{\partial q}{\partial \varepsilon} \right\} dt
\end{aligned}$$

Integrating by parts the second term of the integral

$$\delta \int_{t_1}^{t_2} \widetilde{L}(p,q)dt = \sum_k \frac{\partial \widetilde{L}}{\partial p_k} w_k \bigg|_{t_1}^{t_2} + \int_{t_1}^{t_2} \sum_k \left[-\frac{d}{dt} \frac{\partial \widetilde{L}}{\partial p_k} + \sum_{i,j} c_{ik}^j \frac{\partial \widetilde{L}}{\partial p_j} p_i + v_k(\widetilde{L}) \right] w_k dt$$

Now, we use the fact that

$$Q^T \delta q = \sum_k Q^T v_k w_k$$

to obtain

$$\begin{aligned}
\int_{t_1}^{t_2} \left\{ \delta \widetilde{L}(p,q) + Q^T \delta q \right\} dt &= \sum_k \frac{\partial \widetilde{L}}{\partial p_k} w_k \bigg|_{t_1}^{t_2} \\
&+ \int_{t_1}^{t_2} \sum_k \left[-\frac{d}{dt} \frac{\partial \widetilde{L}}{\partial p_k} + \sum_{i,j} c_{ik}^j \frac{\partial \widetilde{L}}{\partial p_j} p_i + v_k(\widetilde{L}) + Q^T v_k \right] w_k dt
\end{aligned} \tag{5.15}$$

Since the variations w_k are independent in the interval $t_1 < t < t_2$ and vanish at its end points, we have the desired result.   ∎

Remark 5.133 (Remarks on Poincaré's Equations) *We will make a few general observations about Equations (5.11) and (5.12):*

1. *These equations are referred to as Poincaré's equations Arnold et al [5], Chetaev [6, 7] and Lagrange's equations in quasi-coordinates by Meirovitch [8] and Neimark and Fufaev [9]. They are related to Caplygin's equations and to the Boltzman-Hamel equations [9] and also to the generalized Lagrange equations of Noble (see Kwatny et al [12]).*

2. *Poincaré's equation (5.11) or (5.12) along with (5.6) form a closed system of first order differential equations which may be written in the form*

$$\dot{q} = V(q)p \tag{5.16}$$

$$\dot{p}^t \frac{\partial^2 \widetilde{L}}{\partial p^2} + p^t V^t \frac{\partial^2 \widetilde{L}}{\partial q^t \partial p} - \sum_{j=1}^{m} p_j \frac{\partial \widetilde{L}}{\partial p} U X_j - \frac{\partial \widetilde{L}}{\partial q} V = Q^t V \qquad (5.17)$$

3. If M is a Lie group G and $v_i, i = 1, \ldots, m$ are independent right-invariant vector fields on G, then $c_{ij}^k = $ constant, i.e., they are independent of q. If, in addition, $\widetilde{L}$ is invariant under right translations on G, then $v_k(\widetilde{L}) \equiv 0$ and $\widetilde{L}$ depends only on the quasi-velocities p. Thus, the Poincaré equations form a closed system of differential equations on the Lie algebra **g** of the Lie group G, i.e., in the quasi-velocities p.

4. Notice that $L(\dot{q}, q) = \widetilde{L}(U(q)\dot{q}, q)$. Thus, Lagrange's equations can be written

$$\frac{d}{dt} \frac{\partial L}{\partial \dot{q}} - \frac{\partial L}{\partial q} = \frac{d}{dt}\left(\frac{\partial \widetilde{L}}{\partial p} U(q) \right) - \frac{\partial \widetilde{L}}{\partial p} \frac{\partial U(q) \dot{q}}{\partial q} - \frac{\partial \widetilde{L}}{\partial q} = Q^t$$

from which we can derive:

$$\frac{d}{dt} \frac{\partial \widetilde{L}}{\partial p} - \sum_{j=1}^{m} p_j \frac{\partial \widetilde{L}}{\partial p} U X_j - \frac{\partial \widetilde{L}}{\partial q} V = Q^t V$$

Thus, formally, we can derive Poincare's equations from lagrange's equations.

Example 5.134 (Rotating Rigid Body) *A classic example of the application of Poincaré's equations is a rigid body with one point O fixed in space so that the body is free to rotate about O. The configuration of the body at any time t can be associated with the rotation matrix $L(t) \in SO(3)$ which characterizes the relative angular orientation of a body fixed frame with origin at O with respect to a space fixed frame with origin also at O. The velocity of rotation $\dot{L}(t)$ may be thought of as a tangent vector to the group $SO(3)$ at the point $L(t)$. It is commonplace to translate this vector to the tangent space to the group at the identity, therby associating the velocity with an element of the Lie algebra $so(3)$. As described in the previous chapter, we do this with left translations so that the skew symmetric matrix $\widetilde{\omega}_b = L^{-1}(t)\dot{L}(t)$ represents the angular velocity of the the body in the body frame. Recall that the matrix $\widetilde{\omega}_b$ can be associated with the vector ω_b via the relation*

$$\widetilde{\omega}_b = \begin{bmatrix} 0 & -\omega_{b3} & \omega_{b2} \\ \omega_{b3} & 0 & -\omega_{b1} \\ -\omega_{b2} & \omega_{b1} & 0 \end{bmatrix}$$

The mapping $f : L^{-1}\dot{L} \mapsto \omega$ defines an isomorphism of the Lie algebra $so(3)$ to R^3.

Notice also that a basis for the tangent space to SO(3) at the identity is

$$A_1 = \begin{bmatrix} 0 & 0 & 0 \\ 0 & 0 & -1 \\ 0 & 1 & 0 \end{bmatrix}, \ A_2 = \begin{bmatrix} 0 & 0 & 1 \\ 0 & 0 & 0 \\ -1 & 0 & 0 \end{bmatrix}, \ A_3 = \begin{bmatrix} 0 & -1 & 0 \\ 1 & 0 & 0 \\ 0 & 0 & 0 \end{bmatrix} \qquad (5.18)$$

We can regard these as a basis for the Lie algebra **so**(3).

If $R(t)$ is the position vector, in the space frame, of a point fixed in the body, then $R(t) = L(t)R(0)$. Thus,

$$V(t) = \dot{R}(t) = \dot{L}(t)R(0) = L(t)\widetilde{\omega}(t)R(t) = \omega_s \times R(t) \qquad (5.19)$$

where ω_s is the angular velocity vector represented in the space frame. Thus the abstract characterization of the rotational velocity of a rigid body does indeed coincide with the conventional notion of angular velocity. Similarly, if $r(t)$ is the inertial position vector of the body fixed point, represented in the body frame, then $r(t) = L^{-1}(t)R(t)$, and

$$v(t) = L^{-1}(t)V(t) = L^{-1}\omega_s \times R(t) = L^{-1}(t)\omega_s \times L(t)r(t) = \widetilde{\omega}_b(t)r(t) = \omega_b(t) \times r(t)$$

which is the body frame equivalent of (5.19).

Suppose that $I_b = \mathrm{diag}(I_1, I_2, I_3)$ is the inertia tensor in principle (orthogonal) body coordinates and suppose e_1, e_2, e_3 denote unit vectors of the principle axes, indexed in the usual way to provide a right hand system: $e_1 \times e_2 = e_3$, $e_2 \times e_3 = e_1$, $e_3 \times e_1 = e_2$. Let v_1, v_2, v_3 denote the preimages of e_1, e_2, e_3 under the isomorphism $f : \mathbf{so}(3) \to R^3$. Then v_1, v_2, v_3 are left-invariant vector fields on SO(3) and they satisfy.[3]

$$[v_1, v_2] = v_3, \ [v_2, v_3] = v_1, \ [v_3, v_1] = v_2 \qquad (5.20)$$

Thus, we can define quasi-velocities in terms of these vector fields as in (5.6) and (5.7). Let

$$\omega_b = \omega_{b1}e_1 + \omega_{b2}e_2 + \omega_{b3}e_3 \qquad (5.21)$$

so that the kinetic energy can be written

$$T(\omega_b) = \tfrac{1}{2}\left\{ I_1\omega_{b1}^2 + I_2\omega_{b2}^2 + I_3\omega_{b3}^2 \right\} = \tfrac{1}{2}\omega_b^T I_b \omega_b \qquad (5.22)$$

The potential energy is zero, so we have $\widetilde{L} = T(\omega_b)$. Since $\widetilde{L}$ is independent of the coordinates and in the absence of external forces Poincaré's equations reduce to

$$\frac{d}{dt}\frac{\partial \widetilde{L}}{\partial \omega_b} - \sum_{j=1}^{3} \omega_{bj}\frac{\partial \widetilde{L}}{\partial \omega_b}UX_j = 0 \qquad (5.23)$$

Recall that (think local) $X_j = [[v_j, v_1]\ [v_j, v_2]\ [v_j, v_3]]$, from which we compute

$$X_1 = [0 \ v_3 \ -v_2], \ X_2 = [-v_3 \ 0 \ v_1], \ X_3 = [v_2 \ -v_1 \ 0]$$

---

[3]This can also be verified by computing the commutators of the basis elements (5.18) via AB-BA

which can be expressed

$$X_1 = [v_1 \; v_2 \; v_3] \begin{bmatrix} 0 & 0 & 0 \\ 0 & 0 & -1 \\ 0 & 1 & 0 \end{bmatrix} \tag{5.24}$$

$$X_2 = [v_1 \; v_2 \; v_3] \begin{bmatrix} 0 & 0 & 1 \\ 0 & 0 & 0 \\ -1 & 0 & 0 \end{bmatrix} \tag{5.25}$$

$$X_3 = [v_1 \; v_2 \; v_3] \begin{bmatrix} 0 & -1 & 0 \\ 1 & 0 & 0 \\ 0 & 0 & 0 \end{bmatrix} \tag{5.26}$$

Since $U = [v_1 \; v_2 \; v_3]^{-1}$ *we have*

$$\sum_{j=1}^{3} \omega_{bj} \frac{\partial \widetilde{L}}{\partial \omega_b} U X_j = \omega_b^T I_b \left\{ \omega_{b1} \begin{bmatrix} 0 & 0 & 0 \\ 0 & 0 & -1 \\ 0 & 1 & 0 \end{bmatrix} \right.$$

$$\left. + \omega_{b2} \begin{bmatrix} 0 & 0 & 1 \\ 0 & 0 & 0 \\ -1 & 0 & 0 \end{bmatrix} + \omega_{b3} \begin{bmatrix} 0 & -1 & 0 \\ 1 & 0 & 0 \\ 0 & 0 & 0 \end{bmatrix} \right\}$$

$$= \omega_b^T I_b \widetilde{\omega}_b^T$$

and finally,

$$I_b \dot{\omega}_b + \widetilde{\omega}_b I_b \omega_b = 0 \tag{5.27}$$

These are recognized as Euler's equations.

Example 5.135 (Submerged Rigid Body) *Consider a rigid body free to translate and rotate in an fricitionless, incompressible fluid of density ρ and infinite extent. The configuration manifold is the group of rotations and translations of R^3, $SE(3)$. As discussed in the previous chapter, the rigid body configuration may be regarded as the matrix*

$$X = \begin{bmatrix} L^T & R \\ 0 & 1 \end{bmatrix}$$

and the corresponding velocity is an element in the corresponding Lie algebra **se(3)**,

$$p = X^{-1} \dot{X} = \begin{bmatrix} \widetilde{\omega}_b & v_b \\ 0 & 0 \end{bmatrix} \leftrightarrow \begin{bmatrix} \omega_b \\ v_b \end{bmatrix}$$

Recall that $SE(3)$ is the product of the rotation group $SO(3)$ and the translation group R^3. Its Lie algebra **se(3)** *has basis vectors:*

$$A_1 = \begin{bmatrix} 0 & 0 & 0 & 0 \\ 0 & 0 & -1 & 0 \\ 0 & 1 & 0 & 0 \\ 0 & 0 & 0 & 0 \end{bmatrix}, \; A_4 = \begin{bmatrix} 0 & 0 & 0 & 1 \\ 0 & 0 & 0 & 0 \\ 0 & 0 & 0 & 0 \\ 0 & 0 & 0 & 0 \end{bmatrix}$$

$$A_2 = \begin{bmatrix} 0 & 0 & 1 & 0 \\ 0 & 0 & 0 & 0 \\ -1 & 0 & 0 & 0 \\ 0 & 0 & 0 & 0 \end{bmatrix}, \quad A_5 = \begin{bmatrix} 0 & 0 & 0 & 0 \\ 0 & 0 & 0 & 1 \\ 0 & 0 & 0 & 0 \\ 0 & 0 & 0 & 0 \end{bmatrix}$$

$$A_3 = \begin{bmatrix} 0 & -1 & 0 & 0 \\ 1 & 0 & 0 & 0 \\ 0 & 0 & 0 & 0 \\ 0 & 0 & 0 & 0 \end{bmatrix}, \quad A_6 = \begin{bmatrix} 0 & 0 & 0 & 0 \\ 0 & 0 & 0 & 0 \\ 0 & 0 & 0 & 1 \\ 0 & 0 & 0 & 0 \end{bmatrix}$$

Let $v_1, v_2, v_3, v_4, v_5, v_6$ denote the corresponding left-invariant vector fields. Then we easily compute the commutator relations. The nontrivial ones are:

$$[v_1, v_2] = v_3, \quad [v_1, v_3] = -v_2, \quad [v_1, v_5] = v_6, \quad [v_1, v_6] = -v_5, \quad [v_2, v_3] = v_1,$$

$$[v_2, v_4] = -v_6, \quad [v_2, v_6] = v_4, \quad [v_3, v_4] = v_5, \quad [v_3, v_5] = -v_4$$

Thus, we can express each of the X_j in the form

$$X_j = [v_1 v_2 v_3 v_4 v_5 v_6] \Lambda_j = V \Lambda_j$$

where each Λ_j is a 6×6 column 'reordering' matrix. These are

$$\Lambda_1 = \begin{bmatrix} 0 & 0 & 0 & 0 & 0 & 0 \\ 0 & 0 & 0 & 0 & 0 & 0 \\ 0 & 0 & 1 & 0 & 0 & 0 \\ 0 & -1 & 0 & 0 & 0 & 0 \\ 0 & 0 & 0 & 0 & 0 & 1 \\ 0 & 0 & 0 & 0 & -1 & 0 \end{bmatrix}, \quad \Lambda_2 = \begin{bmatrix} 0 & 0 & 0 & 0 & 0 & 0 \\ 0 & 0 & -1 & 0 & 0 & 0 \\ 0 & 0 & 0 & 0 & 0 & 0 \\ 1 & 0 & 0 & 0 & 0 & -1 \\ 0 & 0 & 0 & 0 & 0 & 0 \\ 0 & 0 & 0 & 1 & 0 & 0 \end{bmatrix}$$

$$\Lambda_3 = \begin{bmatrix} 0 & 1 & 0 & 0 & 0 & 0 \\ -1 & 0 & 0 & 0 & 0 & 0 \\ 0 & 0 & 0 & 0 & 0 & 0 \\ 0 & 0 & 0 & 0 & 1 & 0 \\ 0 & 0 & 0 & -1 & 0 & 0 \\ 0 & 0 & 0 & 0 & 0 & 0 \end{bmatrix}, \quad \Lambda_4 = \begin{bmatrix} 0 & 0 & 0 & 0 & 0 & 0 \\ 0 & 0 & 0 & 0 & 0 & -1 \\ 0 & 0 & 0 & 0 & 1 & 0 \\ 0 & 0 & 0 & 0 & 0 & 0 \\ 0 & 0 & 0 & 0 & 0 & 0 \\ 0 & 0 & 0 & 0 & 0 & 0 \end{bmatrix}$$

$$\Lambda_5 = \begin{bmatrix} 0 & 0 & 0 & 0 & 0 & -1 \\ 0 & 0 & 0 & 0 & 0 & 0 \\ 0 & 0 & 0 & 1 & 0 & 0 \\ 0 & 0 & 0 & 0 & 0 & 0 \\ 0 & 0 & 0 & 0 & 0 & 0 \\ 0 & 0 & 0 & 0 & 0 & 0 \end{bmatrix}, \quad \Lambda_6 = \begin{bmatrix} 0 & 0 & 0 & 0 & 1 & 0 \\ 0 & 0 & 0 & -1 & 0 & 0 \\ 0 & 0 & 0 & 0 & 0 & 0 \\ 0 & 0 & 0 & 0 & 0 & 0 \\ 0 & 0 & 0 & 0 & 0 & 0 \\ 0 & 0 & 0 & 0 & 0 & 0 \end{bmatrix}$$

Any motion of an ideal fluid can be characterized by a velocity potential ϕ that satisfies the partial differential equation $\nabla^2 \phi = 0$ in the coordinates (x, y, z) of a body fixed frame. The velocity of at any point within the fluid or on the body surface is given by $u = -\nabla\phi$ and the pressure by $P = \rho\phi$. Khirchoff showed that *the if the fluid is at rest at infinity ($\nabla \phi = 0$ at infinity), the potential function is a linear function of the rigid body velocity, i.e.,*

$$\phi(x, y, z) = a(x, y, z)p$$

the coefficient row vector can be explicitly computed for simple rigid body shapes, such as an elipsoid, or it can be computed via finite element approximaton for shapes of arbitrary complexity.

The kinetic energy of the fluid can be expressed as the integral of the fluid pressure times the normal velocity over the surface of the body. The result is that the fluid kinetic energy is a quadratic function of the body velocity

$$T_f = \tfrac{1}{2}p^T M_f p, \; M_f^T = M_f = \begin{bmatrix} M_{f11} & M_{f12} \\ M_{f12}^T & M_{f22} \end{bmatrix}$$

For convenience, we fix the origin of the body frame at the center of bouyancy and suppose that in this frame the center of mass is located at r_{com}. If the body has mas m_b and inertia matrix J_b (about the center of mass) the the body kinetic energy is

$$T_b = \tfrac{1}{2}p^T M_b p, \; M_b = \begin{bmatrix} m_b I & -m_b \widetilde{r}_{com} \\ m_b \widetilde{r}_{com} & J_b \end{bmatrix}$$

Thus, we have

$$T(p) = \tfrac{1}{2}p^T M p, \; M = M_f + M_b$$

The system potential energy arises from the gravitational field. If we assume that the body has neutral bouyancy (displace fluid mass is the same as the vehicle mass) and that the center of bouyancy and center of mass coincide, then the potential energy function is identically zero. The Lagrangian, then, is $\widetilde{L} = T(p)$. Now, we compute

$$\sum_{j=1}^{6} p_j \frac{\partial \widetilde{L}}{\partial p} U X_j = (p^T M) \sum_{j=1}^{6} p_j \Lambda_j = p^T M \begin{bmatrix} \widetilde{\omega}_b^T & \widetilde{v}_b^T \\ 0 & \widetilde{\omega}_b^T \end{bmatrix}$$

and, finally

$$M \begin{bmatrix} \dot{\omega}_b \\ \dot{v}_b \end{bmatrix} + \begin{bmatrix} \widetilde{\omega}_b & \widetilde{v}_b \\ 0 & \widetilde{\omega}_b \end{bmatrix} M \begin{bmatrix} \omega_b \\ v_b \end{bmatrix} = 0$$

Notice that if we define the momentum

$$\begin{bmatrix} \Pi \\ P \end{bmatrix} = M \begin{bmatrix} \omega_b \\ v_b \end{bmatrix}$$

Then the equations can be expressed (recall, $\widetilde{a}b = a \times b$, $a, b \in R^3$)

$$\dot{\Pi} + \omega_b \times \Pi + v_b \times P = 0$$

$$\dot{P} + \omega_b \times P = 0$$

These can be compared with those given by Leonard [13].

5.3   Chain and Tree Configurations

In general a multibody system can be viewed in terms of an underlying tree structure upon which is imposed additional algebraic and/or differential constraints. In this section we describe the assembly of dynamical equations for the tree. In later sections we show how these equations are modified to accommodate any additional constraints. A tree can be defined in terms of a set of chains, each beginning at the root body. We describe in some detail the process of modeling a chain. Extending the process to a tree requires is merely a book keeping process.

5.3.1   Kinetic Energy and Poincaré's Equations

The key issue in developing Lagrange's or Poincaré's equations is the formulation of the kinetic energy function and we focus on that construction. It is necessary to define a spatial inertia tensor. Recall the recursive velocity relation from the previous Chapter.

$$V^k(k) = \phi(k, k-1)V^{k-1}(k-1) + H^k(k)\beta^k(k), \quad k = 1, \ldots, K, \quad V^0(0) = 0$$
$$(5.28)$$

Consider the kth rigid link and let $I_{cm}(k)$ denote the inertia tensor about the center of mass in coordinates F^k, $m(k)$ denote the mass, and $a(k)$ denote the position vector from the center of mass to an arbitrary point O. The spatial inertia about the center of mass, M_{cm}, and about O, M_o, are

$$M_{cm}(k) = \begin{bmatrix} I_{cm} & 0 \\ 0 & mI \end{bmatrix}$$

$$M_o(k) = \phi^*(a)M_{cm}(k)\phi(a) = \begin{bmatrix} I_o & m\widetilde{a} \\ -m\widetilde{a} & mI \end{bmatrix} \qquad (5.29)$$

where I_o is the inertia tensor about O.

The spatial velocity and spatial inertia matrix and, hence, the kinetic energy function for the entire chain can now be conveniently constructed. Let us define the chain spatial velocity and joint quasi-velocity

$$V = [V^T(1), \ldots, V^T(K)]^T, \quad \beta = [\beta^T(1), \ldots, \beta^T(k)]^T \qquad (5.30)$$

so that we can write

$$V = \Phi \mathcal{H} \beta \qquad (5.31)$$

where

$$\Phi = \begin{bmatrix} I & 0 & \ldots & 0 \\ \phi(2,1) & I & \ldots & 0 \\ \vdots & \vdots & \vdots & \vdots \\ \phi(K,1) & \phi(K,2) & \ldots & I \end{bmatrix}$$

$$\mathcal{H} = \begin{bmatrix} H(1) & 0 & \ldots & 0 \\ 0 & H(2) & \ldots & 0 \\ \vdots & \vdots & \vdots & \vdots \\ 0 & 0 & \ldots & H(K) \end{bmatrix} \tag{5.32}$$

$$\phi(i,j) = \phi(i, i-1)\ldots\phi(j+1, j), \quad i = 2, .., K \quad \text{and} \quad j = 1, .., K-1$$

The following result is easily verified.

Proposition 5.136 *The kinetic energy function for the chain consisting of links 1 through K is*

$$K.E._{chain} = \tfrac{1}{2}\beta^T \mathcal{M}\beta \tag{5.33}$$

where the chain inertia matrix is

$$\mathcal{M} = \mathcal{H}^T \Phi^T M \Phi \mathcal{H}, \quad M = diag\{M_o(1), \ldots, M_o(K)\} \tag{5.34}$$

Remark 5.137 (The Structure of Poincaré's Equations) *The above definitions and constructions provide the kinetic energy function in the form*

$$\widetilde{T}(q, p) = (1/2)p^T M(q)p$$

Hence, we reduce Poincaré's Equations to the form:

$$M(q)\dot{p} + C(q, p)p + F(q) = Q_p \tag{5.35}$$

where

$$C(q, p) = -\left[\frac{\partial M(q)p}{\partial q} V(q)\right] + \frac{1}{2}\left[\frac{\partial M(q)p}{\partial q} V(q)\right]^T + \left[\sum_{j=1}^{m} p_j X_j^T\right] V^{-T} \tag{5.36}$$

$$F(q) = V^T(q)\frac{\partial \mathcal{U}(q)}{\partial q^T}, \quad Q_p = V^T(q)Q \tag{5.37}$$

$\mathcal{U}(q)$ *is the potential energy function. Notice that Q_p denotes the generalized forces represented in the p-coordinate frame whereas Q denotes the generalized forces in the $\dot{q}$-coordinate frame (aligned with q).*

Remark 5.138 (Remark on Computations) *The key point to be noted is that the matrix Φ (and hence the product $\Phi\mathcal{H}$) can be recursively computed. Thus, we can compute the spatial velocity of any or all of the bodies via (5.30) and the inertia matrix using (5.34). Once this is done, we compute $C(q, p)$, $F(q)$, and Q_p explicitly using equations (5.36) and (5.37), assuming that the potential energy function $P(q)$ and the generalized force vector Q are available. In general, both P and Q are defined in terms of coordinates and velocities (in the case of Q) other than the configuration coordinates q and the quasi-velocities p. Thus, it is necessary to develop any transformations required to* *obtain P and Q in terms of q and p. We will illustrate this process below. For now, we note that velocity transformations are recursively constructed using relations like (5.28) or (5.31), and coordinate transformations are built up from the usual sequential multiplications of configuration matrices. Assembly of the system gravitational potential energy and end effector position and orientation, needed below, require constructions of this type.*

Remark 5.139 (Poincaré vs. Lagrange Equations) *Notice that the kinetic energy can be expressed in terms of $\dot{q}$ rather than p,*

$$T(q,p) = (1/2)\dot{q}^T \left\{ V^{-T}(q)M(q)V(q)^{-1} \right\} \dot{q},$$

and hence we have the essential data to construct Lagrange's equations rather than Poincaré's equations. However, Poincaré's equations may have important advantages. An obvious and practical one is the relative simplicity of the inertia matrix. However, there is an important theoretical consideration as well. Lagrange's equations fundamentally constitute a local representation whenever local coordinates are introduced, whereas Poincaré's equations may still admit a global description of the dynamics. This is easily seen by comparing the Lagrange and Poincaré formulations for the dynamics of a rotating rigid body. We do this below.

Example 5.140 (Double Pendulum) *As a simple example we consider the double pendulum.*

Define Joint Data

```
In[79]:= r1={1};H1={{1},{0},{0},{0},{0},{0}};
         q1={a1x};p1={w1x};
         r2={1};H2={{1},{0},{0},{0},{0},{0}};
         q2={a2x};p2={w2x};
         JointLst={{r1,H1,q1,p1},{r2,H2,q2,p2}};
```

Define Body Data

```
In[80]:= com1={0,0,-l}; mass1=m; out1={2,{0,0,-l}};
         Inertia1=DiagonalMatrix[{0,0,0}];
         com2={0,0,-l}; mass2=m; out2={3,{0,0,-l}};
         Inertia2=DiagonalMatrix[{0,0,0}];
         BodyLst={{com1,{out1},mass1,Inertia1},{com2,{out2},mass2,Inertia2}};
```

Define Interconnection Structure

```
In[81]:= TreeLst={{{1,1},{2,2}}};
```

Define Potential Energy

In this case only gravity contributes to the potential energy. The only generalized forces are external torques acting at the two joints.

```
In[82]:= g=gc; PE=0; Q={T1,T2};
```

In[83]:= {V,X,H,M,Cp,Fp,p,q}=CreateModel[JointLst,BodyLst,TreeLst,g,PE,Q];

We can look at some results.

In[84]:= V
Out[84]= {{{1}},{{1}}}

Recall that V returns as a list of kinematic matrices - one for each joint. Hence in this case we get two 1×1 matrices. This can be assembled into a single block diagonal matrix but it is more efficient to retain the list form.

In[85]:= M
Out[85]= $\{\{2 \ l^2 \ m + l^2 \ m \ \text{Cos}[a2x] +$
$l \ \text{Cos}[a2x] \ (l \ m + l \ m \ \text{Cos}[a2x]) + l^2 \ m \ \text{Sin}[a2x]^2, l^2 \ m + l^2 \ m \ \text{Cos}[a2x]\},$
$\{l^2 \ m + l^2 \ m \ \text{Cos}[a2x], l^2 \ m\}\}$

In[86]:= Fp//MatrixForm
Out[86]= $\begin{pmatrix} -T1 - \text{gc} \ l \ m \ (-2 \ \text{Sin}[a1x] - \text{Sin}[a1x + a2x]) \\ -T2 + \text{gc} \ l \ m \ \text{Sin}[a1x + a2x] \end{pmatrix}$

We will return to this example below.

Example 5.141 (Thin Disk) *As another example of the application of these functions let us consider a thin disk free to rotate about its center of mass in space without any external or gravitational forces. The single joint defining relative motion between the space frame and the body frame is considered as a simple spherical joint.*

In[87]:= $r1 = \{3\}$;
$H1 = Join[IdentityMatrix[3], DiagonalMatrix[\{0, 0, 0\}]]$;
$q = \{q1, q2, q3\}; \quad p = \{w1, w2, w3\}$;
$JointLst = \{\{r1, H1, q, p\}\}$;
$m1 = 5; R1 = 2; I1 =$
$DiagonalMatrix[\{(1/4) * m1 * R1\hat{\ }2, (1/4) * m1 * R1\hat{\ }2, (1/2) * m1 * R1\hat{\ }2\}]$;
$cm1 = \{0, 0, 0\}; oc1 = \{2, \{1, 0, 0\}\}$;
$BodyLst = \{\{cm1, \{oc1\}, m1, I1\}\}$;
$TreeLst = \{\{\{1, 1\}\}\}; Q = \{0, 0, 0\}$;
$\{V, X, H, M, Csys, Fsys, psys, qsys\} =$
$CreateModel[JointLst, BodyLst, TreeLst, g, 0, Q]$;

We summarize the results as follows. The Euclidean configuration matrix $X(q)$ is given in Example (4.127). The kinematic equations are

$$\dot{q} = V(q)p$$

$$V(q) = \begin{bmatrix} 1 & \sin q1 \tan q2 & \cos q1 \tan q2 \\ 0 & \cos q1 & -\sin q1 \\ 0 & \sec q2 \sin q1 & \cos q1 \sec q2 \end{bmatrix},$$

and the dynamic equations

$$M(q)\dot{p} + F(q, p) = 0$$

$$M(q) = \begin{bmatrix} 5 & 0 & 0 \\ 0 & 5 & 0 \\ 0 & 0 & 10 \end{bmatrix}, \quad F(t,w) = C(q,p)p = \begin{bmatrix} 5\,w1\,w2 \\ -5\,w1\,w2 \\ 0 \end{bmatrix}$$

Poincaré's equations are recognizable as Euler's equations.

It is interesting to repeat this calculation with the simple spherical joint replaced by a compound 3 dof universal joint. The only change required in the above Mathematica code is to replace the definition of r1={3} *by* r1={1,1,1}. *As noted above, the parameterization of the configuration of the rigid body is the same as that of the simple joint, i.e., $X(q)$ is unchanged and $V(q) = I_3$. The other relevant results are as follows*

$M(q) =$

$$\begin{bmatrix} 5 & 0 & -5\sin q2 \\ 0 & 5/2(3 - \cos q1) & -5/2\cos q2\sin 2q1 \\ -5\sin q2 & -5/2\cos q2\sin 2q1 & 10\cos q1^2\cos q2^2 + 5\cos q2^2\sin q1^2 + 5\sin q2^2 \end{bmatrix}$$

$$
\begin{aligned}
F_1(q,w) &= w3(-5w2\cos 2q1\cos q2 - 5w3\cos q2^2\sin 2q1)/2 + w2(5u3\cos q2 \\
&+ \quad (-5w3\cos 2q1\cos q2 + 5w2\sin 2q1)/2) \\
F_2(q,w) &= 1(-5w3\cos q2/2 + 5w3\cos 2q1\cos q2 - 5w2\sin 2q1) \\
&- \quad 5w2w3\sin 2q1\sin q2/4 \\
&+ \quad (w3(-5w1\cos q2 + 5w2\sin 2q1\sin q2/2 - 5w3\cos q1^2\sin 2q2))/2 \\
F_3(q,w) &= w1(5w2\cos 2q1\cos q2 + 5w3\cos q2^2\sin 2q1) \\
&+ \quad w2(5w1\cos q2 - 5w2\sin 2q1\sin q2/2 + 5w3\cos q1^2\sin 2q2)
\end{aligned}
$$

These equations are, of course, Lagrange's equations – a consequence of the fact that the joint formulation commits us to velocity coordinates alligned with the configuration coordinates. The simplicity of the kinematic matrix is more than offset by the complexity of the dynamical equations. Notice also that the dynamical equations in the previous case are independent of the configuration parameters. They are globally valid equations, whereas the latter are not.

5.3.2   Generalized Force Calculations

Building models not only requires the construction of the kinematic relations and kinetic energy function but it is also necessary to characterize the forces that act on the system. This is normally accomplished through the definition of a potential energy function, a dissipation function and/or the specification of generalized forces. The *ProPac* function CreateModel accepts each of these as arguments. Several computational tools are provided in *ProPac* to assist in the development of these quantities.

Potential Energy Constructions

The potential energy function $\mathcal{U}(q)$ is typically used to characterize forces due to gravity and elastic storage elements. The associated generalized force is

$$F(q) = -\frac{\partial \mathcal{U}(q)}{\partial q}$$

or

$$F(q) = -V(q)^T \frac{\partial \mathcal{U}(q)}{\partial q}$$

in the p-frame. Computing the gravitational potential energy for a multibody mechanical system is a straightforward but tedious task, since it is necessary to locate the center of mass for each body in space. For convenience, this calculation is automated in CreateModel for a uniform gravitational field acting in the negative z-direction. To use a coordinate system with z-axis pointing downward requires specification of the gravitational constant as $-g$.

The potential energy associated with elastic components can also involve complicated geometry. Two functions in *ProPac* can be helpful in performing these calculations. SpringPotential computes the potential energy expression in terms of the configuration coordinates for a spring connected between two nodes located anywhere in the system. It is assumed that the spring potential energy is known as a function of the spring length.

LeafPotential is designed to facilitate computing the potential energy of components (such as tires) that interact elastically with the space frame. The potential energy function is presumed known in terms of the spatial location of a contact node. The potential energy function is computed as a function of the generalized coordinates.

Dissipation Functions

Dissipation functions are generally specified in terms of the configuration coordinates and velocities in the form of the Rayleigh dissipation potential

$$\mathcal{R}(\dot{q}, q) = \dot{q}^T A(q)\dot{q} + a^T(q)\dot{q} \tag{5.38}$$

or in the more general more general Lur'e form (see [14])

$$D(q, p) = \sum_i f_i(q)v_i(p) \tag{5.39}$$

where p denotes the quasi-velocities. The generalized force associated with the Rayleigh dissipation function (in the p-frame) is

$$F(p, q) = -V^T(q)\frac{\partial R}{\partial \dot{q}}(V(q)p, q)$$

The generalized force associated with the Lur'e potential is

$$F(p,q) = -\frac{\partial \mathcal{D}(p,q)}{\partial p}$$

The function `CreateModel` accepts Lur'e type dissipation potential as an argument and computes the generalized force. The calculation required for the Rayleigh function is performed by the function `RayleighDissipationForce`.

A *ProPac* function, `DamperPotential`, can be used to construct the dissipation potential associated with a damper connected between two nodes in the system. The function is analogous to `SpringPotential`. It requires that the damper can be characterized in terms of a dissipation function depending on the relative velocity across the damper. Then a Lur'e type potential function is constructed as a function of the system coordinates and quasi-velocities.

Applied Force

One way to construct the generalized force associated with an external force applied at a specific node in the system is to view the node as an energy port. Suppose a force $F_A \in R^6$, composed of three torques and three forces, acts at a node A in a multibody system. Let V_A denote the spatial velocity at A. Both F_A and V_A are represented in the coordinates of a body fixed frame at node A. The instantaneous power flowing into the system is $\mathcal{P} = F_A^T V_A$. In general, the application of Equations (4.48) and (4.54) of the last chapter, as in the derivation of Equation (5.31), leads to the representation of the spatial velocity at node A in terms of the system coordinates and quasi-velocities. If F_A is defined as a function of system coordinates and quasi-velocities, then $\mathcal{P}$ can be expressed in terms of these variables. Once $\mathcal{P}(p,q)$ is constructed we obtain

$$Q_p = \frac{\partial \mathcal{P}(p,q)}{\partial p} \tag{5.40}$$

To facilitate computing Q_p *ProPac* provides several computational tools. One of these, `GeneralizedForce`, is based on the construction described above. The force F_A is assumed to be given as an expression involving the spatial velocity components of a body fixed frame at the node of force application. The function `GeneralizedForce` then computes the generalized force.

Impact

The Hertz model of impact incorporates a simple characterization of the force interaction between two elastic bodies during the contact phase of a collision. During a collision the two actual colliding bodies deform. However, Hertz introduces a parameter that defines the relative position of two nondeforming virtual bodies (labeled A and B), x (see Figure (5.1)). Then a force (on body A) displacement relationship that applies during the contact phase is introduced:

$$f(x) = -K^{-3/2}x^{3/2}$$

where K is a constant that depends on the material properties and (local) geometry of the colliding bodies:

$$K = \tfrac{4}{3}\left\{q/\left(Q_1 + Q_2\sqrt{a+b}\right)\right\}$$

As an example, consider a sphere of radius r colliding with a plane surface. In

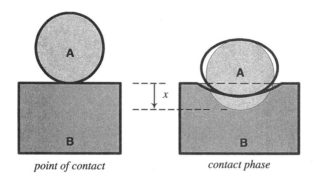

point of contact   contact phase

Figure 5.1: During the contact phase of a collision the two actual bodies deform. The relative position (sometimes called relative approach) of the two bodies is an indicator of the relative location of two (virtual) undeformed bodies.

this case,

$$A = B = 1/2r, \; q = \pi^{1/3}, \; Q_1 = (1 - \mu_1^2)/E_1\pi, \; Q_2 = (1 - \mu_2^2)/E_2\pi$$

where E_i and μ_i denote Young's modulus and Poisson's ratio for the respective bodies.

Notice that the force interaction can be completely characterized by a potential energy function, e.g. for the interaction described above, it has the form:

$$V(x) = \begin{cases} (2/5)K^{-3/2}x^{5/2} & x > 0 \\ 0 & x \le 0 \end{cases}$$

The interaction force is then recovered by $f(x) = -\partial V/\partial x$. In summary, a Hertz impact model consists of: a potential function $\mathcal{V}(x) = \pi(x)u(x)$, where π is a differentiable map $\pi : R \to R$ with $\pi(0) = 0$, and an associated force function $f(x) = -\mathcal{V}_x(x) = -\pi_x(x)u(x)$, where $u(x)$ is the unit step function.

Backlash

Suppose that a symmetric backlash element with dead zone parameter ε has a smooth force function $f(x)$ during the contact phase defined by a potential energy function $\mathcal{V}(x)$. Then the backlash mechanism can be characterized by a potential function:

$$\mathcal{V}_b = \mathcal{V}(|x| - \varepsilon) = \pi(|x| - \varepsilon)u(|x| - \varepsilon)$$

so that the backlash force is given by

$$f_b(x) = \begin{cases} f(x - \varepsilon) & x > \varepsilon \\ 0 & -\varepsilon \leq x \leq \varepsilon \\ -f(-x + \varepsilon) & x < -\varepsilon \end{cases}$$

Working with backlash is facilitated by two functions in *ProPac*

    BacklashPotential,

and BacklashForce.

The former constructs the backlash potential given a Hertz impact potential function and a backlash parameter. The latter returns the associated force. Note that the backlash potential can be included as part of the potential energy function.

Friction

Friction, particularly in joints, is an important factor in many situations. A basic characterization of friction as a static function of contact velocity should include viscous, coulomb and static (Stribeck) effects. In order to do this efficiently with large scale multibody models we can use a dissipation potential of Lur'e type (5.39).

Suppose the friction depends on a single velocity variable, v, that can ultimately be expressed as a function of the system coordinates and quasi-velocities. Potential functions giving rise to viscous, coulomb and static effects are:

- viscous: $\frac{1}{2}c_v v^2$

- coulomb: $c_c v \operatorname{sgn}(v)$

- static: $\frac{1}{2}c_s v_s \sqrt{\pi}\operatorname{erf}\left(v/v_s\right)\operatorname{sgn}(v)$

The friction parameters used above are:

c_v, viscous friction coefficient

c_c, coulomb friction coefficient

c_s, static friction coefficient

v_s, Stribeck velocity

The dissipation potential associated with a joint can be assembled with the function `JointFrictionPotential`.

Example 5.142 (Two masses with backlash and friction) *Consider the system illustrated in Figure (5.2). The system is composed of two bodies and*

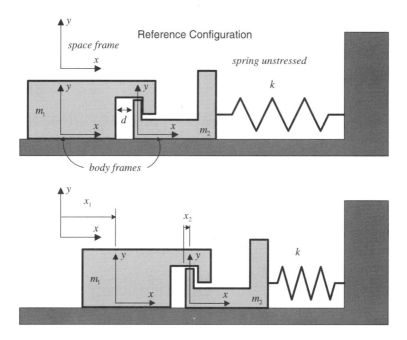

Figure 5.2: Two masses interact through a 'loose' joint exhibiting backlash.

two joints. Body one translates relative to the space frame and body two translates relative to body one. Thus, the joint definitions are as follows.

```
In[88]:=  r1 = {1}; H1 = Transpose[{{0, 0, 0, 1, 0, 0}}];
          q1 = {x1}; p1 = {v1};
          r2 = {1}; H2 = Transpose[{{0, 0, 0, 1, 0, 0}}];
          q2 = {x2}; p2 = {v2};
          JointLst = {{r1, H1, q1, p1}, {r2, H2, q2, p2}};
```

Now, we define the body data. The masses can be treated as point masses so we define the inertia matrices to be zero.

In[89]:= $com1 = \{0, 0, 0\}$; $mass1 = m1$; $out1 = \{2, \{0, 0, 0\}\}$;
$Inertia1 = \{\{0, 0, 0\}, \{0, 0, 0\}, \{0, 0, 0\}\}$;
$com2 = \{0, 0, 0\}$; $mass2 = m2$; $out2 = \{3, \{0, 0, 0\}\}$;
$Inertia2 = \{\{0, 0, 0\}, \{0, 0, 0\}, \{0, 0, 0\}\}$;
$BodyLst =$
  $\{\{com1, \{out1\}, mass1, Inertia1\}, \{com2, \{out2\}, mass2, Inertia2\}\}$;
$TreeLst = \{\{\{1, 1\}, \{2, 2\}\}\}$;

*Bodies one and two interact through a backlash potential function constructued
as follows. A simple linear material compliance is assumed. The backlash parameter is $d/2$.*

In[90]:= $PEBack[x\_] := (k1 * x\hat{}2)/2$;
$PE1 = BacklashPotential[PEBack, d/2, x2]$

Out[90]= $\dfrac{1}{2} \ k1 \ \left(-\dfrac{d}{2} + Abs[x2]\right)^2 \ UnitStep\left[-\dfrac{d}{2} + Abs[x2]\right]$

*The spring is assumed linear so a simple quadratic potential energy function is
used.*

In[91]:= $PE2 = (1/2) * k * (x1 + x2)\hat{}2$;
$PE = PE1 + PE2$;

*The sliding friction of body one is assumed to be significant so a dissipation
potential function is constructed.*

In[92]:= $DissPot1 = JointFrictionPotential[v1, cv, cc, cs, vs]$

Out[92]= $\dfrac{cv \ v1^2}{2} + cc \ Abs[v1] + \dfrac{1}{2} \ (-cc + cs) \ \sqrt{\pi} \ vs \ Erf\left[\dfrac{v1}{vs}\right] \ Sign[v1]$

*Allowing for forces FF_1 and FF_2 to act on bodies one and two, respctively, the
model is assembled as follows.*

In[93]:= $Q = \{FF1, FF2\}$;
$\{JV, JX, JH, MM, Cp, Fp, pp, qq\} =$
  $CreateModel[JointLst, BodyLst, TreeLst, -g, PE, DissPot1, Q]$;

`Computing Joint Kinematics`

`Computing joint 1 kinematics`

`Computing joint 2 kinematics`

`Computing Potential Functions`

`Computing Inertia Matrix`

`Computing Poincare Function`

Let us examine the F_p function.

In[94]:= Fp

Out[94]=

$$\left\{ -FF1 + cv\ v1 + k\ (x1 + x2) + (-cc + cs)\ e^{-\frac{v1^2}{vs^2}}\ Sign[v1] + \right.$$
$$cc\ (-UnitStep[-v1] + UnitStep[v1]), -FF2 + k\ (x1 + x2) + k1$$
$$\left.\left(-\frac{d}{2} + Abs[x2] \right)\ (-UnitStep[-x2] + UnitStep[x2])\ UnitStep\left[-\frac{d}{2} + Abs[x2] \right] \right\}$$

The expression is mixed in sign *and unit step functions. It may be convert to all* sign *functions.*

In[95]:= UnitStep2Sign[Fp]
Out[95]=

$$\left\{ -FF1 + cv\ v1 + k\ x1 + k\ x2 + e^{-\frac{v1^2}{vs^2}}\left(cs + cc\ \left(-1 + e^{\frac{v1^2}{vs^2}} \right) \right)\ Sign[v1], \right.$$

$$-FF2 + k\ x1 + k\ x2 + \frac{k1\ x2}{2} - \frac{1}{4}\ d\ k1\ Sign[x2] + \frac{1}{2}\ k1\ x2\ Sign\left[-\frac{d}{2} + Abs[x2] \right] -$$
$$\left. \frac{1}{4}\ d\ k1\ Sign[x2]\ Sign\left[-\frac{d}{2} + Abs[x2] \right] \right\}$$

Example 5.143 (Automobile directional stability) *As a somewhat more complex example we consider building a simplified automobile model often used as a basis for investigating directional stability (Figure 5.3 and Figure 5.4).*

For simplicity of the resulting equations, we neglect the rotational energy of each wheel around its axle, that is $I_{yy} \ll 1$. Also, we will assume that $s, t,$ and δ are sufficiently small that linear approximation in these variables is adequate. None of these assumptions are necessary, by any means, but are useful for expository purposes. For tire cornering force and alignment torque constitutive equations we take:

$$F_{corner} = \kappa\beta, \quad T_{align} = 0$$

These are applied at the tire contact point.

Joint definitions

There are three joints. The 3 dof main body joint between the space frame and the automobile body and the two front wheel spindles

In[96]:= r1 = {3}; q1 = {θ, x, y}; p1 = {ω, vx, vy}; (* main body *)
    H1 = {{0, 0, 0}, {0, 0, 0}, {1, 0, 0}, {0, 1, 0}, {0, 0, 1}, {0, 0, 0}};
    r2 = {1}; q2 = {delta2}; p2 = {wdel2}; (* right spindle *)
    H2 = Transpose[{{-sr, -tr, 1, 0, 0, 0}}];
    r3 = {1}; q3 = {delta3}; p3 = {wdel3}; (* left spindle *)
    H3 = Transpose[{{-sl, tl, 1, 0, 0, 0}}];
    JointLst = {{r1, H1, q1, p1}, {r2, H2, q2, p2}, {r3, H3, q3, p3}};

We compute the joint parameters with

In[97]:= {V, X, H} = Joints[JointLst];

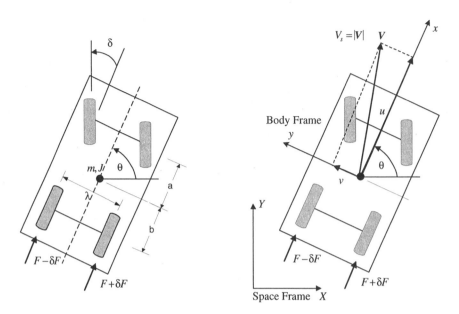

Figure 5.3: These figures define the dimensions and variables of an automobile that moves in the $X - Y$ plane. The $x - y$ frame denotes a reference frame fixed in the body. κ_f, κ_r are the front and rear tire coefficients, respectively. The system is composed of the vehicle body and its four wheels. The front wheels are used for steering with spindles aligned to provide small amounts of caster and camber.

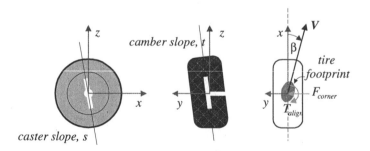

Figure 5.4: The tire rotation involves a revolute joint with rotation axis defined by the caster and camber angles. Tire cornering force and alignment torque are functions of the sideslip angle.

Body data

The body data includes body inertial properties including center of mass locaton in the body frame and outboard node locations.

$In[98]:=$ $cm1 = \{0,0,0\};$
        $out1 = \{\{2, \{a, -\ell/2, 0\}\},$ (* *front left tire spindle**)
          $\{3, \{a, \ell/2, 0\}\},$ (* *front right tire spindle**)
          $\{4, \{-b, -\ell/2, -R\}\},$ (* *rear left tire ground contact point* *)
          $\{5, \{-b, \ell/2, -R\}\}\};$ (* *rear right tire ground contact point* *)
        $I1 = \{\{Jxx, 0, Jxz\}, \{0, Jyy, 0\}, \{Jxz, 0, Jzz\}\};$
        $cm2 = \{0,0,0\}; out2 = \{\{6, \{sr\ R, tr\ R, -R\}\}\};$
        $I2 = DiagonalMatrix[\{Ixx, Iyy, Izz\}];$
        $cm3 = \{0,0,0\}; out3 = \{\{7, \{sl\ R, -tl\ R, -R\}\}\};$
        $I3 = DiagonalMatrix[\{Ixx, Iyy, Izz\}];$

$In[99]:=$ $BodyLst = \{\{cm1, out1, m1, I1\}, \{cm2, out2, m2, I2\}, \{cm3, out3, m2, I3\}\};$
        $TreeLst = \{\{\{1, 1\}, \{2, 2\}\}, \{\{1, 1\}, \{3, 3\}\}\};$
        $q = \{\theta, x, y, delta2, delta3\};$
        $p = \{\omega, vx, vy, wdel2, wdel3\};$

Notice that the system has a tree structure. The tree is composed of two chains: 1) main body and right tire, 2) main body and left tire.

Tire forces

We need to compute the tire generalized forces. This requires four similar, and somewhat complicated, calculations. To do this we use the function

GeneralizedForce.

Furthermore, to reduce computation time (by a factor of ten in this case), we use the function

KinematicReplacements.

GeneralizedForce *computes the generalized force associated with a force applied at any defined node in the system. It is usually the case that the applied force is characterized in the reference frame of the body in which the node is specified. The simplest usage is:*

GeneralizedForce[TerminalNode,TreeLst,BodyLst,
                        X,H,q,p,Force,VelNames]

TerminalNode *is the node number at which the force is applied.* F *is a list of 6 expressions which defines the external torques (first three) and forces (last three) in terms of body velocities (velocities of a body fixed frame at the terminal node). VelNames is a list of (6) names of the velocities used in the expressions F. There must be six names - the first 3 corresponding to the angular velocity and the last 3 to the linear velocity.* KinematicReplacements[V,X,H] *returns* {Vnew,Xnew,Hnew,rules} *where repeated groups of expressions in* V,X,H *are replaced by temporary variables to produce* Vnew,Xnew,Hnew. *The original forms are recovered by applying the rule list "rules". It is often convenient to use the syntax* KinematicReplacements[V,X,H,q], *which returns* {Vnew,Xnew,Hnew,rules1,rules2}. *In this usage, rules is divided into two sets. The set "rules1" depends on the coordinates q. The set "rules2" do not. rules2 involves expressions that depend only on system parameters. The application of rules1 must occur before using* CreateModel *but rules2 can be applied at any time.*

$In[100]:=$ $\{V, X, H, rules1, rules2\} = KinematicReplacements[V, X, H, q];$

$Force = \Big\{0, 0, 0, 0, -kappaf \arctan\Big[\dfrac{v6y}{v6x}\Big], 0\Big\}; (* \ right \ front \ *)$

$VelNames = \{w6x, w6y, w6z, v6x, v6y, v6z\};$

$TerminalNode = 6;$

$Q1 = GeneralizedForce[$
$\qquad TerminalNode, TreeLst, BodyLst, X, H, q, p, Force, VelNames];$

Similar calculations yield the remaining three tire forces. Then we proceed to assemble the model.

$In[101]:=$ $Q = Q1 + Q2 + Q3 + Q4;$

$\{V, X, H, Q\} = Chop[\{V, X, H, Q\}/.rules1, 0.001];$

$\{V, X, H, M, Csys, Fsys, psys, qsys\} =$
$\qquad CreateModel[JointLst, BodyLst, TreeLst, g, 0, Q, V, X, H];$

$\{V, X, H, M, Csys, Fsys, psys, qsys\} =$
$\qquad \{V, X, H, M, Csys, Fsys, psys, qsys\}/.rules2;$

For analysis purposes we create a model with the following two features. First, we assume that delta2 and delta3 are inputs rather than coordinates determined by the dynamics. Thus, we eliminate two degrees of freedom. Second, we choose to ignore any steering imperfections and assume delta2 and delta3 are equal and call them both δ.

$In[102]:=$ $qred = qsys[[\{1, 2, 3\}]];$

$pred = psys[[\{1, 2, 3\}]];$

$Mred =$
$\qquad M[[\{1, 2, 3\}, \{1, 2, 3\}]]/.Inner[Rule, \{delta2, delta3\}, \{0, 0\}, List];$

$Cred =$
$\qquad Csys[[\{1, 2, 3\}, \{1, 2, 3\}]]/.Inner[Rule, \{delta2, delta3\}, \{0, 0\}, List];$

$Fred = Simplify[$
$\qquad (Fsys[[\{1, 2, 3\}]]/.Inner[Rule, \{delta2, delta3\}, \{\delta, \delta\}, List])$
$\qquad /.\{sl-> s, sr-> s, tl-> t, tr-> t, kappaf-> \kappa, kappar-> \kappa\}];$

$Fred = Truncate[Fred, \{s, t\}, 1];$

$Vred = V[[\{1\}]];$

The results are summarized below.

$$
\begin{bmatrix} \dot\theta \\ \dot x \\ \dot y \end{bmatrix} = \begin{bmatrix} 1 & 0 & 0 \\ 0 & \cos\theta & -\sin\theta \\ 0 & \sin\theta & \cos\theta \end{bmatrix} \begin{bmatrix} \omega \\ v_x \\ v_y \end{bmatrix}
$$

$$
\begin{bmatrix} J_{zz} + 2I_{zz} + 2a^2 m_2 + m_2\ell^2/2 & 0 & 2am_2 \\ 0 & m_1 + 2m_2 & 0 \\ 2am_2 & 0 & m + 2m_2 \end{bmatrix} \begin{bmatrix} \dot\omega \\ \dot v_x \\ \dot v_y \end{bmatrix}
$$

$$
+ \begin{bmatrix} 2am_2 v_x\omega \\ -(m_1 + 2m_2)v_y\omega - 2am_2\omega^2 \\ (m_1 + 2m_2)v_x\omega \end{bmatrix} + f(\omega, v_x, v_y, \delta, F, \delta F) = 0
$$

Because of its complexity, we display f only up to second order in v_y, ω:

$$
f = \begin{bmatrix} 2\kappa(-bv_y + Rsv_y + (a^2 + b^2)\omega + a(v_y + 2Rs\omega)/v_x \\ 0 \\ 2\kappa(2v_y + (a - b + Rs)\omega)/v_x \end{bmatrix} + \begin{bmatrix} 0 \\ 2 \\ 0 \end{bmatrix} F'
$$

$$
+ \begin{bmatrix} \ell \\ 0 \\ 0 \end{bmatrix} \delta F + \begin{bmatrix} -2(a + Rs)\kappa \\ -2\kappa(v_y + a\omega + Rs\omega)/v_x \\ -2\kappa \end{bmatrix} \delta
$$

5.4 Systems with Constraints

The above constructions apply to systems interconnected by simple and compound joints and which have a tree structure. Recall that simple and compound joints as we have defined them impose holonomic constraints on the relative motion between two bodies. We wish to generalize the class of systems to include those with closed loops and nonholonomic differential constraints.

5.4.1 Differential Constraints

Consider a system with m-dimensional configuration manifold M and state space TM. Suppose that additional (differential) constraints are imposed on the motion of the system in the form:

$$F(q)\dot q = 0 \tag{5.41}$$

where F is an $r \times m$ matrix of (constant) rank r, or equivalently

$$A(q)p = 0 \tag{5.42}$$

where $A(q) = F(q)V(q)$. We will examine how the imposition of differential constraints affects the equations of motion. Differential constraints may be *adjoined* to the equations of motion via the introduction of Lagrange multipliers or *embedded* which avoids the addition of any auxiliary variables [14]. We consider the latter approach.

Recall that a virtual displacement from an admissible configuration q was required to belong to the tangent space $T_q M$, that is the space to which velocities $\dot{q}$ naturally belong. When differential constraints such as (5.41) and (5.42) apply, velocities are further constrained to lie in the subspace of $T_q M$, ker $F(q)$. Accordingly, virtual displacements are restricted by the same requirement: $F(q)\delta q = 0$.

Proposition 5.144 *Suppose the Lagrangian system of Proposition (5.132) is subject to the constraint $A(q)p = 0$, with $\dim \ker A(q) = m - r$ (a constant). Then the dynamical equations of motion are*

$$\dot{q} = V(q)T(q)\hat{p} \tag{5.43}$$

$$\left\{ \dot{p}^t \frac{\partial^2 \widetilde{L}}{\partial p^2} + p^t V^t \frac{\partial^2 \widetilde{L}}{\partial q^t \partial p} - \sum_{j=1}^m p_j \frac{\partial \widetilde{L}}{\partial p} U X_j - \frac{\partial \widetilde{L}}{\partial q} V \right\} T(q) = Q^t V T(q) \tag{5.44}$$

$$p = T(q)\hat{p} \tag{5.45}$$

where $T(q)$ is an $m \times (m - r)$ matrix whose columns span $\ker A(q)$.

Proof: The calculations in the proof of Proposition (5.132) that lead to equation (5.15) remain true even when the constraint (5.41) applies. However, in this event, the variation w is not arbitrary. When (5.41) obtains, it is necessary that w satisfy $A(q)w = 0$ (recall that $\delta q = Vw$), so that we can write

$$w = T(q)\alpha \tag{5.46}$$

where the columns of $T(q)$ span $\ker[A(q)]$ and $\alpha \in R^{m-r}$ is arbitrary. Rewrite (5.15) as

$$\int_{t_1}^{t_2} \left\{ \delta \widetilde{L}(p, q) + Q^T \delta q \right\} dt = \sum_k \frac{\partial \widetilde{L}}{\partial p_k} T\alpha_k \Big|_{t_1}^{t_2}$$

$$+ \int_{t_1}^{t_2} \sum_k \left[-\frac{d}{dt} \frac{\partial \widetilde{L}}{\partial p_k} + \sum_{i,j} c_{ik}^j \frac{\partial \widetilde{L}}{\partial p_j} p_i + v_k(\widetilde{L}) + Q^T v_k \right] T\alpha_k dt \tag{5.47}$$

Now, we can invoke the independence of the variations to obtain the dynamical equations

$$\left[\frac{d}{dt} \frac{\partial \widetilde{L}}{\partial p_k} - \int_{i;j=1}^m c_{jk}^i \frac{\partial \widetilde{L}}{\partial p_i} p_j - v_k(\widetilde{L}) - Q^t v_k \right] T(q) = 0 \tag{5.48}$$

Equation (5.48) is to be solved along with (5.42) and (5.16).

Remark 5.145 (Constrained dynamics)   *1. In application, we use (5.45) to replace p after the expression in curly brackets is evaluated. Then (5.43) and (5.44) form a closed system of equations in the dependent variables $(\widehat{p}, q) \in R^{2m-r}$.*

2. Notice that when the unconstrained dynamical equations are of the form

$$M(q)\dot{p} + C(q,p)p + F(q) = Q_p$$

then the constrained dynamics (5.44) are:

$$\left[T^t(q)M(q)\right]\dot{p} + T^t(q)C(q,p)p + T^t(q)F(q) = T^t(q)Q_p$$

or

$$\left[T^t(q)M(q)T(q)\right]\dot{\widehat{p}} + T^t(q)\left[C(q,T(q)\widehat{p})T(q) + M(q)\frac{\partial T(q)\widehat{p}}{\partial q}V(q)\right]\widehat{p}$$

$$+T^t(q)F(q) = T^t(q)Q_p \qquad (5.49)$$

First, let us examine a simple example.

Example 5.146 (Sleigh on a horizontal plane) *Consider a sleigh that moves on a horizontal plane as shown in Figure (5.5). Neimark and Fufaev [15] study the more general problem of a sleigh on an inclined plane (see also Problem (5.155)). The knife edge does not admit sideslip. Thus it imposes a simple differential constraint, $v_y = 0$. First, we formulate the equations of*

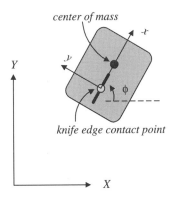

Figure 5.5: A sleigh on a horizontal plane.

motion without the differential constraint imposed by the knife edge. Define the single (main body) joint:

$In[103]:=$ $r1 = \{3\}; q1 = \{\phi, x, y\}; p1 = \{\omega, vx, vy\}; (* \text{ main body } *)$
$H1 = \{\{0, 0, 0\}, \{0, 0, 0\}, \{1, 0, 0\}, \{0, 1, 0\}, \{0, 0, 1\}, \{0, 0, 0\}\};$
$JointLst = \{\{r1, H1, q1, p1\}\};$

Define the body data:

$In[104]:=$ $cm1 = \{d, 0, 0\}; out1 = \{\{2, \{d, 0, 0\}\}\};$
$I1 = \{\{Jxx, 0, Jxz\}, \{0, Jyy, 0\}, \{Jxz, 0, Jzz\}\};$

$BodyLst = \{\{cm1, out1, m1, I1\}\};$
$TreeLst = \{\{\{1, 1\}\}\};$
$q = \{\phi, x, y\};$
$p = \{\omega, vx, vy\};$

Suppose a drive force F acts along the knife edge axis and a torque T acts about the z-axis. Now, set up the generalized force and obtain the (unconstrained) model:

$In[105]:=$ $Q = \{T, F, 0\};$
$\{V, X, H, M, Csys, Fsys, psys, qsys\} =$
$CreateModel[JointLst, BodyLst, TreeLst, g, 0, Q];$

*** Dynamics successfully loaded ***

Computing Joint Kinematics

Computing joint 1 kinematics

Computing Potential Functions

Computing Inertia Matrix

Computing Poincare Function

Finally, add the constraint.

$In[106]:=$ $\{Mm, Cm, Fm, Vm, Trans, phat\} =$
$DifferentialConstraints[M, Csys, Fsys, V, \{vy\}, p, q, \{3\}];$

Vm

Mm

Cm

Fm

$In[107]:=$ Vm
$Out[107]=$ $\{\{1, 0\}, \{0, \cos[\phi]\}, \{0, \sin[\phi]\}\}$

$In[108]:=$ Mm
$Out[108]=$ $\{\{Jzz + d^2 \ m1, 0\}, \{0, m1\}\}$

$In[109]:=$ Cm
$Out[109]=$ $\{\{d \ m1 \ vx, 0\}, \{-d \ m1 \ \omega, 0\}\}$

$In[110]:=$ Fm

Out[110]= $\{-T, -F\}$

Thus, the dynamical equations of motion are

$$\frac{d}{dt}\begin{bmatrix} \phi \\ x \\ y \end{bmatrix} = \begin{bmatrix} 1 & 0 \\ 0 & \cos\phi \\ 0 & \sin\phi \end{bmatrix}\begin{bmatrix} \omega \\ v_x \end{bmatrix}$$

$$\begin{bmatrix} J_{zz} + d^2 m_1 & 0 \\ 0 & m_1 \end{bmatrix}\frac{d}{dt}\begin{bmatrix} \omega \\ v_x \end{bmatrix} + \begin{bmatrix} 0 & d\,m_1\omega \\ -d\,m_1\omega & 0 \end{bmatrix}\begin{bmatrix} \omega \\ v_x \end{bmatrix} + \begin{bmatrix} -T \\ -F \end{bmatrix} = 0$$

Here is another example, only slightly more complicated.

Example 5.147 (Driven Planar Vehicle) *Consider the 3-wheeled planar vehicle shown in Figure (5.6). It illustrates the calculations required to assemble a model involving multiple differential constraints. This system is also useful for illustrating basic properties of nonlinear system controllability.*

The system is assumed to be composed of a main body with two rear wheels and one front wheel. The front wheel is both the steering and drive wheel. The rear wheels rotate freely about an axle fixed in the body. The assumption of pure rolling imposes a sideslip constraint, but they play no other essential role in the system behavior. Thus, we consider them to have no mass or inertia. The front wheel, on the other hand is assumed to have nontrivial inertia properties and both steering and drive torques are applied to it. It is also assumed to undergo pure rolling.

In summary, the model is composed of two bodies: (1) the main vehicle body including the rear wheels and (2) the and the front wheel. The model has two joints: (1) the main body joint – a three degree of freedom (two displacements and a rotation) joint characterizing the motion of the body in a space fixed frame, and (2) the main body/front wheel joint – a two degree of freedom (steering column and front axle rotations) joint that characterizes the relative motion of the front wheel relative to the main body.

Joint data

In[111]:= $r1 = \{3\}; q1 = \{\phi, x, y\}; p1 = \{\omega b, vx, vy\};$
$H1 = \{\{0,0,0\}, \{0,0,0\}, \{1,0,0\}, \{0,1,0\}, \{0,0,1\}, \{0,0,0\}\};$

In[112]:= $r2 = \{1,1\}; q2 = \{\theta f, \theta\}; p2 = \{\omega f, \omega\};$
$H2 = \{\{0,0\}, \{1,0\}, \{0,1\}, \{0,0\}, \{0,0\}, \{0,0\}, \{0,0\}\};$

In[113]:= $JointLst = \{\{r1, H1, q1, p1\}, \{r2, H2, q2, p2\}\};$

Body data

In[114]:= $cm1 = \{-d/2, 0, 0\}; out1 = \{\{2, \{0,0,0\}\}, \{3, \{-d, 0, -R\}\}\};$
$I1 = \{\{Jxx, 0, Jxz\}, \{0, Jyy, 0\}, \{Jxz, 0, Jzz\}\};$
$cm2 = \{0,0,0\}; out2 = \{\{4, \{0, R\sin[\theta f], -R\cos[\theta f]\}\}\};$
$I2 = DiagonalMatrix[\{Ix, J, Ix\}];$

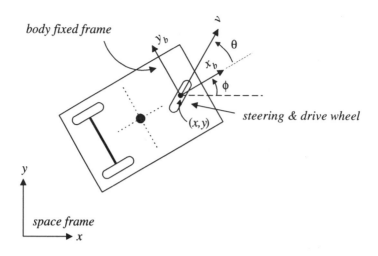

Figure 5.6: This simple vehicle can be driven around the plane using steering and drive torques, T_s and T_d, respectively, that are applied to the front wheel. The wheelbase is denoted d and the front wheel radius R.

$In[115]:=$ $BodyLst = \{\{cm1, out1, m1, I1\}, \{cm2, out2, m2, I2\}\};$
          $TreeLst = \{\{\{1, 1\}, \{2, 2\}\}\};$
          $q = Flatten[\{q1, q2\}]$
          $p = Flatten[\{p1, p2\}]$

$Out[115]=$ $\{\phi, x, y, \theta f, \theta\}$

$Out[115]=$ $\{\omega b, vx, vy, \omega f, \omega\}$

Unconstrained Model Assembly

$In[116]:=$ $Q = \{0, 0, 0, Ts, Td\};$

          $\{V, X, H, M, Csys, Fsys, psys, qsys\} =$
             $CreateModel[JointLst, BodyLst, TreeLst, g, 0, Q];$

Computing Joint Kinematics

Computing joint 1 kinematics

Computing joint 2 kinematics

Computing Potential Functions

Computing Inertia Matrix

Computing Poincare Function

Adding Constraints

The rolling assumption implies that the wheel contact point velocity is zero. We compute the velocities for the unconstrained model at each contact point using the function NodeVelocity. *Because of our assumptions, some components are identically zero but the remaining velocity constraints must be enforced. In the case of the rear wheels we need to enforce a single sideslip constraint (local y-direction), and for the front wheel we need to enforce the sideslip and tangential constraints (local x and y-directions). To do this we use the function* DifferentialConstraints.

$In[117] :=$ $Vrear = NodeVelocity[3, TreeLst, BodyLst, X, H, q, p];$
           $Vfront = NodeVelocity[4, TreeLst, BodyLst, X, H, q, p];$
           $c1 = Vrear[[5]]$
           $c2 = FullSimplify[Vfront[[5]]/.\{\theta f \to 0\}]$
           $c3 = FullSimplify[Vfront[[4]]/.\{\theta f \to 0\}]$

$Out[117] =$ vy $- d\omega b$

$Out[117] =$ vy $\text{Cos}[\theta] - $ vx $\text{Sin}[\theta]$

$Out[117] =$ $-R\omega f + $ vx $\text{Cos}[\theta] + $ vy $\text{Sin}[\theta]$

$In[118] :=$ $\{Mm, Cm, Fm, Vm, Trans, phat\} =$
           $Simplify[DifferentialConstraints[$
                $M, Csys, Fsys, V, \{c1, c2, c3\}, p, q, \{1, 2, 3\}]];$

Vm

Mm

Cm

Fm

$In[119] :=$ $Vm // MatrixForm$

$$Out[119] = \begin{pmatrix} \dfrac{R \ \text{Sin}[\theta]}{d} & 0 \\ R \ \text{Cos}[\theta + \phi] & 0 \\ R \ \text{Sin}[\theta + \phi] & 0 \\ 1 & 0 \\ 0 & 1 \end{pmatrix}$$

$In[120] :=$ q

$Out[120] =$ $\{\phi, x, y, \theta f, \theta\}$

$In[121] :=$ $Simplify[Mm]$

$$Out[121] = \left\{ \left\{ \frac{4 (Ix + Jzz) R^2 + d^2 (8 J + (5 m1 + 8 m2) R^2)}{8 d^2} \right.\right.$$
$$- \frac{(4 Ix + 4 Jzz - 3 d^2 m1) R^2 \ \text{Cos}[2 \theta]}{8 d^2},$$
$$\left. \frac{Ix R \ \text{Sin}[\theta]}{d} \right\}, \left\{ \frac{Ix R \ \text{Sin}[\theta]}{d}, Ix \right\} \right\}$$

$In[122] :=$ $(Cm.phat + Fm) // MatrixForm$

$$Out[122] = \begin{pmatrix} -Ts + \dfrac{\left(4 Ix + 4 Jzz - 3 d^2 m1\right) R^2 \ \omega \ \omega f \ \text{Cos}[\theta] \ \text{Sin}[\theta]}{4 d^2} \\ -Td + \dfrac{Ix R \ \omega \ \omega f \ \text{Cos}[\theta]}{d} \end{pmatrix}$$

Suppose now, that the two control torques can be used to precisely regulate the two remaining quasi-velocities ω, the steering angle rate, and ω_f, the drive wheel angular velocity. Then the problem of moving the vehicle around the plane becomes purely a kinematic one in which these two quasi-velocities can be specified to steer the vehicle along a desired path. Normally, the drive wheel rotation angle is not a coordinate of interest, and since it does not entire into any of the elements of V_m, we can ignore it. Accordingly, eliminating the $\dot{\theta}_f$ equation from the kinematics and reordering the states for convenience we obtain:

$In[123]:=$ $Vm[[\{2,3,1,5\}]]//MatrixForm$

$Out[123]=$ $\begin{pmatrix} R\ \text{Cos}[\theta+\phi] & 0 \\ R\ \text{Sin}[\theta+\phi] & 0 \\ \dfrac{R\ \text{Sin}[\theta]}{d} & 0 \\ 0 & 1 \end{pmatrix}$

Now, introduce the drive velocity $v = R\omega_f$ to replace ω_f and take $d = 1$ to obtain the kinematic equations in the final form:

$$\frac{d}{dt}\begin{bmatrix} x \\ y \\ \phi \\ \theta \end{bmatrix} = \begin{bmatrix} \cos(\theta+\phi) & 0 \\ \sin(\theta+\phi) & 0 \\ \sin\theta & 0 \\ 0 & 1 \end{bmatrix}\begin{bmatrix} v \\ \omega \end{bmatrix}$$

We will use this model in the next chapter (Example (6.165)) to illustrate certain important aspects of the controllability of nonlinear systems.

Here is a more complex example.

Example 5.148 (Rolling Disk) *As an illustration we take another example from Neimark and Fufaev [9]. Consider a disk that rolls without slipping on the x-y plane (z is up as in Figure (5.7)). Assume that the disk is of mass m and radius R. One approach is to ignore the rolling constraints and formulate the equations of motion for the free disk in space, and then add the required constraints.*

A body frame is established with origin at the center of the disk. The six degree of freedom, simple joint is defined by:

```
In[124]:= r={6};
          H=IdentityMatrix[6];
          q={psi,theta,phi,x,y,z};
          p={wx,wy,wz,vx,vy,vz};
          JointList={{r,H,q,p}};
```

In this setup the x, y, z coordinates locate the center of the disk in the space frame and ψ, θ, ϕ are Euler angles in the $z - y - x$ (or 3-2-1) convention. The body data is:

```
In[125]:= Mass=m; CenterOfMass={0,0,0};
          OutboardNode={2,{0,-R*Sin[psi],-R*Cos[psi]}};
          InertiaMatrix=DiagonalMatrix[{J,Iy,Iy}];
          BodyList={{CenterOfMass,{OutboardNode},Mass,InertiaMatrix}};
```

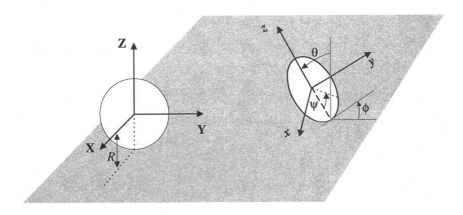

Figure 5.7: The rolling disk.

and the remaining data:

In[126]:= TreeList={{{1,1}}};
*   PE=0; Q={0,0,0,0,0,0};*

The unconstrained disk model is obtained with

In[127]:= {V,X,H,M,Cp,Fp,p,q}=CreateModel[JointList,BodyList,TreeList,g,PE,Q];

*Now, we formulate the differential constraints. Rolling without slipping implies
that the velocity of the disk contact point must be zero. An expression for the
(angular and translational) velocity at the outboard node as a function of the configuration variables is easily obtained using the function* EndEffectorVelocity
or NodeVelocity.

In[128]:= {ChainList}=TreeList;
*   VCont=NodeVelocity[ChainList,BodyList,X,H,p]*
Out[128]= {wx,wy,wz,vx − R wy Cos[psi] + R wz Sin[psi],vy + R wx Cos[psi],
*   vz − R wx Sin[psi]}*

In[129]:= {Mm,Cm,Fm,Vm,T,phat}=
*   DifferentialConstraints[M,Cp,Fp,V,VCont[[Range[4,6]]],p,q,{4,5,6}];*

*The dynamics are reduced to three dimensions and the original six quasi-velocities
are reduced to three, in fact, we have:* phat={wx,wy,wz}. *The set of configuration coordinates is not reduced by the function* DifferentialConstraints.
*In general, a set of differential constraints may not admit any such reduction.
Such would be the case if the constraints were completely nonholonomic. In the
present case, however, the constraints are 'partially' integrable and from basic
geometry one can see that* $height_{com} = R(1 − \cos\theta)$. *Using this relationship,
the coordinate z can be eliminated from the equations. Because the translation
parameters x, y, z are, in fact, the space coordinates, z is precisely the height of
the center of mass. Inspection shows that only Mm and Vm depend on z so we
define:*

```
In[130]:= {Sols}=Solve[{X[[1]][[3,4]]==R*(Cos[theta]-1)},{z}];
          Mmo=Simplify[Mm/.Sols];
          Vmo=Simplify[Vm[[ {1,2,3,4,5} ]]/.Sols];
```

Now, we assemble the governing equations.

```
In[131]:= Eqns=MakeODEs[phat,q[[{1,2,3,4,5}]],Vmo,Mmo,Cm,Fm,t]
```

Consider the equations of motion as given by the list Eqns. *Careful inspection of the equations suggests that representation of the angular velocity in a frame that does not rotate about the body-x axis may simplify them. Thus, we transform the angular velocity coordinates,* wy, wz, *via the relations:*

```
In[132]:= Trans = {wy[t] → cos[psi[t]] wyy[t] + sin[psi[t]] wzz[t]
          wz[t] → - sin[psi[t]] wyy[t] + cos[psi[t]] wzz[t]};
```

The transformed equations are obtained with the function **StateTransformation***:*

```
In[133]:= StateVars={psi,theta,phi,x,y,ux,uy,uz};
          TrEqns=StateTransformation[Eqns,StateVars,Trans,{wyy,wzz},t]
```

to obtain:

$$\dot{\theta} = w_{yy}$$

$$\dot{\phi} = \sec(\theta)w_{zz}$$

$$\dot{\psi} = w_x + \tan(\theta)w_{zz}$$

$$\dot{x} = R(\sin(\phi)w_x + \cos(\phi)\cos(\theta)w_{yy})$$

$$\dot{y} = -R\cos(\phi)w_x + R\cos(\theta)\sin(\phi)w_{yy}$$

$$\dot{w}_x = \frac{mR^2 w_{yy}w_{zz}}{J + mR^2}$$

$$\dot{w}_{zz} = \frac{w_{yy}(Jw_x + I_y\tan(\theta)w_{zz})}{I_y}$$

$$\dot{w}_{yy} = \frac{gmR\sin(\theta) - (J + mR^2)w_x w_{zz} - I_y\tan(\theta)w_{zz}^2}{I_y + mR^2}$$

These equations are obviously equivalent to the disk equations given by Neimark and Fufaev [9].

Another derivation is given by Meirovitch [8]. To compare our equations with his, it is necessary to reduce them to second order form by eliminating the quasi-velocities (thus, we get Lagrange's equations), and to perform a minor transformation of angular coordinates:

```
In[134]:= Rules1=Flatten[Solve[TrEqns[[{1,2,8}]]],{ux[t],wyy[t],wzz[t]}]];
```

```
In[135]:= Rules2={ux'[t]->D[ux[t]/.Rules1,t],
                  wyy'[t]->D[wyy[t]/.Rules1,t],
                  wzz'[t]->D[wzz[t]/.Rules1,t]};
```

```
In[136]:= Rules3={theta[t]->th[t]-Pi/2,
                 theta'[t]->th'[t],
                 theta''[t]->th''[t]};
In[137]:= Rules4={phi[t]->-pi/2+ph[t],
                 phi'[t]->ph'[t],
                 phi''[t]->ph''[t]}
In[138]:= LagEqns=Simplify[TrEqns[[{5,6,7}]]/.Rules1/.Rules2/.Rules3/.Rules4]
```

$$Out[138]= \left\{ \text{Cos}[\text{th}[t]]\, \text{ph}''[t] + \text{psi}''[t] == \frac{(J + 2\,m\,R^2)\,\text{Sin}[\text{th}[t]]\,\text{ph}'[t]\,\text{th}'[t]}{J + m\,R^2}, \right.$$

$$\frac{(2\,\text{Iy} - J)\,\text{Cos}[\text{th}[t]]\,\text{ph}'[t]\,\text{th}'[t] - J\,\text{psi}'[t]\,\text{th}'[t] + \text{Iy}\,\text{Sin}[\text{th}[t]]\,\text{ph}''[t]}{\text{Iy}}$$

$$== 0,$$

$$\text{th}''[t] == \frac{1}{\text{Iy} + m\,R^2}(-g\,m\,R\,\text{Cos}[\text{th}[t]] + \text{Iy}\,\text{Cos}[\text{th}[t]]\,\text{Sin}[\text{th}[t]]\,\text{ph}'[t]^2 -$$

$$\left. (J + m\,R^2)\,\text{Sin}[\text{th}[t]]\,\text{ph}'[t]\,(\text{Cos}[\text{th}[t]]\,\text{ph}'[t] + \text{psi}'[t])) \right\}$$

These equations are easily confirmed to be equivalent to those given by Meirovitch [8], p. 163.

5.4.2 Holonomy and Integrability

If the distribution $\Delta = \ker F(q)$ is completely integrable (in the sense of Frobenius) then any motion of the constrained system is confined to an integral manifold of a submanifold of M. Thus, the differential constraint (5.41) can be replaced by a configuration constraint of the form

$$f(q) = 0 \tag{5.50}$$

Such differential constraints are called *holonomic*. Of course, an integrable distribution will have an infinity of integral manifolds and the one on which motion takes place is determined by the initial conditions of the system. Since the distribution is nonsingular and of dimension r all of the integral manifolds are of dimension r. Thus, in principle, it is possible to reduce the configuration space by r coordinates. If Δ is not completely integrable, then the constraint is said to be nonholonomic. A nonholonomic constraint may be 'partially integrable.' Recall that the distribution Δ is completely integrable if and only if it is involutive. Suppose this is not the case. Then construct the smallest involutive distribution that contains Δ. Denote this distribution Δ^*. Now let $G(q)$ be a matrix whose columns $\{g_1(q), \ldots, g_{m-r}(q)\}$span $\Delta(q)$. Then (5.41) is satisfied if and only if $q(t)$ satisfies the differential equation:

$$\dot{q} = G(q)v(t) \tag{5.51}$$

where $v(t)$ is arbitrary (except for any conditions necessary to achieve smoothness requirements on $q(t)$). By considering (5.51) to be a control system, we can apply known results on nonlinear system controllability to reach the following conclusion. If Δ^* is nonsingular with dimension $m - r^*$, $0 \leq r^* \leq r$, then there is an $m - r^*$-dimensional integral manifold of Δ^* passing through any point $q_0 \in M$, say S_{q_0}, and all points reachable from q_0 via an admissible motion (one that satisfies (5.41) or, equivalently, (5.51)) belong to S_{q_0}.

Consequently, the differential constraints restrict the motion to an $m - r^*$ dimensional submanifold of the space M. If $r^* = r$, then we have the holonomic case which yields the maximally restricted configuration space. If $r^* = 0$, the configuration space is the entire space M, there is no restriction of the accessible configurations. In this case the constraints are said to be completely nonholonomic.

A standard procedure that can be used to construct Δ^* is described in Chapter 2, Algorithm (3.91). Define

$$
\begin{aligned}
\Delta_0 &= \Delta \\
\Delta_{k+1} &= \Delta_k + [\Delta, \Delta_k]
\end{aligned}
$$

and terminate at $k = k^*$ when $\Delta_{k^*+1} = \Delta_{k^*}$. The integer k^* is sometimes referred to as the 'degree of nonholonomy' [15]. It represents the number of Lie bracket operations necessary to achieve integrability by expansion of the distribution.

5.4.3   Configuration Constraints

Suppose that a set of constraints of the form of (5.50) are imposed on a Lagrangian system with configuration manifold M. Assume further that

$$
\text{rank } [\partial f / \partial q] = r \text{ on } S = \{q \mid f(q) = 0\}
$$

Then, S is a regular submanifold of M of dimension $n - r$ and S is the configuration manifold for the constrained system. There are a number of ways to formulate the equations of motion for the constrained system. The two most commonly used are:

- If (5.50) can be solved for r coordinates in terms of the remaining coordinates, then these relations can be used to eliminate r coordinates from the unconstrained equations of motion.

- It is always possible to differentiate (5.50) to obtain

$$
J(q)\dot{q} = 0, \quad J(q) := \frac{\partial f}{\partial q} \tag{5.52}
$$

Now proceed to use the procedure described above for differential constraints. The resulting equations describe the motion in terms of the original n coordinates but the manifold S can be shown to be an invariant manifold so that if the initial conditions satisfy

$$
f(q_0) = 0, \quad J(q_0)\dot{q}_0 = J(q_0)V(q_0)p_0 = 0 \tag{5.53}
$$

then the resultant motion evolves in S.

Example 5.149 (Two bar linkage) *Consider a planar two bar linkage in which the lower end of bar 1 is constrained by a revolute joint on the y-axis at $y = 0$ and the upper end of bar two slides on the z-axis. The upper end of bar 1 and the lower end of bar 2 are connected by a revolute joint. Figure (5.8) illustrates the system.*

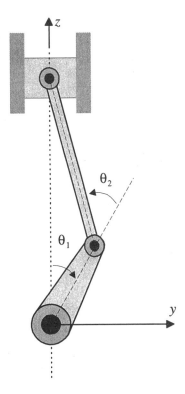

Figure 5.8: A two bar assembly is illustrative of a closed chain configuration.

```
In[139]:= r1={1};
          H1=Transpose[{{1,0,0,0,0,0}}];
          q1={theta1};p1={omega};

          r2={1};
          H2=Transpose[{{1,0,0,0,0,0}}];
          q2={theta2};p2={omega2};

          JointList={{r1,H1,q1,p1},{r2,H2,q2,p2}};
```

In[140]:= Mass1=m1; CenterOfMass1={0,0,L1/2}; OutboardNode1={2,{0,0,L1}};
InertiaMatrix1=DiagonalMatrix[{m1*L1^2/12,m1*L1^2/12,0}];

Mass2=m2; CenterOfMass2={0,0,L2/2}; OutboardNode2={3,{0,0,L2}};
InertiaMatrix2=DiagonalMatrix[{m2*L2^2/12,m2*L2^2/12,0}];

BodyList={{CenterOfMass1,{OutboardNode1},Mass1,InertiaMatrix1},
{CenterOfMass2,{OutboardNode2},Mass2,InertiaMatrix2}};

In[141]:= TreeList={{{1,1},{2,2}}};
PE=0; Q={0,0};

In[142]:= {V,X,H,M,Cp,Fp,p,q}=CreateModel[JointList,BodyList,TreeList,g,PE,Q];

*We have the model for the unconstrained tree structure. Now, we formulate
the constraint.*

In[143]:= ChnBodyList={{CenterOfMass1,{0,0,L1},Mass1,InertiaMatrix1},
{CenterOfMass2,{0,0,L2},Mass2,InertiaMatrix2}};

In[144]:= EndPos=EndEffector[ChnBodyList,X];

In[145]:= G=EndPos[[{2},4]]
Out[145]= {−L1 Sin[theta1]+
L2 (−Cos[theta2] Sin[theta1] − Cos[theta1] Sin[theta2])}

*The relation $G = 0$ is not solvable for either theta1 or theta2 for all values of
L1 and L2 - although, as we will describe below, it can be solved for particular
values of these parameters. Thus, we will not try to eliminate any configuration
coordinate. The constrained dynamics are obtained with:*

In[146]:= {Mm,Cm,Fm,Vm,phat,qhat}=AlgebraicConstraints[M,Cp,Fp,V,G,p,q]

*These parameters define the constrained system dynamics. Now, let's consider
the special case: L1=L2=L. This is the only situation where G=0 can be solved
for either theta1 or theta2, so that either angle can be used to parameterize the
system configuration. For simplicity, also set m1=m2=m. Since L1=L2=L it
is easy to show that theta2=-2 theta1. Hence we can make this replacement. In
this case the system matrices are:*

In[147]:= {Mm,Cm,Fm,Vm}=Simplify[{Mm,Cm,Fm,Vm}/.{L1->L,L2->L,m1->m,m2->m}]

It is convenient to assemble the corresponding differential equations.

In[148]:= SpEqns=
Simplify[MakeODEs[phat,qhat,Vm,Mm,Cm,Fm,t]/.theta2[t]->-2*theta1[t]]

The result is:

$$\dot{\theta}_1 = -L\cos\theta_1\, w_1$$

$$\dot{\theta}_2 = 2L\cos\theta_1\, w_1$$

$$(7\cos\theta_1 - 3\cos 3\theta_1)\,\dot{w}_1 = -\frac{12}{L^2}g\sin\theta_1 - L\,(2\sin 2\theta_1 - 3\sin 4\theta_1)\,w_1^2$$

*Notice that the last two equations consitute a closed system of two first order
differential equations in θ_1 and w_1. Let us define some replacement rules to
replace these by a single second order equation in θ_1, i.e., Lagrange's equation.*

In[149]:= $Rules = \{w1[t] \rightarrow D[theta1[t], t]/(-L \cos[theta1[t]]),$
$\qquad\qquad D[w1[t], t] \rightarrow D[D[theta1[t], t]/(-L \cos[theta1[t]]), t]\}$

In[150]:= $Simplify[SpEqns[[3]]/.Rules]$

Out[150]= $0 == \dfrac{1}{3} L^2 \ m \ Cos[theta1[t]] \ (-6 \ g \ Sin[theta1[t]] +$
$\qquad\qquad 3 \ L \ Sin[2 \ theta1[t]] \ theta1'[t]^2 - L \ (-5 + 3 \ Cos[2 \ theta1[t]]) \ theta1''[t]$

*Simple trigonometric identities can be used to verify that these equations are
equivalent to those given by Ginsberg [16], p.275 for this example.*

5.5 Simulation

The equations developed above can be further manipulated symbolically, for
example, they can be put into state variable form or linearized or transformed
in other ways of interest to control systems engineers. However, it sometimes
desired to perform numerical computations or simulations with these models.
It may be convenient to perform simulations within Mathematica, or it may
desirable to employ other standard simulation packages. We will describe and
illustrate both approaches.

5.5.1 Computing with *Mathematica*

It is easy to construct a simulation within Mathematica. First, assemble the
system parameter matrices as computed above into a system of ordinary differential equations using the *ProPac* function MakeODEs. MakeODEs[p,q,V,M,C,F,t]
builds and returns a list of ordinary differential equations in *Mathematica* syntax. They can be integrated using the built in differential equaton solver
NDSolve. The process is illustrated with a simple example.

Example 5.150 (Double Pendulum Revisited) *Let us revisit Example
5.140. All of the information required to invoke the function* MakeODEs *has
been assembled.*

In[151]:= $Equations=MakeODEs [p,q,V,M,Cp,Fp,t];$

*Before numerical computation can proceed it is necessary to replace parameter
symbols by numbers and set up initial conditions.*

In[152]:= $DataReplacements=\{m->1, l->1, gc->1, T1->0, T2->0\};$

In[153]:= $Equations1=Simplify [Equations/.DataReplacements]$

Out[153]= $\{ a1x'[t] == w1x[t],$
$\qquad a2x'[t] == w2x[t],$
$\qquad 2 \ Sin[a1x[t]]+Sin[a1x[t]+a2x[t]]-2 \ Sin[a2x[t]] \ w1x[t] \ w2x[t]-Sin[a2x[t]] \ w2x[t]^2 +$
$3 \ w1x'[t] + 2 \ Cos[a2x[t]] \ w1x'[t] + w2x'[t] + Cos[a2x[t]] \ w2x'[t] == 0,$
$\qquad Sin[a1x[t] + a2x[t]] + Sin[a2x[t]] \ w1x[t]^2 + (1 + Cos[a2x[t]]) \ w1x'[t] + w2x'[t] == 0\}$

Set up and join the initial conditions:

```
In[154]:= InitialConditions={w1x[0] == 0, w2x[0] == 0, a1x[0] == .1, a2x[0] == .2};
          Equations2=Join[Equations1,InitialConditions];
```

Finally, we are ready to integrate the equations using the Mathematica function NDSolve *and plot the results.*

```
In[155]:= sols=NDSolve[Equations2,vars,{t,0,40}];
```

```
In[156]:= Plot[Evaluate[a1x[t]/.sols],{t,0,40},AxesLabel-> {t,a1x}]
```

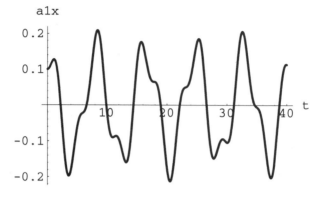

5.5.2   Computing with SIMULINK

It may be desirable to use the system model in an external program. Matlab/SIMULINK is especially popular among control systems engineers. Simulink provides a convenient block diagram environment for building and running simulations. *ProPac* provides functions to create C-Code that compiles as a MEX-File for use as an S-function in SIMULINK. The code is computationally optimized. An S-function may have inputs and outputs so that it can be interconnected with other subsystems within SIMULINK and it can have parameters that can be defined from within SIMULINK. The main tool for building MEX-files is the function CreateModelMEX. It has the calling syntax:

```
CreateModelMEX[p,q,Inputs,Outputs,PassedParameters,
        PassedParametersDimensions,V,C,Fp,M,MEXFilename]
```

Example 5.151 (Double Pendulum Revisited) *In this example, we define the joint torques as inputs and the (y, z) coordinates of mass 2 as the outputs. Parameters include the two masses, the two lengths and the gravitational constant.*

The ProPac function EndEffector *is employed to define the output expressions.*

```
In[157]:= ChainLst={{1,1},{2,2}};
          TerminalNode=3;
          Xout=EndEffector[ChainLst,TerminalNode,BodyLst,X];
          yout={Xout[[2,4]],Xout[[3,4]]}
```

Out[157]= {1 Sin[a1x] − 1 (− Cos[a2x] Sin[a1x] − Cos[a1x] Sin[a2x]),
  −1 Cos[a1x] − 1 (Cos[a1x] Cos[a2x] − Sin[a1x] Sin[a2x])}

Now, the data for `CreateModelMEX` *is set up and the function is executed.*

```
In[158]:= Inputs = Q;
          Outputs = yout;
          MEXFilename = "dbl_{}pend.c";
          paramvec ={m1,l1,m2,l2,gc};
          PassedParams = {"X0","m1","l1","m2","l2","gc"};
          PassedParamsDimensions = {{2*Length[p],1},{1,1},{1,1},{1,1},{1,1},{1,1}};

          CreateModelMEX[p,q,Inputs,Outputs,PassedParams,PassedParamsDimensions,
                         V,Cp,Fp,M,MEXFilename];
```

...*Generating Header Code*

...*Generating Initial Condition Function Code*

...*Generating State Derivative Function Code*

...*Collecting all function terms*

...*Generating temp variable declarations*

...*Converting Expressions to C form*

...*Generating Output Function Code*

MEX File created with name: dbl_pend.c

Detailed information on compiling and using MEX files with Matlab/SIMULINK can be found in the appropriate MATLAB references. CreateModelMEX (and also CreateControllerMEX) assemble C source code that needs to be compiled. During the compilation process the compiled code will be linked with additional code segments and libraries provided with *ProPac* or MATLAB or the compiler. It is necessary that this code be available at the time of compilation. The easiest way to proceed is to perform the compilation from within MATLAB using scripts provided with MATLAB (either cmex with MATLAB 4 or mex with MATLAB 5). This should automatically define the locations of all required MATLAB or compiler code segments. In addition to the file created by *ProPac* (e.g, dbl_pend.c) you will need the files linsolv.c, f2c.h and trigfun.h - all provided with *ProPac*.

One way to compile dbl_pend.c is to place the files dbl_pend.c, linsolv.c, f2c.h and trigfun.h in a common directory, for example C:\DoublePend. Then in the MATLAB command window set the current directory to C:\DoublePend to insure that the four required files are in the search path:

cd C:\DoublePend

then use the command

mex dbl_pend.c linsolv.c

in MATLAB 5, or

!cmex dbl_pend.c linsolv.c

in MATLAB 4.2.

If you have followed the MATLAB installation instructions as appropriate for your compiler this should work fine. Otherwise, you can reinstall MATLAB 5 and follow the instructions (which setsup the appropriate paths in mex), or in MATLAB 4.2 you can edit cmex as described therein.

5.6   Problems

Problem 5.152 (Reconaissance robot, continued) *Reconsider the robot of Problem (4.130). Suppose the vehicle has mass m and inertia J_{zz} about the verticle axis, at its center of mass. The radar is mounted at the vehicle center of mass. It has mass m_r and it rotates about its center of mass with body fixed inertias, I_{zz}, and I_{yy}. The drive force T and steering angle δ are inputs to the system. Assume perfect rolling without side slip and derive the equations of the motion for the system.*

Problem 5.153 (Overhead crane, continued) *Consider the overhead crane of Problem (4.131). Assume the cable is massless and treat the payload as a point mass. Assign inertial parameters and dimensions as required and assemble the equations of motion for this system. Consider the three following controllable inputs to be applied to the system: a drive force, F, applied to the cart, a joint torque, T_a, applied to the revolute upper arm joint, and a cable tension force, T_c.*

Problem 5.154 (Synchronous motor) *Consider a three phase synchronous motor with the variables and parameters defined in Table (5.1). The load torque T_L is an exogenous disturbance and the voltages v_1, v_2, v_3, v_f are control inputs. The generalized coordinates are $q = [\theta \quad q_1 \quad q_2 \quad q_3 \quad q_f]$. Define the Blondel transformation matrix*

$$B = \sqrt{\frac{2}{3}} \begin{bmatrix} \cos\theta & \cos\left(\theta - \frac{2\pi}{3}\right) & \cos\left(\theta + \frac{2\pi}{3}\right) \\ \sin\theta & \sin\left(\theta - \frac{2\pi}{3}\right) & \sin\left(\theta + \frac{2\pi}{3}\right) \\ \frac{1}{\sqrt{2}} & \frac{1}{\sqrt{2}} & \frac{1}{\sqrt{2}} \end{bmatrix}$$

Note that $B^{-1} = B^T$. Define the quasi-velocities, $p^T = [\omega \quad i_d \quad i_q \quad i_0 \quad i_f]$, via $\dot{q} = V(q)p$, in this case

$$\begin{bmatrix} \dot{\theta} \\ \dot{q}_1 \\ \dot{q}_2 \\ \dot{q}_3 \\ \dot{q}_f \end{bmatrix} = \begin{bmatrix} 1 & 0 & 0 \\ 0 & B^T & 0 \\ 0 & 0 & 1 \end{bmatrix} \begin{bmatrix} \omega \\ i_d \\ i_q \\ i_0 \\ i_f \end{bmatrix}$$

Symbol	Definition
θ	rotor angle relative inertial reference
ω	motor speed
v_f	field winding voltage
$v_i,\, i = 1, 2, 3$	stator winding voltages
i_f	field current
$i_i,\, i = 1, 2, 3$	stator winding currents
q_f	field winding charge
$q_i,\, i = 1, 2, 3$	stator winding charges
J	rotor inertia
M	inertia matrix in rotor (Blondel) frame
T_L	load torque
$I_{ii} = L_1,\, i = 1, 2, 3$	stator windings self inductances
$I_{ij} = -L_3,\, i \neq j,\, i = 1, 2, 3$	stator windings mutual inductances
$I_{ff} = L_4$	field windings self inductances
$I_{f1} = L5 \cos\theta$ $I_{f2} = L_5 \cos(\theta - 2\pi/3)$ $I_{f3} = L_5 \cos(\theta + 2\pi/3)$	field/stator mutual inductances
r	stator winding resistance
r_f	field winding resistance
R	dissipation matrix

Table 5.1: **AC motor nomenclature**

The currents i_d, i_q, i_0 are called the Blondel currents. It is also convenient to define the Blondel voltages v_d, v_q, v_0,

$$\begin{bmatrix} v_d \\ v_q \\ v_0 \end{bmatrix} = B \begin{bmatrix} v_1 \\ v_2 \\ v_3 \end{bmatrix}$$

The kinetic energy of the system can be expressed in terms of the quasivelocities, the potential energy is trivial and the generalized force can be obtained from a (Lur'e) potential function, $Q = \partial\mathcal{D}/\partial p$

$$\mathcal{T}(p, q) = \tfrac{1}{2} p^T M p$$

$$\mathcal{U}(q) = 0$$

$$\mathcal{D}(p, q) = \tfrac{1}{2} p^T R p + [\, -T_L \quad v_d \quad v_q \quad v_0 \quad v_f \,] p$$

$$
M = \begin{bmatrix}
J & 0 & 0 & 0 & 0 \\
0 & L_1 + L_3 & 0 & 0 & \sqrt{\tfrac{3}{2}}L_5 \\
0 & 0 & L_1 + L_3 & 0 & 0 \\
0 & 0 & 0 & L_1 - 2L_3 & 0 \\
0 & \sqrt{\tfrac{3}{2}}L_5 & 0 & 0 & L_4
\end{bmatrix}
$$

$$
R = \mathrm{diag}\{0, r, r, r, r_f\}
$$

Show that Poincaré's equations are

$$
\begin{bmatrix}
J & 0 & 0 & 0 & 0 \\
0 & L_1 + L_3 & 0 & 0 & \sqrt{\tfrac{3}{2}}L_5 \\
0 & 0 & L_1 + L_3 & 0 & 0 \\
0 & 0 & 0 & L_1 - 2L_3 & 0 \\
0 & \sqrt{\tfrac{3}{2}}L_5 & 0 & 0 & L_4
\end{bmatrix}
\frac{d}{dt}
\begin{bmatrix}
\omega \\ i_d \\ i_q \\ i_0 \\ i_f
\end{bmatrix}
$$

$$
= - \begin{bmatrix}
0 & 0 & -i_f\sqrt{\tfrac{3}{2}}L_5 & 0 & 0 \\
0 & 0 & \omega(L_1 + L_3) & 0 & 0 \\
i_f\sqrt{\tfrac{3}{2}}L_5 & -\omega(L_1 + L_3) & r & 0 & 0 \\
0 & 0 & 0 & r & 0 \\
0 & 0 & 0 & 0 & r_f
\end{bmatrix}
\begin{bmatrix}
\omega \\ i_d \\ i_q \\ i_0 \\ i_f
\end{bmatrix}
+
\begin{bmatrix}
-T_L \\ v_d \\ v_q \\ v_0 \\ v_f
\end{bmatrix}
$$

Problem 5.155 (Sleigh on inclined plane) *Consider the problem of a sleigh on an inclined plane as described in [15] and illustrated in Figure (5.9). A single knife edge along the centerline represents the runners. Assume that the center of mass lies along a straight line forward of the knife edge a distance d, the sleigh has mass m and moment of inertia J about the center of mass. The sleigh is constrained by the knife edge so that it can not side slip. Supose that a drive force F acts along the centerline and a turning torque T acts about the z-axis.*

(a) *Derive the equatons of motion.*

(b) *Derive an expression for the total energy of the system.*

(c) *Suppose $d = 0$ so that the center of mass is alligned with the knife edge contact point, $m = 1$, $F = 0$, $T = 0$. Verify that the total energy is constant along trajectories. Consider trajectories that begin with only an angular velocity and derive expressions for $x(t), y(t)$ using the fact that total energy is constant. Plot some typical curves in the (x, y)-plane.*

(d) *Continue with the assumptions of (c) and suppose $\alpha = 0$ so that the motion takes place on a horizontal plane. Notice that the potential energy term now vanishes. Show that the sleigh moves with constant angular velocity and constant forward speed, i.e., it moves in a circle of constant radius. What is the radius?*

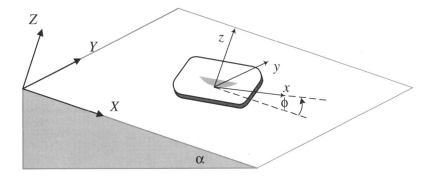

Figure 5.9:

5.7 References

1. Kwatny, H.G. and G.L. Blankenship, Symbolic Construction of Models for Multibody Dynamics. IEEE Transactions on Robotics and Automation, 1995. 11(2): p. 271-281.

2. Leu, M.C. and N. Hemati, Automated Symbolic Derivation of Dynamic Equations for Robotic Manipulators. Journal of Dynamic Systems, Measurement and Control, 1986. 108(September): p. 172-179.

3. Cetinkunt, S. and B. Ittoop, ComputerAutomated Symbolic Modeling of Dynamics of Robotic Manipulators with Flexible Links. IEEE Transactions on Robotics and Automation, 1992. 8(1): p. 94105.

4. Blankenship, G.L., et al., Integrated tools for Modeling and Design of Controlled Nonlinear Systems. IEEE Control Systems, 1995. 15(2): p. 65-79.

5. Arnold, V.I., V.V. Kozlov, and A.I. Neishtadt, Mathematical Aspects of Classical and Celestial Mechanics. Encyclopedia of Mathematical Sciences, ed. V.I. Arnold. Vol. 3. 1988, Heidelberg: SpringerVerlag.

6. Chetaev, N.G., On the Equations of Poincaré. PMM (Applied Mathematics and Mechanics), 1941(5): p. 253262.

7. Chetaev, N.G., Theoretical Mechanics. 1989, New York: SpringerVerlag.

8. Meirovitch, L., Methods of Analytical Dynamics. 1970, New York: McGrawHill, Inc.

9. Neimark, J.I. and N.A. Fufaev, Dynamics of Nonholonomic Systems. Translations of Mathematical Monographs. Vol. 33. 1972, Providence: American Mathematical Society.

10. Gantmacher, F., Lectures in Analytical Mechanics. English Translation ed. 1975, Moscow: Mir.

11. Marsden, J.E. and T. Ratiu, Introduction to Mechanics and Symmetry. 2nd ed. Texts in Applied Mathematics. Vol. 17. 1998, New York: Springer-Verlag.

12. Kwatny, H.G., F.M. Massimo, and L.Y. Bahar, The Generalized Lagrange Equations for Nonlinear RLC Networks. IEEE Trans. on Circuits and Systems, 1982. CAS29(4): p. 220233.

13. Leonard, N.E., Stability of a Bottom-Heavy Underwater Vehicle. Automatica, 1997. 33(3): p. 331-346.

14. Rosenberg, R.M., Analytical Dynamics of Discrete Systems. 1977, New York: Plenum Press.

15. Bloch, A.M., M. Reyhanoglu, and N.H. McClamroch, Control and Stabilization of Nonholonomic Dynamic Systems. IEEE Transactions on Automatic Control, 1992. 37(11): p. 1746-1757.

16. Ginsberg, J.H., Advanced Engineering Dynamics. 1988, New York: Harper and Row.

Chapter 6

Smooth Affine Control Systems

6.1 Introduction

This chapter is concerned with the analysis and design of controls for nonlinear systems that are linear in control. They take the form:

$$
\begin{aligned}
\dot{x} &= f(x) + G(x)u \\
y &= h(x)
\end{aligned}
\tag{6.1}
$$

where $x \in R^n$, $y \in R^p$, $u \in R^m$ and f, G and h are smooth functions. Systems of this type are called *smooth, affine systems*. Consistent with the main theme of this book, the discussion to follow emphasizes computations. We will only summarize the theoretical concepts that underlie the computations. More complete developments can be found elsewhere, notably, in [1] and [2]. More specialized results will be referenced at appropriate points in the sequel.

In Chapter 3 we discussed some basic tools and computations involving vector fields and distributions. This material is an essential prerequisite for what follows. Section 2 of this chapter deals with controllability and Section 3 observability. Section 4 describes a local decomposition of affine nonlinear systems into four parts: controllable and observable, controllable and not observable, not controllable and observable, and not controllable and not observable. A decomposition very much like the Kalman decomposition for linear systems. Then we address exact feedback linearization, input-output linearization, dynamic inversion and dynamic extension, respectively, in Sections 5 through 8. Section 9 describes observers and one approach to observer design. In each instance, we summarize the method, describe the relevant computations and give examples.

6.2   Controllability

The notions of reachability and controllability are fundamental to control system design for nonlinear systems just as they are for linear systems. Perhaps more so, because there are important nuances of nonlinear controllability whose counterparts in the linear context are nonexistant or inconsequential. As we will see, these lead to new paradigms for nonlinear control.

Roughly speaking, reachability is the ability to reach any desired terminal state from a given initial state in finite time and controllability is the ability to reach a given terminal state from any initial state in finite time. For linear systems the two properties are equivalent but this may not be the case for nonlinear systems. The controllability of nonlinear systems has many subtleties that will not be explored here. We shall provide only some basic definitions and results. For more information about the controllability of affine systems along the lines of our discussion below see [1] and [2].

The notion of controllability to be used herein is provided in the following definition. Readers desiring a deeper motivation for these concepts should consult the pioneering paper [3].

Definition 6.156   *1. The state x_f is U-reachable from x_0 if given a neighborhood U of x_0 containing x_f, there exists a time $t_f > 0$ and a piecewise constant control $u(t)$, $t \in [0, t_f]$ such that if the system (6.1) starts in the state x_0 at time 0, it reaches the state x_f at time t_f along a trajectory that remains entirely in U.*

2. The control system (6.1) is said to be locally reachable from x_0 if for each neighborhood U of x_0 the set of states U-reachable from x_0 contains a neighborhood of x_0. If the reachable set contains merely an open set (not a neigborhood) the system is said to be locally weakly reachable from x_0

3. The control system (6.1) is said to be locally controllable on R^n if it is locally reachable from every initial state $x_0 \in R^n$. It is locally weakly controllable on R^n if it is locally weakly reachable from every initial state $x_0 \in R^n$.

Now, let us define two related distributions (sometimes referred to as *controllability distributions*):

$$\Delta_C = \langle f, g_1 \ldots g_m \, | \mathrm{span}\{f, g_1, \ldots, g_m\} \rangle \qquad (6.2)$$

i.e., the smallest involutive distribution containing $\mathrm{span}\{f, g_1 \ldots g_m\}$ and invariant with respect to $f, g_1 \ldots g_m$, and

$$\Delta_{C_0} = \langle f, g_1 \ldots g_m \, | \mathrm{span}\{g_1, \ldots, g_m\} \rangle \qquad (6.3)$$

i.e., the smallest involutive distribution containing span$\{g_1 \ldots g_m\}$ and invariant with respect to $f, g_1 \ldots g_m$. These distributions are clearly closely related. The most important relationships are given by the following Lemma from [1].

Lemma 6.157 *The distributions Δ_C and Δ_{C_0} satisfy*

1. $\Delta_{C_0} + \text{span}\{f\} \subseteq \Delta_C$

2. *if x is a regular point of $\Delta_{C_0}(x) + \text{span}\{f(x)\}$, then $\Delta_{C_0}(x) + \text{span}\{f(x)\} = \Delta_C(x)$.*

3. *if Δ_{C_0} and $\Delta_{C_0} + \text{span}\{f\}$ are of constant dimension, then $\dim \Delta_C - \dim \Delta_{C_0} \leq 1$.*

Proof: see[1], p. 61. ∎

When the distribution Δ_C is of constant dimension on a neighborhood of a state x_0 we can contruct a local coordinate transformation that reveals the controllability properties of the system. We summarize the main result in the following Lemma.

Lemma 6.158 *Suppose Δ_C and Δ_{C_0} are of constant dimension on some open set U of R^n. Furthermore, suppose that Δ_{C_0} is properly contained in Δ_C, so that $\dim \Delta_{C_0} = r - 1$ and $\dim \Delta_C = r$. Then for each point $x_0 \in U$ there exists a neighborhood U_0 of x_0 and a local coordinate transformation $z = \Phi(x)$ on U_0 such that in the new coordinates the system equations take the form:*

$$\dot{\zeta}_1 = f_1(\zeta_1, \zeta_2) + G_1(\zeta_1, \zeta_2)u$$

$$\dot{\zeta}_2 = f_2(\zeta_2)$$

where $\zeta_1 = (z_1, .., z_{r-1})$, $\zeta_2 = (z_r, .., z_n)$

$$f_2(\zeta_2) = \begin{bmatrix} f_r(\zeta_2) \\ 0 \\ \vdots \\ 0 \end{bmatrix}$$

Moreover, if $\Delta_C = \Delta_{C_0}$ (so that $\dim \Delta_{C_0} = r - 1$ and $\dim \Delta_C = r - 1$) then the first component of f_2 vanishes so that $\dot{z}_r = 0$.

Proof: Recall Lemma (3.87). It follows that there exists a local coordinate transformation (matched to Δ_{C_0}) such that each of the vector fields $f, g_1, \ldots, g_m$

have the form

$$\bar{f}(z) = \begin{bmatrix} f_1(z_{r-1}, \ldots z_d, z_r, \ldots, z_n) \\ \cdots \\ f_{r-1}(z_1, \ldots z_{r-1}, z_r, \ldots, z_n) \\ f_r(z_r, \ldots, z_n) \\ \cdots \\ f_n(z_r, \ldots, z_n) \end{bmatrix}$$

Furthermore, since $g_1, \ldots, g_m$ belong to Δ_{C_0}, their last $n - r + 1$ coordinates must vanish in the new coordinate system. Now, in view of the fact that Δ_C is of dimension r and contains Δ_{C_0} as well as f, the last $n - r$ components of f must vanish in the new coordinate system. Finally, in the case $\Delta_C = \Delta_{C_0}$, the same arguments lead to the stated conclusion. ∎

Remark 6.159 *Suppose that* $\dim \Delta_C = r$ *and consider the controlled motion from any initial state* $x_0 \in U$. *Trajectories are resticted to the* r-*dimensional set*

$$S_{x_0} = \{ x \in U_0 \, | \phi_{r+1}(x) = \phi_{r+1}(x_0), \ldots, \phi_n(x) = \phi_n(x_0) \}$$

by virtue of the fact that the derivatives $\dot{z}_{r+1}, \ldots, \dot{z}_r$ *all vanish. Thus a necessary condition for local weak reachability from* x_0 *is that* $\dim \Delta_C = n$.

Remark 6.160 *Suppose that* $\dim \Delta_C = n$ *and* $\dim \Delta_{C_0} = n - 1$ *on* U. *The controlled motion from an initial state* $x_0 \in U$ *is again in the set* S_{x_0} *defined above, with* $r = n$. *But we can still exploit the last equation which is*

$$\dot{z}_r = f_r(z_n)$$

Now, f *can not belong to* Δ_{C_0} *Since* Δ_C *is strictly larger than* Δ_{C_0} *(everywhere on* U). *Thus,* f_n *can not vanish anywhere on* U. *Suppose* $z_n(t)$ *denotes the solution of this differential equation which has the boundary condition* $z_n(0) = \phi_n(x_0)$. *This relation defines a diffeomorphism*

$$\mu : t \mapsto z_n$$

on a time interval $(-\varepsilon, \varepsilon)$ *to its image on the* z_n *axis. Thus, we can take time* $t = \mu^{-1}(z_n)$ *as the new* n^{th} *coordinate. In terms of the transformed states* $(z - 1, \ldots, z_n - 1, t)$, *the points reachable from* x_0 *at precisely time* T *belong to the set*

$$S_{x_0}^T = \{ x \in U_0 \, | \mu^{-1}(\phi_n(x)) = T \}$$

It follows that x_0 *is on the boundary of the reachable set, it can not be in its interior. Consequently, the set of states reachable from* x_0 *is not a neighborhood of* x_0. *It follows that* $\dim \Delta_{C_0} = n$ *is a necessary condition for local reachability.*

Example 6.161 *Consider the following example (Isidori [1], Example 1.8.4), involving the system:*

$$\dot{x} = f(x) + g(x)u = \begin{bmatrix} x_1 x_3 + x_2 e^{x_2} \\ x_3 \\ x_4 - x_2 x_3 \\ x_3^2 + x_2 x_4 - x_2^2 x_3 \end{bmatrix} + \begin{bmatrix} x_1 \\ 1 \\ 0 \\ x_3 \end{bmatrix} u$$

In[159]:= f = {*x1 x3* + *x2* exp[*x2*], *x3, x4* − *x2 x3, x3^2* + *x2 x4* − *x2^2 x3*};
   g = {*x1*, 1, 0, *x3*};
   var = {*x1, x2, x3, x4*};
   del = *Span*[{*g*}]
Out[159]= {{x1, 1, 0, x3}}

First, compute the smallest f, g invriant distribution containing span{*g*}.

In[160]:= Del0 = *SmallestInvariantDistribution*[{*f, g*}, *del, var*]
Out[160]= {{1, 0, 0, 0}, {0, 1, 0, x3}}

Now, augment this distribution with two additional vector fields,

In[161]:= Del = *Join*[*Del0*, {{0, 0, 1, 0}, {0, 0, 0, 1}}]
Out[161]= {{1, 0, 0, 0}, {0, 1, 0, x3}, {0, 0, 1, 0}, {0, 0, 0, 1}}

and check that the distribution does span R^4.

In[162]:= Span[*Del*]
Out[162]= {{1, 0, 0, 0}, {0, 1, 0, 0}, {0, 0, 1, 0}, {0, 0, 0, 1}}

Finally, we generate the transformation and the system equations in the new state variables.

In[163]:= TriangularDecomposition[*f* + *g* ∗ *u, Del, var*, {0, 0, 0, 0}, ∞]
Out[163]= {{z1, z2, z3, z2 z3 + z4}, {x1, x2, x3, −x2 x3 + x4},
   {u z1 + e^{z2} z2 + z1 z3, u + z3, z4, 0}}

So, we obtain the system in triangular form as anticipated:

$$\dot{z} = \begin{bmatrix} z_1 z_3 + z_2 e^{z_2} \\ z_3 \\ z_4 \\ 0 \end{bmatrix} + \begin{bmatrix} z_1 \\ 1 \\ 0 \\ 0 \end{bmatrix} u$$

Now, we are in a position to establish the main result.

Proposition 6.162 *1. A necessary and sufficient condition for the control system (6.1) to be locally weakly controllable on R^n is that* dim $\Delta_C(x_0) = n$ *for all $x_0 \in R^n$.*

2. A necessary and sufficient condition for the control system (6.1) to be locally controllable on R^n is that dim $\Delta_{C_0}(x_0) = n$ *for all $x_0 \in R^n$.*

Proof: Only a sketch of the main ideas will be given. The main argument follows [1].

necessity: Necessity is a conseqence of the remarks following Lemma (6.158).

sufficiency: We will summarize a constructive proof that the condition $\dim \Delta_C = n$ is sufficient for local weak controllability. To do this we will show that from any point $x_0 \in R^n$ we can construct a piecewise continuous control that steers to an arbitrary point in an open set contained in the slice S_{x_0}. First, let us make several preliminary observations.

(a) Suppose that $\dim \Delta_C = r \leq n$ on a neighborhood U_0 of x_0. All trajectories from x_0 are restricted to the r-dimensional slice S_{x_0}. At any point x we can steer the trajectory in the direction

$$\theta_i(x) = f(x) + g_1(x)u_1^i + \cdots + g_m(x)u_m^i \qquad (6.4)$$

where $u_1^i, \ldots u_m^i$ are real numbers. By choosing $k \leq m$ sets of constant controls we can define k vector fields $\theta_i(x)$ of the form (6.4), $i = 1, \ldots, k$. Define the mapping $F : V = (-\varepsilon, \varepsilon)^k \to S_{x_0}$ realized by

$$F(t_1, \ldots, t_k) = \phi_k^{t_k} \circ \cdots \circ \phi_1^{t_1}(x_0) \qquad (6.5)$$

where $\phi_i^{t_i}$ is the flow corresponding to the vector field θ_i. Suppose that the differential at $s_i, \ldots, s_k$, $0 < s_i < \varepsilon$ is of rank k, then for ε sufficiently small,

$$M = \{x \in S_{x_0} \mid x = F(t_1, \ldots, t_k), \ t_i \in (s_i, \varepsilon), \ i = 1, \ldots, k\}$$

is a regular submanifold of S_{x_0}.

(b) For any $x \in M$, $TM_x \subset \Delta_C(x)$. If $k < r$, it can not be true that for all $x \in M$, $f(x) \in TM_x$ and $g_i(x) \in TM_x$, $i = 1, \ldots, k$. Because if it were true, TM_x would define an involutive distribution containing $f, g_1, \ldots, g_m$ that is smaller than Δ_C, a contradiction. But if it is not true, then it is possible to find an $\bar{x} \in M$ and constants $u_1^{k+1}, \ldots u_m^{k+1}$ such that $\theta_{k+1}(\bar{x}) \notin TM_{\bar{x}}$. In fact, $\bar{x}$ can be found arbitrarily close to x (because ε can be taken arbitrarily small). Let $\bar{x} = F(\bar{s}_1, \ldots, \bar{s}_k)$, $\bar{s}_i > s_i$, $i = 1, \ldots, k$. Define the mapping $\bar{F}(t_1, \ldots, t_k, t_{k+1}) = \phi_{k+1}^{t_{k+1}} \circ F(t_1, \ldots, t_k)$. It can be shown that this mapping has rank $k+1$ at the point $(\bar{s}_1, \ldots \bar{s}_k, 0)$ (see [1], p66).

Now, let us turn to the main construction. We can choose constants $u_1^1, \ldots, u_m^1$ to define a vecor field

$$\theta_1 = f + \sum_{i=1}^m g_i u_i^1$$

that is not zero at x_0. Define the map

$$F_1 : (0, \varepsilon) \to S_{x_0}$$

$$F_1(t_1) = \phi_1^{t_1}(x_0)$$

and let M^1 be the image of F_1.

Let $\bar{x}_1 = F_1(s_1^1)$ be a point of M^1 and θ_2 be a vector field

$$\theta_2 = f + \sum_{i=1}^{m} g_i u_i^2$$

such that $\theta_2(\bar{x}_1) \notin TM_{\bar{x}_1}^1$. Define the mapping

$$F_2 : (s_1^1, \varepsilon) \times (0, \varepsilon) \to S_{x_0}$$

$$F_2(t_1, t_2) = \phi_2^{t_2} \circ \phi_1^{t_1}(x_0)$$

By construction this mapping has rank 2 on its domain. Repeating this procedure, at the kth step we choose a point $\bar{x} = F_{k-1}(s_1^{k-1}, \ldots, s_{k-1}^{k-1})$, with $s_i^{k-1} > s_i^{k-2}$ for $i = 1, \ldots, k-2$, $s_{k-1}^{k-1} > 0$, and constants u_i^k such that the vector field $\theta_k \notin TM_{\bar{x}}^{k-1}$. Then construct the map

$$F_k : (s_1^{k-1}, \varepsilon) \times \cdots \times (s_{k-1}^{k-1}, \varepsilon) \times (0, \varepsilon) \to S_{x_0}$$

$$F_k(t_1, \ldots, t_k) = \phi_k^{t_k} \circ \cdots \circ \phi_1^{t_1}(x_0)$$

The procedure stops with $k = r$.

Notice that any point $x = F_r(t_1, \ldots, t_r)$ in the image of M_r can be reached from x_0 with the piecewise control

$$\begin{aligned} u_i(t) &= u_i^1 & t \in [0, t_1) \\ u_i(t) &= u_i^k & t \in [t_1 + \cdots + t_{k-1}, t_1 + \cdots + t_k) \end{aligned}$$

Now, by construction, each mapping F_k parametrically defines regular submanifolds of S_{x_0} of dimension k. Thus, the images under F_r of open sets

$$V_r = (s_1^{r-1}, \varepsilon) \times \cdots \times (s_{r-1}^{r-1}, \varepsilon) \times (0, \varepsilon)$$

are open sets of S_{x_0} of dimension r. ∎

Example 6.163 (Linear system controllability) *Let us consider the controllability of a linear system*

$$\dot{x} = Ax + Bu, \ x \in R^n, \ u \in R^m$$

by computing the distributions Δ_C and Δ_{C_0} using algorithm (3.91). In this case the relevant vector fields are $f(x) = Ax$ and $g_i(x) = b_i$, $i = 1, \ldots, m$. Notice that

$$[Ax, b_i] = -\frac{\partial Ax}{\partial x} b_i = -Ab_i$$

and

$$[b_j, b_i] = 0, \ [Ax, Ax] = 0$$

To compute Δ_{C_0} *begin with* $\Delta = \text{span}\{B\}$ *and apply the algorithm.*

$$\Delta_0 = \text{span}\{B\}$$
$$\Delta_1 = \text{span}\{B \quad AB\}$$
$$\vdots$$
$$\Delta_k = \text{span}\{B \quad AB \quad \cdots \quad A^{k-1}B\}$$

In view of the Caley-Hamilton theorem, we may as well stop at $k = n$. *Thus, we find that the linear system is locally controllable if and only if*

$$\text{rank}\,[B \quad AB \quad \cdots \quad A^{k-1}B] = n \tag{6.6}$$

To compute Δ_C *we begin the process with* $\Delta = \text{span}\{Ax, B\}$ *to obtain*

$$\Delta_C = \text{span}\{Ax \quad B \quad AB \quad \cdots \quad A^{n-1}B\}$$

If $\text{rank}\,\Delta_{C_0} = n-1$ *it is possible that* $\text{rank}\,\Delta_C = n$ *at points other than* $x = 0$, *so weak local reachability around points* $x_0 \neq 0$ *is possible. However, controllability still requires (6.6).*

ProPac provides functions that construct the distributions Δ_C and Δ_{C_0} and implement the test for controllability. The following example exploits these calculations to illustrate the distinction between local controllability and weak local controllability.

Example 6.164 (Bilinear System Controllability) *Consider the following bilinear, scalar input system (Example 7.35 in Vidyasagar [4]):*

$$\dot{x} = \begin{bmatrix} 0 & 0 & -14 \\ 0 & 0 & 0 \\ 0 & 0 & -19 \end{bmatrix} x + \begin{bmatrix} 1 & 2 & 4 \\ 0 & 2 & 0 \\ 0 & 0 & 3 \end{bmatrix} x\,u$$

We use the ProPac package to compute the control distributions. First, the distribution

$$\langle f, g_1 \ldots g_m \,|\, f, g_1 \ldots g_m \rangle$$

In[164]:=　$A = \{\{0, 0, -14\}, \{0, 0, 0\}, \{0, 0, -1\}\};$
　　　　$B = \{\{1, 2, 4\}, \{0, 2, 0\}, \{0, 0, 3\}\};$
　　　　$x = \{x1, x2, x3\};$
　　　　$f = A.x;$
　　　　$g = B.x;$
　　　　$ControlDistribution[f, g, x]//MatrixForm$

Out[164]=　$\begin{pmatrix} 1 & 0 & 0 \\ 0 & 1 & 0 \\ 0 & 0 & 1 \end{pmatrix}$

The control distribution has constant dimension 3 (=n) and thus the system is weakly locally controllable.

It does not always make sense to rely on the drift term, f, to steer a system from one state to another. So an option to compute the control distribution without the drift term is made available, i.e.,

$$\langle f, g_1 \cdots g_m \, | g_1 \cdots g_m \rangle$$

In[165]:= $ControlDistribution[f, g, x, IncludeDrift \to False] // MatrixForm$

Out[165]= $\begin{pmatrix} 1 & 0 & 0 \\ 0 & 2\,x2 & 3\,x3 \end{pmatrix}$

In this case the generic rank is 2. Thus without accounting for the drift term, the bilinear system is not controllable, i.e., it is not locally controllable.

The following example illustrates another important and distinctive aspect of nonlinear system controllability.

Example 6.165 (Parking) *A classic problem in control system analysis is the 'parking problem' [5]. The simplified equations of motion of the vehicle to be parked are (see Example (5.147)):*

$$\frac{d}{dt} \begin{bmatrix} x \\ y \\ \phi \\ \theta \end{bmatrix} = \begin{bmatrix} \cos(\phi + \theta) & 0 \\ \sin(\phi + \theta) & 0 \\ \sin(\theta) & 0 \\ 0 & 1 \end{bmatrix} \begin{bmatrix} v \\ w \end{bmatrix}$$

where x, y and ϕ represent the planar location and orientation of the center of mass, and θ represents the steering angle. The controls v and w represent the drive velocity and the angular velocity of the steering angle, respectively. The equations represent the kinematics of the vehicle motion. It is assumed that the velocities can be changed instantaneously. There are two control actions defined by the vector fields:

$$drive = \begin{bmatrix} \cos(\phi + \theta) \\ \sin(\phi + \theta) \\ \sin(\theta) \\ 0 \end{bmatrix}, \quad steer = \begin{bmatrix} 0 \\ 0 \\ 0 \\ 1 \end{bmatrix}$$

First, let us compute the controllability distribution. Note that since the system is drift free ($f = 0$), $\Delta_C = \Delta_{C_0}$.

In[166]:= $f = \{0, 0, 0, 0\};$
$G = \{\{\cos[\phi + \theta], \sin[\phi + \theta], \sin[\theta], 0\}, \{0, 0, 0, 1\}\};$
$var = \{x, y, \phi, \theta\};$
$ControlDistribution[f, Transpose[G], var] // MatrixForm$

Out[166]= $\begin{pmatrix} 1 & 0 & 0 & 0 \\ 0 & 1 & 0 & 0 \\ 0 & 0 & 1 & 0 \\ 0 & 0 & 0 & 1 \end{pmatrix}$

The control distribution has constant dimension 4 (=n) so that the system is locally controllable. Thus, the vehicle can be moved from any position and orientation to any other position and orientation in finite time. Notice, however, the linearization of the system at any point is not controllable. To better appreciate the information contained in the controllability distibution Δ_C (or Δ_{C_0}) let us consider the details of its construction. Since $f = 0$, we begin with span$\{drive, steer\}$ and expand this distribution by taking Lie brackets of its component vector fields until we achieve a distribution of maximum dimension.

As it turns out we need to add two vector fields to reach the maximum dimension of 4. These are called wriggle and slide.

In[167]:= drive $= \{\cos[\phi + \theta], \sin[\phi + \theta], \sin[\theta], 0\}$;
           steer $= \{0, 0, 0, 1\}$;
           wriggle $= LieBracket[steer, drive, \{x, y, \phi, \theta\}]$;
           wriggle//MatrixForm

Out[167]= $\begin{pmatrix} -\operatorname{Sin}[\theta + \phi] \\ \operatorname{Cos}[\theta + \phi] \\ \operatorname{Cos}[\theta] \\ 0 \end{pmatrix}$

In[168]:= slide $= Simplify[LieBracket[wriggle, G[[1]], \{x, y, \phi, \theta\}]]$;

           slide//MatrixForm

Out[168]= $\begin{pmatrix} -\operatorname{Sin}[\phi] \\ \operatorname{Cos}[\phi] \\ 0 \\ 0 \end{pmatrix}$

These two new control vector fields enable complete configuration control of the vehicle. Experienced drivers will recognize that maneuvering a vehicle in and out of parking a space requires the control action generated by wriggle. The wriggle vector is the Lie bracket of the steer and the drive vector fields. To actually move the vehicle in the direction of this vector field, at least approximately, would require infinitesimal excursions along the steer vector field, then the drive vector field, then the reverse of the steer vector field, then the reverse of the drive vector field. Thus, the name wriggle. Movement along higher order bracket directions (e.g. slide) involves more complicated switching schemes.

The following example shows that, in contrast to linear systems, local controllability does not imply asymptotic stabilizability via simple state feedback control.

Example 6.166 (Sleigh on a horizontal plane, continued) *Let us return to the sleigh of Example (5.146). We will first show that the system is locally controllable. To do this, we need to put the equations obtained previously into state space form.*

In[169]:= f1 $= Vm.phat$;
           f2 $= Inverse[Mm].(-Cm.phat - Fm)$;
           f $= Join[f1, f2]/.\{T \to 0, F \to 0\}$

$Out[169]=$ $\left\{\omega, \text{vx } \cos[\phi], \text{vx } \sin[\phi], -\dfrac{\text{d m1}^2 \text{ vx } \omega}{\text{Jzz m1} + \text{d}^2 \text{ m1}^2}, \dfrac{\text{d m1 (Jzz} + \text{d}^2 \text{ m1}) \omega^2}{\text{Jzz m1} + \text{d}^2 \text{ m1}^2}\right\}$

$In[170]:= \quad G = Simplify[\,Transpose[\,Map[\,Coefficient[\,Join[f1, f2], \#]\&, \{T, F\}]]]$

$Out[170]= \left\{\{0, 0\}, \{0, 0\}, \{0, 0\}, \left\{\dfrac{1}{\text{Jzz} + \text{d}^2 \text{ m1}}, 0\right\}, \left\{0, \dfrac{1}{\text{m1}}\right\}\right\}$

$In[171]:= \quad var = Join[qsys, phat]$

$Out[171]= \{\phi, x, y, \omega, vx\}$

In summary, the state equations are:

$$\frac{d}{dt}\begin{bmatrix} \phi \\ x \\ y \\ \omega \\ v_x \end{bmatrix} = \begin{bmatrix} \omega \\ v_x \cos\phi \\ v_x \sin\phi \\ -\frac{d m_1}{J_{zz}+d^2 m_1} v_x \omega \\ d\omega^2 \end{bmatrix} + \begin{bmatrix} 0 & 0 \\ 0 & 0 \\ 0 & 0 \\ \frac{1}{J_{zz}+d^2 m_1} & 0 \\ 0 & \frac{1}{m_1} \end{bmatrix} \begin{bmatrix} T \\ F \end{bmatrix}$$

Now, we can apply the controllability test.

$In[172]:= \quad Controllability[f, G, var, LocalControllability \to True]$

$Out[172]= \text{True}$

Thus we confirm that the system is locally controllable. We can obtain more details about controllability by computing the controllability distribution Δ_{C_0}. In the following calculation we display intermediate results. That is, beginning with span $\{g_1, g_2\}$, each time the distribution is expanded by addition of a new vector field arising from Lie bracket operations, the new distribution is displayed.

$In[173]:= \quad ControlDistribution[f, G, var, IntermediateResults \to True,$
$\qquad ControlDrift \to False]$

Intermediate distribution is:

$\{\{\omega, \text{vx } \cos[\phi], \text{vx } \sin[\phi], 0, 0\}, \{0, 0, 0, 1, 0\}, \{0, 0, 0, 0, 1\}\}$

Intermediate distribution is:

$\{\{1, 0, 0, 0, 0\}, \{0, 1, \tan[\phi], 0, 0\}, \{0, 0, 0, 1, 0\}, \{0, 0, 0, 0, 1\}\}$

Intermediate distribution is:

$\left\{\begin{array}{l} \{1, 0, 0, 0, 0\}, \{0, 1, 0, 0, 0\}, \{0, 0, 1, 0, 0\}, \{0, 0, 0, 1, 0\}, \\ \{0, 0, 0, 0, 1\} \end{array}\right\}$

$Out[173]= \{\{1, 0, 0, 0, 0\}, \{0, 1, 0, 0, 0\}, \{0, 0, 1, 0, 0\}, \{0, 0, 0, 1, 0\},$
$\qquad \{0, 0, 0, 0, 1\}\}$

Since the system is locally controllable consider the posibility of asymptotically stabilizing the origin via smooth state feedback. Suppose $u_1(\phi, x, y, \omega, v_x)$ and $u_2(\phi, x, y, \omega, v_x)$ are arbitrary feedback functions that have continuous first derivatives and $u_1(0, 0, 0, 0, 0) = u_2(0, 0, 0, 0, 0) = 0$. The closed loop dynamics are:

$In[174]:= \quad fcl = f + G.\{u1[\phi, x, y, \omega, vx], u2[\phi, x, y, \omega, vx]\};$

$\qquad fcl//MatrixForm$

$$Out[174] = \begin{pmatrix} \omega \\ vx\ Cos[\phi] \\ vx\ Sin[\phi] \\ -\dfrac{d\ m1^2\ vx\ \omega}{Jzz\ m1 + d^2\ m1^2} + \dfrac{u1[\phi, x, y, \omega, vx]}{Jzz + d^2\ m1} \\ \dfrac{d\ m1\ (Jzz + d^2\ m1)\ \omega^2}{Jzz\ m1 + d^2\ m1^2} + \dfrac{u2[\phi, x, y, \omega, vx]}{m1} \end{pmatrix}$$

Equilibria of the closed loop system occur when $f_{cl} = 0$. Clearly, the origin is an equilibrium point. However, it is not an isolated equilibrium point. To see this, observe that $f_{cl} = 0$ if and only if $\omega = 0$, $v_x = 0$, and $u(\phi, x, y, 0, 0) = 0$, i.e., all points (ϕ, x, y) that satisfy the two equations $u(\phi, x, y, 0, 0) = 0$ are equilibrium points. Notice that

$$\text{rank} \begin{bmatrix} \frac{\partial u_1(0)}{\partial \phi} & \frac{\partial u_1(0)}{\partial x} & \frac{\partial u_1(0)}{\partial y} \\ \frac{\partial u_2(0)}{\partial \phi} & \frac{\partial u_2(0)}{\partial x} & \frac{\partial u_2(0)}{\partial y} \end{bmatrix} \leq 2$$

If the rank is 2, the Implicit Function Theorem establishes that there is are explicit smooth functions expresing two of the variables, ϕ, x, y, as functions of the third and passing through the origin. Consequently, there is a one dimensional set of equilibrium states in every neighborhood of the origin. If the rank is 1, then there is a two dimensional set and if it is zero, there is a three dimensional set (all values of ϕ, x, y are equilibria).

Thus, the origin is certainly not an isolated equilibrium point. It follows from Lemma (2.16) that the origin can not be asymptotically stable. We conclude that even though the system is locally controllable it can not be asymptotically stabilized via smooth state feedback.

In order to better appreciate the relationship of controllability for nonlinear and linear systems we consider two additional distributions.

$$\Delta_L = \text{span}\{f, ad_f^k g_i, 1 \leq i \leq m, 0 \leq k \leq n - 1\} \tag{6.7}$$

and

$$\Delta_{L_0} = \text{span}\{ad_f^k g_i, 1 \leq i \leq m, 0 \leq k \leq n - 1\} \tag{6.8}$$

Note that $\Delta_L \subseteq \Delta_C$ and $\Delta_{L_0} \subseteq \Delta_{C_0}$. What is missing in these distributions, in relation to Δ_C and Δ_{C_0}, are the Lie brackets among the control vector fields, g_i. These new distributions have obvious connections to the Kalman test for controllability of linear systems.

Example 6.167 (Linear system controllability, revisited) *For the linear system of Example (6.163) we have $f(x) = Ax$, $G = B$, so it is easy to compute*

$$\Delta_C = \Delta_L = \text{span}\left\{Ax, B, AB, \ldots, A^{n-1}B\right\}$$

$$\Delta_{C_0} = \Delta_{L_0} = \text{span} \left\{ B, AB, \ldots, A^{n-1}B \right\}$$

With linear systems the control vector fields are constant so that the missing Lie brackets contribute nothing to the controllability distributions.

Now, let us state the following obvious results:

Proposition 6.168   *1. A sufficient condition for the control system (1) to be locally weakly reachable around x_0 is that* dim $\Delta_L(x_0) = n$.

  2. A sufficient condition for the control system (1) to be locally weakly controllable on R^n is that dim $\Delta_L(x_0) = n$ *for all $x_0 \in R^n$.*

The relationships between the various distributions, weak local controllability and local controllability are summarized in the following diagram.

$$
\begin{array}{ccccc}
weak\ local\ controllability & \Leftrightarrow & \dim \Delta_C = n & \Leftarrow & \dim \Delta_L = n \\
\Uparrow & & \Uparrow & & \Uparrow \\
local\ controllability & \Leftrightarrow & \dim \Delta_{C_0} = n & \Leftarrow & \dim \Delta_{L_0} = n
\end{array}
$$

6.3 Observability

Like controllability, observability is a fundamental property of nonlinear control systems just as it is for linear systems. Our treatment will be similar in many ways to the previous discussion of controllability. While observability does not have precisely the dual structure to controllability as it has in linear system theory, there are interesting parallels. Moreover, we will see that nonlinear observability has important nuances that distinguish it from the linear counterpart.

Consider the nonlinear affine system described by Equation (6.1). We denote by $y(\cdot; x_0, u)$ the entire output response $y(t; x_0, u)$, $\forall t > 0$ corresponding to the initial state x_0 and control $u(t)$, $\forall t > 0$.

Definition 6.169   *1. Let U be an open set in R^n. Two states $x_1, x_2 \in U$ are said to be U-distinguishable if there exists a control $u(t)$, $\forall t > 0$, whose trajectories from both x_1, x_2 remain in U, such that $y(\cdot; x_1, u) \neq y(\cdot; x_2, u)$. Otherwise they are U-indistinguishable.*

  2. The control system (6.1) is said to be strongly locally observable at $x_0 \in R^n$ if for every neighborhood U of x_0, every state in U other than x_0 is U-distinguishable from x_0. It is said to be locally observable at $x_0 \in R^n$ if there exists a neighborhood W of x_0 such that for every neighborhood U of x_0 contained in W every state in U other than x_0 is U-distinguishable from x_0.

3. *The control system (6.1) is said to be strongly locally observable if it is strongly locally observable at x_0 for every $x_0 \in R^n$. It is said to be locally observable if it is locally at observable x_0 for every $x_0 \in R^n$.*

In essence, local observability at x_0 requires only that x_0 be distinguishable from its immediate neighbors. More insight into the distinction between strong local observability and local observability can be found in [3] (where they are termed, respectively, local observability and weak local observability).

We can establish an observability triangular decomposition similar to the controllability decomposition of Lemma (6.158). First, we define the *observability codistribution*

$$\Omega_O = \langle f, g_1, \ldots, g_m | \text{ span } \{dh_1, \ldots, dh_p\} \rangle \qquad (6.9)$$

and its kernel

$$\Delta_O = \Omega_O^\perp \qquad (6.10)$$

This distribution is invariant with respect to $f, g_1, \ldots, g_m$ and it is contained in the kernel of span $\{dh_1, \ldots, dh_p\}$. If it is nonsingular, it is also involutive.

Proposition 6.170 *Suppose Δ_O is of constant dimension r on some open set U of R^n.*

(i) *Then for each point $x_0 \in U$ there exists a neighborhood U_0 of x_0 and a local coordinate transformation $z = \Phi(x)$ on U_0 such that in the new coordinates the system equations take the form:*

$$\dot{\zeta}_1 = f_1(\zeta_1, \zeta_2) + G_1(\zeta_1, \zeta_2)u$$

$$\dot{\zeta}_2 = f_2(\zeta_2) + G_2(\zeta_2)u$$

$$y = h(\zeta_2)$$

where $\zeta_1 = (z_1, \ldots, z_r)$ and $\zeta_2 = (z_{r+1}, \ldots, z_n)$.

(ii) *Moreover,*

(a) *any two initial states x_1 and x_2 in U_0 such that*

$$\phi_i(x_1) = \phi_i(x_2), \ i = r + 1, \ldots, n$$

produce identical output functions for any input that keeps the trajectory in U_0.

(b) *any initial state $x \in U_0$ that cannot be distinguished from $x_0 \in U_0$ under piecewise constant input functions belongs to the slice*

$$S_{x_0} = \{x \in U_0 | \ \phi_i(x) = \phi_i(x_0), \ i = r + 1, \ldots, n\}$$

Proof: For part (i), Again, recall Lemma (3.87) from which it follows that there exists a local coordinate transformation (matched to Δ_O) such that each of the vector fields $f, g_1, \ldots, g_m$ have the form

$$\bar{f}(z) = \begin{bmatrix} f_1(z_r, \ldots z_d, z_{r+1}, \ldots, z_n) \\ \cdots \\ f_r(z_1, \ldots z_r, z_{r+1}, \ldots, z_n) \\ f_{r+1}(z_{r+1}, \ldots, z_n) \\ \cdots \\ f_n(z_{r+1}, \ldots, z_n) \end{bmatrix}$$

In the new coordinates the covector fields $dh_1, \ldots, dh_2$ must belong to $\Omega_O = \Delta_O^\perp$, so that

$$\frac{\partial h_i}{\partial z_j} = 0$$

for all $0 \le j \le r$ and $1 \le i \le p$.

For part (ii), we provide a sketch of the proof, along the lines of [1]. Consider the following points.

(a) Consider a piecewise constant control function (recall the proof of Proposition (6.162))

$$\begin{aligned} u_i(t) &= u_i^1 & t \in [0, t_1) \\ u_i(t) &= u_i^k & t \in [t_1 + \cdots + t_{k-1}, t_1 + \cdots + t_k), \ k > 1 \end{aligned}$$

for $i = 1, \ldots, m$, and define the vector field

$$\theta_k = f + \sum_{i=1}^m g_i u_i^k$$

Let ϕ_k^t denote its flow. The state reached at time t_k from x_0 at time $t = 0$ is given by the composition

$$x(t_k) = \phi_k^{t_k} \circ \cdots \circ \phi_1^{t_1}(x_0)$$

The output at time t_k is

$$y(t_k) = h(x_{t_k})$$

Accordingly, we may define an output map $Y^{x_0} : (-\varepsilon, \varepsilon)^k \to R^p$

$$Y^{x_0}(t_1, \ldots, t_k) = h \circ \phi_k^{t_k} \circ \cdots \circ \phi_1^{t_1}(x_0)$$

If two arbitrarily close initial states x_1 and x_2 produce identical outputs for any possible piecewise input, we have

$$Y^{x_1}(t_1, \ldots, t_k) = Y^{x_2}(t_1, \ldots, t_k)$$

for all possible $(t_1, \ldots, t_k)$, $t_i \in [0, \varepsilon)$. From this we can verify by direct computation that

$$L_{\theta_1} \ldots L_{\theta_k} h_i(x_1) = L_{\theta_1} \ldots L_{\theta_k} h_i(x_2)$$

(b) Since θ_j, $j = 1, \ldots, k$ depends on the constants $(u_1^j, \ldots, u_m^j)$, and the equality of (a) must hold for all possible choices of $(u_1^j, \ldots, u_m^j) \in R^m$, it can be verified that

$$L_{v_1} \ldots L_{v_k} h_i(x_1) = L_{v_1} \ldots L_{v_k} h_i(x_2)$$

for any set of vector fields, $v_1, \ldots, v_k$ belong to $\{f, g_1, \ldots, g_m\}$.

(c) Recall that the distribution Δ_O is invariant with respect to $f, g_1, \ldots, g_m$ and contains the kernel of span $\{dh_1, \ldots, dh_p\}$. consequently, in view of (b), we can conclude that x_2 belongs to a set that is contained in the maximal integral manifold of Δ_O that passes through x_1, i.e., it belongs to S_{x_1}.

■

An immediate consequence of the above theorem is the following.

Corollary 6.171 *If Ω_O (equivalently, Δ_O) is of constant dimension on some open set U then the system (6.1) is locally observable on U if and only if the observability codistribution Ω_O has dimension n, or equivalently, its kernel Δ_O has dimension 0.*

As a first example of the necessary computations consider the observability of a linear system.

Example 6.172 (Linear System Observability) *Let us consider the observability of a linear system*

$$\dot{x} = Ax + Bu$$

$$y = Cx$$

We will compute the codistribution Ω_O. To do this, we apply the algorithm (3.96) and the formula for the directional derivative of a covector field as defined in (3.72). For the linear system the necessary vector fields are $f(x) = Ax$ and $g_i(x) = b_i$, $i = 1, \ldots m$. Also, $h(x) = Cx$ so that codistribution $dh = \text{span } C$. Notice that for any constant covector field, c_j, using (3.72),

$$L_{Ax}c_j = c_jA, \text{ and } L_{b_i}c_j = 0$$

Thus, compute

$$\Omega_0 = \text{span}\,\{C\},\ \Omega_1 = \text{span}\left\{\begin{array}{c} C \\ CA \end{array}\right\}, \ldots, \Omega_k = \text{span}\left\{\begin{array}{c} C \\ CA \\ \vdots \\ CA^{k-1} \end{array}\right\}$$

From the Caley-Hamilton theorem, we may as well stop at $k = n$. Consequently, the observability necessary condition reduces to the familiar

$$\text{rank} \begin{bmatrix} C \\ CA \\ \vdots \\ CA^{n-1} \end{bmatrix} = n$$

The following example illustrates the decomposition claimed in Proposition (6.170).

Example 6.173 *We consider a modification of Example (6.161) in which a single output equation is added:*

$$y = x_3$$

First, define the system

In[175]:= $\quad f = \{x1 \ x3 + x2 \ \exp[x2], x3, x4 - x2 \ x3, x3\char`\^2 + x2 \ x4 - x2\char`\^2 \ x3\};$
$\qquad\quad g = \{x1, 1, 0, x3\};$
$\qquad\quad h = \{x3\};$
$\qquad\quad var = \{x1, x2, x3, x4\};$

Now, compute the distribution Δ_O:

In[176]:= $\quad DelO = LargestInvariantDistribution[\{f, g\}, h, var]$
Out[176]= $\{\{0, 1, 0, x3\}, \{1, 0, 0, 0\}\}$

Since Δ_O is not empty, the system is not observable. Proceed to obtain the transformation by appending a set of independent vector fields to Δ_O to obtain a distribution of rank 4.

In[177]:= $\quad Del = Join[DelO, \{\{0, 0, 1, 0\}, \{0, 0, 0, 1\}\}]$
Out[177]= $\{\{0, 1, 0, x3\}, \{1, 0, 0, 0\}, \{0, 0, 1, 0\}, \{0, 0, 0, 1\}\}$

Check the rank,

In[178]:= $\quad Span[Del]$
Out[178]= $\{\{1, 0, 0, 0\}, \{0, 1, 0, 0\}, \{0, 0, 1, 0\}, \{0, 0, 0, 1\}\}$

and compute the transformed system equations.

In[179]:= $\quad TriangularDecomposition[f + g * u, h, Del, var, \{0, 0, 0, 0\},$
$\qquad\qquad \infty]$
Out[179]= $\big\{\{z2, z1, z3, z1 \ z3 + z4\}, \{x2, x1, x3, -x2 \ x3 + x4\},$
$\qquad\qquad \big\{u + z3, e^{z1} \ z1 + z2 \ (u + z3), z4, 0\big\}, \{z3\}\big\}$

Thus, the equations are in the anticipated form with $\zeta_1 = (\ z_1 \quad z_2\ )$ and $\zeta_2 = (\ z_3 \quad z_4\ )$.

$$\dot{z} = \begin{bmatrix} z_3 \\ e^{z_1}z_1 + z_2z_3 \\ z_4 \\ 0 \end{bmatrix} + \begin{bmatrix} 1 \\ z_2 \\ 0 \\ 0 \end{bmatrix} u$$

$$y = z_3$$

The next example shows that nonlinear system observability does provide some new twists not evident in linear systems.

Example 6.174 *This example is from Vidyasagar [4] (Example 65, Section 7.3). Consider the system*

$$\begin{bmatrix} \dot{x}_1 \\ \dot{x}_2 \\ \dot{x}_3 \end{bmatrix} = \begin{bmatrix} x_2 \\ x_3 \\ 0 \end{bmatrix} + \begin{bmatrix} 0 \\ x_1 \\ 0 \end{bmatrix} u, \quad y = x_2$$

Based on linear system results it would be anticipated that this system is not observable. However, let us compute $\Omega_O^\perp$.

$In[180]:=$ $f = \{x2, x3, 0\};$
           $g = \{0, x1, 0\};$
           $h = x2;$
           $var = \{x1, x2, x3\};$

$In[181]:=$ $LargestInvariantDistribution[\{f, g\}, \{h\}, var]$

$Out[181]=$ $\{\}$

Thus, we conclude that the system is indeed observable. The reason for this is that the state x_1 can be easily ascertained by observing the response to specified control signals.

The question of observability can be answered directly with the ProPac test **Observability***, and the observability codistribution can be obtained with the function* **ObservabilityCodistribution***. Here are the calculations with and without the control input.*

$In[182]:=$ $Observabiity[f, g, h2, var]$

$Out[182]=$ True

$In[183]:=$ $Observability[f, \{\{\}\}, \{h2\}, var]$

$Out[183]=$ False

$In[184]:=$ $ObservabilityCodistribution[f, Transpose[\{g\}], \{h2\}, var]$

$Out[184]=$ $\{\{1, 0, 0\}, \{0, 1, 0\}, \{0, 0, 1\}\}$

$In[185]:=$ $ObservabilityCodistribution[f, \{\{\}\}, \{h2\}, var]$

$Out[185]=$ $\{\{0, 0, 1\}, \{0, 1, 0\}\}$

Similarly to the case of controllability, it is of interest to introduce the codistribution

$$\Omega_L = \text{span} \left\{ L_f^k(dh_i), \ 1 \le i \le p, \ 0 \le k \le n - 1 \right\} \tag{6.11}$$

Clearly, Ω_L is a subdistribution of Ω_O. Thus, a sufficient condition for local observability around x_0 is $\dim \Omega_L(x_0) = n$. Moreover, for a linear system it is easy enough to verify that $\Omega_O = \Omega_L$, so that in a crude sense $\dim \Omega_L(x_0) = n$ establishes an observability that is linear-like. Indeed, what is missing in Ω_L as compared to Ω_O are: 1) the Lie derivatives of the covector fields dh_i with respect to the control input vector fields and 2) the higher order ($> n - 1$) lie derivatives. Thus, if $\dim \Omega_L(x_0) = n$, observability is achieved without the need to exploit the control input.

We summarize the main observability relationships in the following diagram.

$$strong\,local\,observability$$
$$\Downarrow$$
$$local\,observability \qquad \Leftrightarrow \quad \dim \Omega_O = n \quad \Leftarrow \quad \dim \Omega_L = n$$

6.4 Local Decompositions

In linear system theory, the *Kalman decomposition* utilizes a set of coordinates that explicitly reveals the controllability/observability structure of a control system. Thus, new state variables are identified such that the system equations are in the form

$$\begin{bmatrix} \dot{z}_1 \\ \dot{z}_2 \\ \dot{z}_3 \\ \dot{z}_4 \end{bmatrix} = \begin{bmatrix} A_{11} & A_{12} & A_{13} & A_{14} \\ 0 & A_{22} & 0 & A_{24} \\ 0 & 0 & A_{33} & A_{34} \\ 0 & 0 & 0 & A_{44} \end{bmatrix} \begin{bmatrix} z_1 \\ z_2 \\ z_3 \\ z_4 \end{bmatrix} + \begin{bmatrix} B_1 \\ B_2 \\ 0 \\ 0 \end{bmatrix} u$$

$$y = \begin{bmatrix} 0 & C_2 & 0 & C_4 \end{bmatrix} \begin{bmatrix} z_1 \\ z_2 \\ z_3 \\ z_4 \end{bmatrix} + Du$$

The coordinates z_1, z_2 correspond to the controllable subspace and the coordinates z_2, z_4 correspond to the observable subspace. A similar decomposition is achievable for affine nonlinear systems.

Proposition 6.175 (Local Decomposition) *Consider the system (6.1). Suppose that the controllability distribution Δ_{C_0} and the observability codistribution Ω_O as well as $\Delta_{C_0} + \Omega_O^\perp$ are all of constant dimension on a neigborhood U of $x_0 \in R^n$. Then there exists a local diffeomorphism Ψ on U such that the system* *equations in the new coordinates are:*

$$
\begin{bmatrix} \dot{\zeta_1} \\ \dot{\zeta_2} \\ \dot{\zeta_3} \\ \dot{\zeta_4} \end{bmatrix} = \begin{bmatrix} f_1(\zeta_1, \zeta_2, \zeta_3, \zeta_4) \\ f_2(\zeta_2, \zeta_4) \\ f_3(\zeta_3, \zeta_4) \\ f_4(\zeta_4) \end{bmatrix} + \begin{bmatrix} G_1(\zeta_1, \zeta_2, \zeta_3, \zeta_4) \\ G_2(\zeta_2, \zeta_4) \\ 0 \\ 0 \end{bmatrix} u
$$

$$
y = h(\zeta_2, \zeta_4)
$$

Moreover, the system restricted to $\zeta_3 = 0, \zeta_4 = 0$ is locally controllable and the system restricted to $\zeta_1 = 0, \zeta_3 = 0$ is locally observable.

Proof: The theorem is proved in [2]. We outline a constructive proof, an implementation of which is described later. It proceeds as follows.

1. Compute the controllability distribution Δ_{C_0}

2. Compute its complement $\Delta_{\bar{C}_0}$

3. Compute the observability codistribution Ω_O and its annihilator $\Omega_O^\perp$

4. Compute the intersection $\Delta_{C\bar{O}}$ of Δ_{C_0} and $\Omega_O^\perp$

5. Compute the complement Δ_{CO} of $\Delta_{C\bar{O}}$ in Δ_{C_0}

6. Compute the intersection $\Delta_{\bar{C}\bar{O}}$ of $\Delta_{\bar{C}_0}$ and $\Omega_O^\perp$

7. Compute the complement $\Delta_{\bar{C}O}$ of $\Delta_{\bar{C}\bar{O}}$ in $\Delta_{\bar{C}_0}$

8. Compute and apply the transformation based on $\Delta = \Delta_{C\bar{O}} + \Delta_{CO} + \Delta_{\bar{C}\bar{O}} + \Delta_{\bar{C}O}$

Of course there are technical arguments required at various stages in the process. ∎

Example 6.176 *Let us revisit Examples (6.161) and (6.173). In those instances we computed the controllable and observable decompositions. Now, the ProPac function* LocalDecomposition *is used to compute the complete local decomposition.*

```
In[186]:=  f = {x1 x3 + x2 exp[x2], x3, x4 − x2 x3, x3^2 + x2 x4 − x2^2 x3};
           G = Transpose[{{x1, 1, 0, x3}}];
           h = {x3};
           var = {x1, x2, x3, x4};
```

```
In[187]:=  LocalDecomposition[f, G, h, var, {u}, ∞]
Out[187]=  {{z2, z1, z3, z1 z3 + z4}, {x2, x1, x3, −x2 x3 + x4},
           {z3, e^{z1} z1 + z2 z3, z4, 0}, {{1}, {z2}, {0}, {0}}, {z3}}
```

Notice that the equations turn to be exactly those of Example (6.173).

6.5 Input–Output Linearization

When confronted with a nonlinear control design problem, it is reasonable to ask if it is transformable into a linear one. The earliest investigations of this question considered the possibility of using a state transformation to do this. Of course, the set of transformable systems turns out to be quite limited. The idea of using feedback to accomplish linearization is generally attributed to Brockett [6]. As a matter of fact, many practical control system designs already used feedback to accomplish linearization. We will consider a constructive process for linearizing the input-output dynamics of a given nonlinear system using state transformations and feedback. When this is possible, a reasonable approach for controller design is a two level strategy that implements first the linearizing control and then a linear feedback that regulates the linearized system. *ProPac* contains the constructions required to implement this process. We describe the essentials in the following paragraphs.

A system is exactly linearizable or input-state linearizable if the state equations are linearizable by a combination of a state transformation and state feedback. If a system is not exactly feedback linearizable, it may still be linearizable in an input-output sense. In this event, we can find a state transformation and a nonlinear state feedback control such that the input-output behavior is described by a linear dynamical system. However, in this case there remain residual nonlinear dynamics, called the internal dynamics, which are decoupled from the output. Hence the input-output behavior is linear even though the entire state dynamics are not. In this section we consider input-output linearization and in the next section we consider input-state linearization.

6.5.1 SISO Case

First, we consider the single-input single-output case:

$$\begin{aligned}
\dot{x} &= f(x) + g(x)u \\
y &= h(x)
\end{aligned} \tag{6.12}$$

where $x \in R^n$, $u \in R$ and $y \in R$. Now, let us differentiate $y = h(x)$ with respect to time to obtain

$$\dot{y} = L_f h(x) + L_g h(x)u$$

If $L_g h(x) \neq 0$ we stop, if $L_g h(x) = 0$, we differentiate again to obtain

$$\ddot{y} = L_f^2 h(x) + L_g L_f h(x)u$$

Again, the coefficient of u vanishes or it does not. If not, we continue to differentiate until after r steps we have

$$y^{(r)} = L_f^r h(x) + L_g L_f^{r-1} h(x)u \tag{6.13}$$

with $L_g L_f^{r-1} h(x) \neq 0$ and the process is terminated. Assume that the process does stop in a finite number of steps.

Definition 6.177 *Consider the system (6.12). Let U be a neighborhood of x_0 and suppose there is a finite integer r such that*

$$L_g L_f^k h(x) = 0, \ \forall x \in U, \ k = 0, \ldots, r-2 \ \ L_g L_f^{r-1} h(x_0) \neq 0 \qquad (6.14)$$

Then r is the relative degree of (6.12). If the sequence specified in (6.14) does not terminate in finite steps the system relative degree is $r = \infty$.

Example 6.178 (Linear system relative degree) *Consider a SISO linear system*

$$\dot{x} = Ax + bu$$

$$y = cx$$

We make the associations with (6.12): $f(x) = Ax$, $g(x) = b$ and $h(x) = cx$. It is easy enough to verify that $L_f^k h = cA^k x$ and $L_g L_f^k h = cA^k b$. Thus, the conditions expressed in (6.14) for a system of finite relative degree r is

$$cb = 0, \ cAb = 0, \ldots, cA^{r-2} b = 0, \ cA^{r-1} b \neq 0$$

Define the functions

$$z_i(x) = L_f^{i-1} h(x) \qquad (6.15)$$

We intend to show, that these functions define a partial state transformation that reduces the system to an important *normal form* from which a linearizing feedback control is obvious. To do this we need to establish two essential facts. First, if the process terminates in finite steps, it does so with $r \leq n$. Second, the functions $z_i(x)$ are independent. Independence will be considered first. However, we will need the following identity.

Lemma 6.179 *Suppose the system (6.12) has finite relative degree r. Then*

$$L_{ad_f^i g} L_f^k h = \begin{cases} 0 & if \ i + k \leq r-1 \\ (-1)^i L_g L_f^{r-1} h & if \ i + k = r-1 \end{cases} \qquad (6.16)$$

Proof: Let us fix k and prove the claim by induction on i. First note that for $i = 0$, the claim (6.16) reduces to the definition of relative degree (6.14). Now, assume that the (6.16) is true for $i = 0, \ldots, p-1$. We will prove that it is true for $i = p$. Recall that $ad_f^i g = [f, ad_f^{i-1} g]$ and compute

$$L_{ad_f^p g} L_f^k h = L_f L_{ad_f^{p-1} g} L_f^k h - L_{ad_f^{p-1} g} L_f L_f^k h$$

In view of the induction hypothesis this reduces to

$$L_{ad_f^p g} L_f^k h = -L_{ad_f^{p-1} g} L_f L_f^k h = -L_{ad_f^{p-1} g} L_f^{k+1} h$$

Now, $p + k < r - 1$ implies $(p - 1) + (k + 1) < r - 1$ so that by the induction hypothesis

$$L_{\mathrm{ad}_f^p g} L_f^k h = 0, \; p + k < r - 1$$

If $p + k = r - 1$ the induction hypothesis allows the sequential reduction

$$L_{\mathrm{ad}_f^p g} L_f^k h = -L_{\mathrm{ad}_f^{p-1} g} L_f^{k+1} h = \cdots (-1)^p L_g L_f^{r-1} h$$

■

Lemma 6.180 *Consider the system (6.12) and suppose it has finite relative degree r. Then the covectors $\{dz_1, dz_2, \ldots, dz_r\}$ associated with the functions $z_i(x)$ defined in (6.15) are independent and $r \leq n$.*

Proof: We will show that the only set of constants $a_1, \ldots, a_r$ for which the relation

$$\sum_{i=1}^{r} a_i dz_i(x_0) = 0 \tag{6.17}$$

is satisfied is the trivial set $a_i = 0$ for $i = 1, \ldots, r$. To do this consider the scalar function

$$\alpha(x) = \sum_{i=1}^{r} a_i z_i(x) \tag{6.18}$$

First we show that $a_r = 0$. Suppose that (6.17) is true, which means that $d\alpha(x_0) = 0$. Then

$$L_g \alpha(x_0) = d\alpha(x) \cdot g(x)|_{x=x_0} = 0 \tag{6.19}$$

Now,

$$L_g \alpha(x) = \sum_{i=0}^{r-1} L_g L_f^i h(x) \tag{6.20}$$

But, by assumption $L_g L_f^k h(x) = 0$, $k = 0, \ldots, r - 2$, $L_g L_f^{r-1} \neq 0$, so we have

$$L_g \alpha(x) = a_r L_g L_f^{r-1} h(x) \tag{6.21}$$

Thus, we conclude $a_r = 0$, so that

$$\alpha(x) = \sum_{i=1}^{r-1} a_i z_i(x) \tag{6.22}$$

Now, we show that $a_{r-1} = 0$. Note that

$$L_{\mathrm{ad}_f g} \alpha(x_0) = 0 \tag{6.23}$$

From the previous lemma, we have

$$L_{\mathrm{ad}_f g} \alpha(x) = \sum_{i=0}^{r-2} L_{\mathrm{ad}_f g} L_f^i h(x) = -a_{r-1} L_g L_f^{r-1} h(x) \tag{6.24}$$

so that we must have $a_{r-1} = 0$. Continuing in this way, we show that all $a_i = 0$.
∎

If $r < n$ we can always complete the mapping $x \mapsto z(x)$ to be a transformation
by specifying $n-r$ functions $\xi(x)$ independent of $z(x)$ in the sense that the set of
covectors $d\xi_1(x_0), \ldots, d\xi_{n-r}(x_0), dz_1(x_0), \ldots, dz_r(x_0)$ are linearly independent.
Then the transformed equations are

$$\dot{\xi} = F(\xi, z, u)$$

$$\dot{z} = Az + b\left[\alpha\left(x(\xi, z)\right) + \rho\left(x(\xi, z)\right)u\right]$$

where
$$A = \begin{bmatrix} 0 & 1 & 0 & & \\ 0 & 0 & 1 & 0 & \\ \vdots & & \ddots & \ddots & \\ & & & \ddots & 1 \\ 0 & & & 0 & 0 \end{bmatrix}, \ b = \begin{bmatrix} 0 \\ \vdots \\ \vdots \\ 0 \\ 1 \end{bmatrix}$$

and
$$\alpha(x) = L_f^r h(x), \ \rho(x) = L_g L_f^{r-1} h(x)$$

However, we can actually do more that that. We seek functions $\xi(x)$ with the
property that $L_g \xi_i(x) = 0$ around x_0. That is, $d\xi_1(x), \ldots, d\xi_{n-r}(x) \in \mathcal{G}^\perp$,
where $\mathcal{G} = \text{span}\{g\}$.

Proposition 6.181 *Suppose the system (6.12) has finite relative degree r at
x_0. Then $r \le n$. Moreover, if $r < n$ it is possible to find $n - r$ functions
$\xi_1(x), \ldots, \xi_{n-r}$ such that the mapping*

$$\Phi(x) = \begin{bmatrix} \xi(x) \\ z(x) \end{bmatrix}$$

*is a local coordinate transformation on a neighborhood of x_0. Moreover, it is
always possible to choose $\xi_1(x), \ldots, \xi_{n-r}(x)$ so that*

$$L_g \xi_i(x) = 0, \ 1 \le i \le n - r$$

The transformed equations are

$$\dot{\xi} = F(\xi, z) \tag{6.25}$$

$$\dot{z} = Az + b\left[\alpha\left(x(\xi, z)\right) + \rho\left(x(\xi, z)\right)u\right] \tag{6.26}$$

$$y = cz \tag{6.27}$$

where
$$A = \begin{bmatrix} 0 & 1 & 0 & & \\ 0 & 0 & 1 & 0 & \\ \vdots & & \ddots & \ddots & \\ & & & \ddots & 1 \\ 0 & & & 0 & 0 \end{bmatrix}, \ b = \begin{bmatrix} 0 \\ \vdots \\ \vdots \\ 0 \\ 1 \end{bmatrix}$$

$$c = [1 \quad 0 \quad \cdots \quad \cdots \quad 0]$$

and

$$\alpha(x) = L_f^r h(x), \; \rho(x) = L_g L_f^{r-1} h(x)$$

Proof: By the definition of relative degree $g(x_0)$ is not zero. Thus, the distribution $\mathcal{G} = \text{span}\{g\}$ is nonsingular around x_0. Since it is one dimensional it is also involutive. Thus, the Frobenius theorem implies the existence of $n-1$ functions $\lambda_1, \ldots, \lambda_{n-1}$ defined on a neighborhood of x_0 such that

$$\text{span}\{d\lambda_1, \ldots, d\lambda_{n-1}\} = \mathcal{G}^\perp \tag{6.28}$$

Now, it must be that

$$\dim \left(\mathcal{G}^\perp + \text{span} \left\{ dh, dL_f h, \ldots, dL_f^{r-1} h \right\} \right) = n$$

at x_0. Otherwise,

$$\mathcal{G}(x_0) \cap \ker \left(\text{span} \left\{ dh(x_0), dL_f h(x_0), \ldots, dL_f^{r-1} h(x_0) \right\} \right) \neq \{\} \tag{6.29}$$

In other words, the vector $g(x_0)$ is annihilated by all of the covectors in

$$\text{span} \left\{ dh(x_0), dL_f h(x_0), \ldots, dL_f^{r-1} h(x_0) \right\}$$

But this is a contradiction because $\left\langle dL_f^{r-1} h(x_0), g(x_0) \right\rangle = L_g L_f^{r-1} h(x_0)$ is nonzero by assumption. Since span $\left\{ dh, dL_f h, \ldots, dL_f^{r-1} h \right\}$ has dimension r, it follows from (6.28) and (6.29) that there are $n-r$ covectors in $\mathcal{G}^\perp$, say, $d\lambda_1, \ldots, d\lambda_{n-r}$ so that $dh, dL_f h, \ldots, dL_f^{r-1} h, d\lambda_1, \ldots, d\lambda n - r$ are independent at x_0. Furthermore, by construction, we have $L_g \lambda_i(x) = 0$, $1 \leq i \leq n-r$. Finally, the form of the equations follows from the construction of the functions $z(x)$ and $\xi(x)$. ∎

Remark 6.182 *Equations (6.26) and (6.27) are called the* linearizable dynamics *because we can introduce a new control variable v and define $\alpha + \rho u = v$ to reduce (6.26) and (6.27) to the linear system*

$$\dot{z} = Az + bv, \; y = cx$$

with input v and output y. Equation (6.25) are called the internal dynamics *because they are decoupled from the (linearized) input-output dynamics. Moreover, since the linear system is controllable (it is in controllable form) it can be stabilized by an appropriate linear control of the form $v = Kz$. If this is done, then the overall system is stabilized if and only if the system*

$$\dot{\xi} = F(\xi, 0) \tag{6.30}$$

The system of equations (6.30) are referred to as the zero dynamics *or* zero output dynamics *of (6.12) because they represent the residual dynamical behavior that can take place under the constraint $y(t) \equiv 0$.*

The following is an important property of 'relative degree' is that it is invariant under state transformation and feedback

Lemma 6.183 *Suppose the system (6.12) has finite relative degree r, then r is invariant under state transformation and feedback.*

Proof: Consider a transformation $x \mapsto z$ realized by the mapping $z = \Phi(x)$ and its inverse $x = \Phi^{-1}(z)$. In the new state coordinates the system equations are
$$\dot{z} = d\Phi\left(\Phi^{-1}(z)\right)\left[f\left(\Phi^{-1}(z)\right) + g\left(\Phi^{-1}(z)\right)u\right]$$
$$= \bar{f}(z) + \bar{g}(z)$$
and
$$y = h\left(\Phi^{-1}(z)\right) = \bar{h}(z)$$

Now, let us compute $L_{\bar{g}}\bar{h}(z)$:

$$L_{\bar{g}}\bar{h}(z) = \left.\frac{\partial h}{\partial x}\right|_{x=\Phi^{-1}(z)} \frac{\partial \Phi^{-1}}{\partial z} \left.\frac{\partial \Phi}{\partial x}\right|_{x=\Phi^{-1}(z)} g\left(\Phi^{-1}(z)\right)$$

But, the relation
$$x = \Phi^{-1}(\Phi(x))$$

implies
$$I = \frac{\partial \Phi^{-1}}{\partial z} \frac{\partial \Phi}{\partial x}$$

so that we conclude
$$L_{\bar{g}}\bar{h}(z) = L_g h\left(\Phi^{-1}(z)\right)$$

Identical calculations lead to the result
$$L_{\bar{f}}\bar{h}(z) = L_f h\left(\Phi^{-1}(z)\right)$$

and indeed,
$$L_{\bar{g}}L_{\bar{f}}^k\bar{h}(z) = L_g L_f^k h\left(\Phi^{-1}(z)\right)$$

Thus, in view of the definition of relative degree we have the result that relative degree is invariant under state coordinate transformations.

Now, let is apply state feedback $u = \kappa(x) + v$ so that the system equations become
$$\dot{x} = f(x) + g(x)\kappa(x) + g(x)u = \bar{f}(x) + g(x)u$$
$$y = h(x)$$

Now compute
$$L_{\bar{f}}h(x) = L_f h(x) + L_{g\kappa}h(x)$$

but notice that
$$L_{g\kappa}h(x) = L_g h(x)\,\kappa(x)$$

So we have

$$L_{\tilde{f}}h(x) = L_f h(x) + L_g h(x)\,\kappa(x) = L_f h(x)$$

since $L_g h(x) = 0$. Similarly, we successively compute

$$L_{\tilde{f}}^i h(x) = L_f^i h(x) + L_g L_f^{i-1} h(x)\,\kappa(x) = L_f^i h(x), \quad 1 \le i \le r - 1$$

using the fact that $L_g L_f^{i-1} h(x) = 0$, $1 \le i \le r - 1$. Thus we have

$$L_g L_{\tilde{f}}^i h(x) = L_g L_f^i h(x), \quad 1 \le i \le r - 1$$

so that $L_g L_{\tilde{f}}^i h(x) = 0$, $1 \le i \le r - 2$ and $L_g L_{\tilde{f}}^r h(x) \ne 0$. Consequently, the system system retains relative degree 'r' under feedback. ∎

In the SISO case, the coordinate transformation can always be chosen such that the decoupled (internal) dynamics are independent of the control, [1] Chapter 4.3. This calculation has been implemented in the *ProPac* function SISONormalFormTrans.

Example 6.184 (I–O Linearization and normal forms) *Consider the following example (example 4.1.5 in Isidori [1]):*

$$\begin{bmatrix} \dot{x}_1 \\ \dot{x}_2 \\ \dot{x}_3 \\ \dot{x}_4 \end{bmatrix} = \begin{bmatrix} x_1 x_2 - x_1^3 \\ x_1 \\ -x_3 \\ x_1^2 + x_2 \end{bmatrix} + \begin{bmatrix} 0 \\ 2 + 2x_3 \\ 1 \\ 0 \end{bmatrix} u \qquad (6.31)$$

$$y = x_4$$

Below we perform the required computations. Notice that a control can be chosen such that the output $y = x_4 = z_1$ is identically equal to zero if and only if $z_1 = 0$ and $z_2 = 0$.

When this obtains, the internal dynamics reduce to the zero (output) dynamics

$In[188] :=$ $f29 = \{x1\ x2 - x1^3, x1, -x3, x1^2 + x2\};$
$\qquad g29 = \{0, 2 + 2\ x3, 1, 0\}; h29 = \{x4\};$
$\qquad var29 = \{x1, x2, x3, x4\}; newvar29 = \{z1, z2, z3, z4\};$
$\qquad \{T1, T2\} = SISONormalFormTrans[f29, g29, h29, var29]$
$Out[188] = \{\{x4, x1^2 + x2\}, \{x1, x2 - 2\ x3 - x3^2\}\}$

$In[189] :=$ $Trans = Join[T1, T2];$
$\qquad InvTrans = InverseTransformation[var29, newvar29, Trans];$

$In[190] :=$ $\{Newf, Newg\} =$
$\qquad\qquad TransformSystem[f29, g29, var29, newvar29, Trans, InvTrans];$
$\qquad Newf//MatrixForm$
$\qquad Newg//MatrixForm$

$$Out[190] = \begin{pmatrix} z2 \\ z3 + 2\ z2\ z3^2 - 4\ z3^4 \\ z3\ (z2 - 2\ z3^2) \\ 2 + 2\ z2 + z3 - 2\ z3^2 + 2\ \sqrt{1 + z2 - z3^2 - z4} - 2\ z4 \end{pmatrix}$$

$$Out[190] = \begin{pmatrix} 0 \\ -2 \ \sqrt{1 + z2 - z3^2 - z4} \\ 0 \\ 0 \end{pmatrix}$$

$In[191] := \quad ZeroDyn = Inner[Equal, \partial_t\{z3[t], z4[t]\},$
$\qquad\qquad (Newf[[\{3,4\}]]/.\{z1 \to 0, z2 \to 0, z3 \to z3[t], z4 \to z4[t]\}), List]//$
$\qquad\qquad MatrixForm$

$$Out[191] = \begin{pmatrix} z3'[t] == -2 \ z3[t]^3 \\ z4'[t] == 2 + z3[t] - 2 \ z3[t]^2 + 2 \ \sqrt{1 - z3[t]^2 - z4[t]} - 2 \ z4[t] \end{pmatrix}$$

6.5.2 MIMO Case

Consider the square MIMO case as described by Equation (6.1) with $p = m$. Recall that the kth Lie derivative (directional derivative) of the scalar function $\phi(x)$ with respect to the vector field $f(x)$ is denoted $L_f^k(\phi)$. Then by successive differentiation of the elements of the output vector y we arrive at the following definitions.

$$\begin{aligned} r_i &:= \inf\{k | L_{g_j}(L_f^{k-1}(h_i)) \neq 0 \quad \text{for at least one } j\} \\ \alpha_i(x) &:= L_f^{r_i}(h_i), \quad i = 1, .., m \\ \rho_{ij}(x) &:= L_{g_j}(L_f^{r_i-1}(h_i)), \quad i, j = 1, .., m \end{aligned} \qquad (6.32)$$

Definition 6.185 *Suppose there exists a set of finite integers $\{r_1, \ldots, r_m\}$ as specified in Equation (6.32) with $\det\{\rho(x_0)\} \neq 0$, then $[r_1, .., r_m]$ is called the vector relative degree at x_0.*

An important result is the MIMO generalization of Lemma (6.180).

Lemma 6.186 *Consider the system (6.1) and suppose it has finite vector relative degree $[r_1, .., r_m]$ at x_0. Let $r = r_1 + .. + r_m$ and define the functions z_i, $i = 1, \ldots, r$,*

$$z(x) = \begin{bmatrix} z^1 \\ \vdots \\ \vdots \\ z^m \end{bmatrix} := \begin{bmatrix} h_1 \\ \vdots \\ L_f^{r_1-1}(h_1) \\ \vdots \\ \vdots \\ h_m \\ \vdots \\ L_f^{r_m-1}(h_m) \end{bmatrix} = \begin{bmatrix} y_1 \\ \vdots \\ y_1^{r_1-1} \\ \vdots \\ y_m \\ \vdots \\ y_m^{r_m-1} \end{bmatrix} \qquad (6.33)$$

where $z^i \in R^{r_i}$. Then the covectors $\{dz_1, dz_2, \ldots, dz_r\}$ are independent and $r \leq n$ on some neighborhood of x_0.

Proof: We need to show that the r n-dimensional row vectors

$$\left\{ dL_f^k h_i(x_0) \,|\, 0 \le k \le r_i - 1, \, 1 \le i \le m \right\}$$

are linearly independent. Then it follows imediately that r can not be greater that n.

To establish independence, select real numbers a_{ik}, $0 \le k \le r_i - 1$, $1 \le i \le m$, such that

$$\sum_{i=1}^{m} \sum_{k=1}^{r_i-1} a_{ik} dL_f^k h_i(x_0) = 0 \tag{6.34}$$

We will show that the only set of such constants is the trivial set. Define

$$a(x) = \sum_{i=1}^{m} \sum_{k=1}^{r_i-1} a_{ik} dL_f^k h_i \tag{6.35}$$

Now, the assumption of finite relative degree implies $L_{g_i} L_f^k h_i(x) = 0$ for $k = \ldots, r_i - 2$. Thus, compute

$$L_{g_j} a = \sum_{i=1}^{m} a_{i,r_i-1} L_{g_j} L_f^{r_i-1} h_i = \sum_{i=1}^{m} a_{i,r_i-1} \rho_{ij} \tag{6.36}$$

Now, using (6.34) we can compute

$$L_{g_j} a(x_0) = da \cdot g_j |_{x_0} = 0$$

which implies

$$\left[a_{1,r_i-1}, \ldots, a_{m,r_m-1} \right] \rho(x_0) = 0$$

Since $\rho(x_0)$ is nonsingular we conclude that the m numbers $a_{i,r_i-1} = 0$, for $i = 1, \ldots m$. Thus, $a(x)$ reduces to

$$a(x) = \sum_{i=1}^{m} \sum_{k=1}^{r_i-2} a_{ik} dL_f^k h_i \tag{6.37}$$

Now compute $L_{ad_f g} a$ using (6.37) for each j. An identical argument as above leads to the conclusion that $a_{i,r_i-2} = 0$, for $i = 1, \ldots m$ and $a(x)$ reduces to

$$a(x) = \sum_{i=1}^{m} \sum_{k=1}^{r_i-3} a_{ik} dL_f^k h_i \tag{6.38}$$

The argument is repeated to show that all $a_{ij} = 0$. ∎

Now, we consider the partial state transformation $x \to z \in R^r$, $r = r_1 + .. + r_m \le n$ as defined in equation (6.33). It is a straightforward calculation to verify that

$$\begin{aligned} \dot{z} &= Az + E[\alpha(x) + \rho(x)u] \\ y &= Cz \end{aligned} \tag{6.39}$$

where A, E, and C have the special structure:

- A is of the form

$$A = diag(A_1, ..., A_m), \; A_i = \begin{bmatrix} 0 & I_{r_i-1} \\ 0 & 0 \end{bmatrix} \in R^{r_i \times r_i}$$

- the only nonzero rows of E are the m rows $r_1, r_1 + r_2, \ldots, r$ and these form the identity I_m

- the only nonzero columns of C are the m columns $1, r_1 + 1, r_1 + r_2 + 1, \ldots, r - r_m + 1$ and these form the identity I_m

The remaining part of the transformation can be defined by arbitrarily choosing additional independent coordinates. The condition $\det \rho(x_0) \neq 0$ insures the existence of a local (around x_0) change of coordinates $x \to (\xi, z), \xi \in R^{n-r}, z \in R^r$ such that

$$\dot{\xi} = \widehat{F}(\xi, z, u) \tag{6.40}$$

$$\dot{z} = Az + E[\alpha(x(\xi, z)) + \rho(x(\xi, z))u] \tag{6.41}$$

It is common to call (6.40) the internal dynamics and (6.41) the linearizable dynamics.

Notice that in view of (6.41), we can apply the control

$$u = \rho^{-1}(x) \{-\alpha(x) + v\} \tag{6.42}$$

and reduce (6.41) to a linear system

$$\begin{aligned} \dot{z} &= Az + Ev \\ y &= Cz \end{aligned} \tag{6.43}$$

This justifies the terminology of linearizable dynamics for (6.41). Notice that when the control (6.42) is applied the internal dynamics (6.40) are decoupled from the output (not necessarily the input, as is the case for SISO systems). It is a simple matter to design a linear stabilizing controller for (6.43) – for example a state feedback law of the form $v = Kz$ that insures $z(t) \to 0$ as $t \to \infty$. Such a control does not necessarily stabilize the complete system (6.40) and (6.41), because we need to account for the decoupled internal dynamics (6.40).

Lemma 6.187 *Suppose that $\rho(x)$ has continuous first derivatives with*

$$\det \rho(x) \neq 0 \text{ on } M_0 = \{x | z(x) = 0\}$$

Then $\partial z(x)/\partial x$ is of maximum rank on the set M_0.

Proof: The result follows directly from Lemma (6.186). ∎

The Lemma is extremely important because it relates the invertibility of the decoupling matrix with the geometry of the set M_0. With it, we can obtain several important results, one of which we state here.

Proposition 6.188 *Suppose that $\rho(x)$ has continuous first derivatives with*

$$\det \rho(x) \neq 0 \text{ on } M_0 = \{x | z(x) = 0\}$$

Then M_0 is a regular, $n - r$ dimensional submanifold of R^n and any trajectory segment $x(t), t \in T$, T an open interval of R^1, which satisfies $h(x(t)) = 0$ on T lies entirely in M_0. Moreover, the control that obtains on T is

$$u_0(x) = -\rho^{-1}(x)\alpha(x) \tag{6.44}$$

and every such trajectory segment with boundary condition $x(t_0) = x_0$, $t_0 \in T$ satisfies

$$\dot{x} = f(x) - G(x)\rho^{-1}(x)\alpha(x), \quad z(x(t_0)) = 0. \tag{6.45}$$

Proof: In view of Lemmas (6.186) and (6.187), it follows from $\det \rho(x) \neq 0$ on M_0 that

1. the covectors $\{dz_1, dz_2, \ldots, dz_r\}$ are independent around every point $x_0 \in M_0$,

2. $\partial z(x)/\partial x$ is of maximum rank on the set $M_0 = \{x | z(x) = 0\}$.

This maximal rank condition insures that M_0 is a well defined regular manifold of dimension $n - r$. From the definition of $z(x)$, it follows that y is identically zero on an open time interval if and only if z is zero on that interval. Thus, it follows from (6.39) that the unique (provided $\det \rho \neq 0$) control which must obtain during any motion constrained by $h(x) = 0$ is (6.44). With this control (6.1) reduces to (6.45). ∎

The analysis above leads to the following observations:

- The manifold M_0 is invariant with respect to the dynamics (6.45).

- These equations are equivalent to the output constrained dynamics

$$\begin{aligned} \dot{x} &= f(x) + G(x)u \\ 0 &= h(x) \end{aligned} \tag{6.46}$$

  hence they are called the zero dynamics.

- The proposition defines the zero dynamics in global form. An equivalent local form is

$$\begin{aligned} \dot{\xi} &= F(\xi, 0, 0) \\ F(\xi, z, v) &= \widehat{F}(\xi, z, \rho^{-1}(x(\xi, z)))\{-\alpha(x(\xi, z)) + v\}) \end{aligned} \tag{6.47}$$

Let us collect these results in the following proposition that justifies the design procedure depicted in Figure (6.1).

Proposition 6.189 *Suppose the conditions of (6.188) hold, and*

1. *$x_0 \in M_0$ is an equilibrium point of (6.1) which implies $x_0 \to (\xi_0, z_0) = (\xi_0, 0)$*

2. *ξ_0 is a stable equilibrium point of the zero dynamics, and*

3. *$v = Kz$ is a stabilizing controller for (6.43).*

Then

$$u = \rho^{-1}(x)\left\{-\alpha(x) + Kz(x)\right\} \qquad (6.48)$$

is a stabilizing controller for the system (6.1).

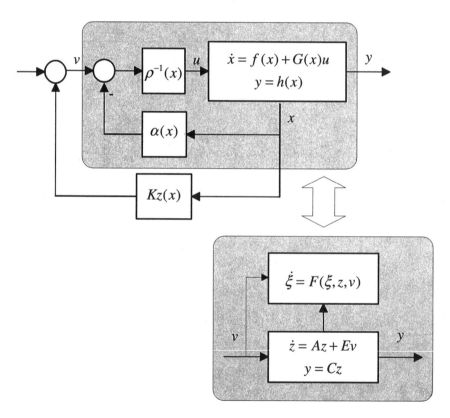

Figure 6.1: The two level controller design process of Proposition (6.189) is depicted in this diagram.

Before proceeding with examples, let us consider the computation of the local zero dynamics. One approach is to obtain the internal dynamics by completing the local transformation and then to set $z = 0, v = 0$. We will describe an alternative in which the local zero dynamics are computed directly. Note that the functions $z(x)$ can be computed using (6.33). Once they are obtained, we are in a position to compute the local form of the zero dynamics near any point $x_0 \in M_0$ in the following way. Without loss of generality assume $x_0 = 0$. Now, split $z(x)$ into its linear and nonlinear parts:

$$z(x) = Ax + N(x), \quad A = \frac{\partial z}{\partial x}(0) \tag{6.49}$$

We assume that $x_0 = 0$ is a regular point (ρ is nonsingular) so that A is of full rank. Let A^* denote a right inverse of A and define K such that its columns span ker A. Define new coordinates v, w so that

$$x = A^*v + Kw \tag{6.50}$$

Then on the zero dynamics manifold, we have

$$v + N(A^*v + Kw) = 0 \tag{6.51}$$

Clearly, the Implicit Function Theorem guarantees the existence of a local solution to (6.51), $v^*(w)$, that is

$$v^*(w) + N(A^*v^*(w) + Kw) = 0 \tag{6.52}$$

on a neighborhood of $w = 0$, and $v^*(0) = 0$. Furthermore, $v^*(w)$ can be efficiently estimated because the mapping

$$v_{i+1} = -N(A^*v_i + Kw) \tag{6.53}$$

is a contraction. In fact, we have the following result.

Proposition 6.190 *Suppose $z(x)$ is smooth, and A is of full rank. Then*

1. *there exists a smooth function $v^*(w) = 0 + O(\|w\|)$,*

2. *if $v_i(w)$ satisfies $\|v^* - v_i\| = O(\|w\|^k)$, then $v_{i+1}(w)$, obtained via (6.53), satisfies $\|v^* - v_{i+1}\| = O(\|w\|^{2k})$.*

Proof: The first conclusion follows directly from the implicit function theorem and the fact that $v^*(w)$ is smooth. To prove the second, first subtract (6.53) from to obtain

$$v^* - v_{i+1} = -N(A^*v^* + Kw) + N(A^*v_i + Kw) \tag{6.54}$$

Now, consider the function $N_x(x) := \partial N(x)/\partial x$. Since N is smooth, with $\partial N(0)/\partial x = 0$, we have $N_x(0) = 0$ and by continuity of the second derivative of N, we conclude that $\partial N_x(x)/\partial x$ is bounded on a neigborhood of $x = 0$. Let

L be such a bound on an appropriately defined neighborhood, U, so that the usual arguments based on the Mean Value Theorem provide

$$\|N_x(x) - N_x(y)\| \leq L \|x - y\|, \ \forall x, y \in U \tag{6.55}$$

Thus, we can write

$$N(x) - N(x + \delta x) = N_x(x)\delta x + O(\|\delta x\|^2) \tag{6.56}$$

which in view of (6.55) gives

$$\|N(x) - N(x + \delta x)\| = O(\|\delta x\|^2) \tag{6.57}$$

for $x, y = x + \delta x \in U$. In order to apply this result to (6.54), take $x = A^*v^* + Kw$ and $\delta x = A^*(v^* - v_i)$. Then (6.54) and (6.57) yield

$$\|v^* - v_{i+1}\| = O(\|A^*(v^* - v_i)\|^2) = O(\|w\|^{2k}) \tag{6.58}$$

which is the desired conclusion.   ∎

Recall the global form of the zero dynamics:

$$\dot{x} = f(x) - G(x)\rho^{-1}(x)\alpha(x) \tag{6.59}$$

which defines the zero dynamics flow everywhere on M_0. Near x_0 we simply project the flow onto the tangent space to M_0 at x_0. K has a left inverse K^* so that

$$\dot{w} = K^* f(x^*(w)) - K^* G(x^*(w))\rho^{-1}(x^*(w))\alpha(x^*(w)) \tag{6.60}$$

$$x^*(w) = A^* v^*(w) + Kw \tag{6.61}$$

ProPac provides the computations necessary to implement the above control design method. The main functions

- `IOLinearize`

- `NormalCoordinates`

- `LocalZeroDynamics`

are illustrated in the following examples.

Example 6.191 (A basic example) *Consider the following simple single-input single-output example from [1]:*

$$f(x) = \begin{bmatrix} 0 \\ x_1 + x_2^2 \\ x_1 - x_2 \end{bmatrix}, \quad G(x) = \begin{bmatrix} e^{x_2} \\ e^{x_2} \\ 0 \end{bmatrix}, \quad h(x) = x_3$$

Below, we compute the relative degree vector, the decoupling matrix and the feedback linearizing using the function IOLinearize. *Then to compute the zero dynamics, we obtain the partial state transformation, i.e, $z(x)$ and the control with $v_i(t) = 0$, $i = 1, \ldots, m$, and finally we compute the local zero dynamics.*

$In[192] := \quad var2 := \{x1, x2, x3\}$
$\qquad\qquad f2 := \{0, \ x1 \ + \ x2\hat{}2, \ x1 \ - \ x2\}$
$\qquad\qquad g2 := \{\exp[x2], \exp[x2], 0\}$
$\qquad\qquad h2 := \{x3\}$

$In[193] := \quad \{\rho, \alpha, ro, control\} = IOLinearize[f2, g2, h2, var2]$

Computing Decoupling Matrix

Computing linearizing/decoupling control

$Out[193] = \left\{\left\{\{-e^{x2} - 2 \ e^{x2} \ x2\}\right\}, \left\{-2 \ x2 \ (x1 + x2^2)\right\}, \{3\}, \left\{\dfrac{v1 + 2 \ x2 \ (x1 + x2^2)}{-e^{x2} - 2 \ e^{x2} \ x2}\right\}\right\}$

$In[194] := \quad z = NormalCoordinates[f2, g2, h2, var2, ro];$
$\qquad\qquad u0 = control/.\{v1 \to 0\};$
$\qquad\qquad LocalZeroDynamics[f2, g2, h2, var2, u0, z]$

The system is completely linearizable.

There are no zero dynamics.

$Out[194] = \{\}$

The result should have been anticipated. Since $r = 3 = n$, there are no decoupled dynamics.

Example 6.192 (Zero dynamics of a simple vehicle) *Consider a simple wheeled vehicle that moves in the plane as illustrated in the Figure (6.2). The model incorporates two simplifications; $m_2 = 0, s << 1$, so that only first order terms in s are included.*

kinematics:

$$
\begin{bmatrix} \dot{\theta} \\ \dot{x} \\ \dot{y} \\ \dot{\delta} \end{bmatrix} = \begin{bmatrix} 1 & 0 & 0 & 0 \\ 0 & \cos(\theta) & -\sin(\theta) & 0 \\ 0 & \sin(\theta) & \cos(\theta) & 0 \\ 0 & 0 & 0 & 1 \end{bmatrix} \begin{bmatrix} \omega_\theta \\ v_x \\ v_y \\ \omega_\delta \end{bmatrix}
$$

dynamics:

$$
\begin{bmatrix} I_{zz} + J_{zz} & 0 & 0 & I_{zz} \\ 0 & m_1 & 0 & 0 \\ 0 & 0 & m_1 & 0 \\ I_{zz} & 0 & 0 & I_{zz} \end{bmatrix} \begin{bmatrix} \dot{\omega}_\theta \\ \dot{v}_x \\ \dot{v}_y \\ \dot{\omega}_\delta \end{bmatrix} + \begin{bmatrix} 0 \\ m_1 v_y \omega_\theta \\ -m_1 v_x \omega_\theta \\ 0 \end{bmatrix} + \begin{bmatrix} f_1 \\ f_2 \\ f_3 \\ f_4 \end{bmatrix} = 0
$$

The coordinates x and y locate the center of mass of the main body, and θ its orientation. The front wheels rotate an amount δ about an axis of slope s ($s = 0$, results in a vertical axis), s is assumed small as are the tire inertial parameters.

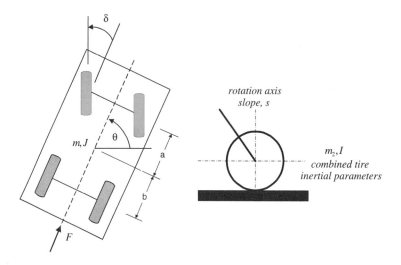

Figure 6.2: The essential parameters of the example are illustrated in this figure.

The functions f_i, which include a description of how the steering torque T and the drive force F enter the model, are omitted for the sake of space.

Our goal is to consider the problem of steering the vehicle along a path of constant radius, and at constant speed V_d. There are several ways of formulating this problem. One common approach is to replace the constant radius condition by the requirement that the angular velocity ω_θ is a constant, say ω_d. This leads to a constant curvature path of radius, with $R = V_d/\omega_d$.. Thus, we introduce two output relations

$$y_1 = v_x^2 + v_y^2 - V_d^2$$

$$y_2 = \omega_\theta - \omega_d$$

We are interested in the zero dynamics relative to these two outputs and the two controls T, F. Notice that in this formulation it is not necessary to retain the kinematic equations which define the vehicle location and orientation in the plane, i.e., θ, x, y. Thus, the system equations include dynamics and only the last equation of the kinematics.

We will compute the zero dynamics corresponding to $V_d =$ constant, $\omega_d = 0$. The equilibrium point is $(x_1, x_2, x_3, x_4, x_5) = (0, V_d, 0, 0, 0)$, so we compute the local zero dynamics near this point. In order to exhibit a complete example without using excessive space we exhibit results for the case $s = 0$.

$In[195] := f27 := \{-\kappa * (a\hat{}2 * x1 + b\hat{}2 * x1 + a * x3 -$
$\qquad b * x3 - a * x2 * x5)/(Jzz * x2),$
$\qquad -(m1 * x1 * x3 - \kappa * (a * x1 + x3) * x5/x2)/m1,$
$\qquad -(-(m1 * x1 * x2) + \kappa * (a * x1 - b * x1 + 2 * x3 - x2 * x5)/x2)/m1,$
$\qquad -(\kappa * (-(a\hat{}2 * x1) - b\hat{}2 * x1 - a * x3 + b * x3 +$
$a * x2 * x5)/(Jzz * x2)),$
$x4\};$
$g27 := \{\{-1/Jzz, 0\}, \{0, 1/m1\}, \{0, 0\}, \{(1/Izz + 1/Jzz), 0\}, \{0, 0\}\};$
$h27 := \{x2\hat{}2 + x3\hat{}2 - Vd\hat{}2, x1 - wd\};$
$var27 := \{x1, x2, x3, x4, x5\};$

$In[196] := \{\rho, \alpha, ro, control\} = IOLinearize[f27, g27, h27, var27];$
$\qquad za = NormalCoordinates[f27, g27, h27, var27, ro]/.\{wd- > 0\};$
$\qquad \{fa, ga, ha, controlsa, za\} =$
$\qquad\qquad \{f27, g27, h27, control, za\}/.\{x2- > x2 + Vd\};$
$\qquad u0 = controlsa/.\{v1- > 0, v2- > 0\};$
$\qquad f0 = LocalZeroDynamics[fa, ga, ha, var27, u0, za, 4];$
$\qquad Simplify[f0]//MatrixForm$

Computing Decoupling Matrix

Computing linearizing/decoupling control

$$Out[196] = \begin{pmatrix} \dfrac{\left(b\ w3\ \left(2\ Vd^2 + w3^2\right) + a\ \left(2\ Vd^3\ w1 - 2\ Vd^2\ w3 - w3^3\right)\right)\ \kappa}{2\ Izz\ Vd^3} \\[1em] \dfrac{\left(Vd^3\ w1 - 2\ Vd^2\ w3 - w3^3\right)\ \kappa}{m1\ Vd^3} \end{pmatrix}$$

*We can test the stability of the equilibrium point, by examining the linearized zero
dynamics. The eigenvalues are readily obtained but they are lengthy functions
of the parameters. Some insight is obtained, however, by examining the special
case, a = b, in which case the eigenvalues simplify to those shown where κ is
the tire coefficient that determines the cornering force. In this case, the zero
dynamics are unstable.*

$In[197] := Anu = Jacobian[f0, \{w1, w2, w3\}]/.\{w1- > 0, w2- > 0, w3- > 0, b- > a + \eta\};$
$\qquad Eigenvalues[Anu/.\{\eta- > 0\}]$

$Out[197] = \{ -\dfrac{\sqrt{a}\ \sqrt{\kappa}}{\sqrt{Izz}}, \dfrac{\sqrt{a}\ \sqrt{\kappa}}{\sqrt{Izz}}, -\dfrac{2\ \kappa}{m1\ Vd} \}$

6.6 Exact (Input-State) Linearization

We begin consideration of the exact linearization problem by considering the
SISO case,

$$\dot{x} = f(x) + g(x)u \tag{6.62}$$

Definition 6.193 *The SISO control system (6.62) is locally exactly feedback linearizable around x_0 if there exists a state transformation $z = \phi(x)$ and non-linear feedback $u = \varphi(x) + \Phi(x)v$, $\Phi(x_0) \neq 0$, all defined on a neighborhood X of $x_0 \in R^n$, that transforms (6.62) into the controllable linear state space system*

$$\dot{z} = Az + bv \tag{6.63}$$

The SISO state space exact feedback linearization problem *is is that of finding the transformation and the feedback, if they exist, given the control system (6.62).*

Recall that any linear controllable SISO system is similar to the canonical system

$$\dot{z} = \begin{bmatrix} 0 & 1 & 0 & \cdots & 0 \\ 0 & 0 & 1 & \ddots & \vdots \\ \vdots & \vdots & 0 & \ddots & 0 \\ \vdots & \vdots & \vdots & \ddots & 1 \\ 0 & 0 & 0 & \cdots & 0 \end{bmatrix} z + \begin{bmatrix} 0 \\ \vdots \\ \vdots \\ 0 \\ 1 \end{bmatrix} u \tag{6.64}$$

Thus, the exact linearization problem is equivalent to achieving the form (6.64) by state transformation and feedback.

Proposition 6.194 *The SISO exact feedback linearization problem is solvable if and only if there exists a function $h(x)$ such the relative degree of the control system*

$$\dot{x} = f(x) + g(x)u$$
$$y = h(x)$$

is n.

Proof: Sufficiency is obvious in view of the input-output linearization result. To establish necessity, assume the the existence of a state transformation and feedback control that transforms (6.62) to (6.64). Then take $y = z_1$, so that (6.64) with this output has relative degree n. The corresponding output map, $h(x)$, is $h(x) = \Phi_1^{-1}(x)$. Now since relative degree is invarient under state transformation and feedback, we conclude that (6.62) with output $y = h(x)$ has relative degree n. ∎

This theorem implies that $h(x)$ must satisfy the system of partial differential equations

$$L_g L_f^i h(x) = 0, \quad 0 \leq i \leq n - 2$$

and the boundary condition

$$L_g L_f^{n-1} h(x_0) \neq 0$$

Using Lemma (6.179) it can be shown that these equations are equivalent to

$$L_{ad_f^i g} h(x) = 0, \quad 0 \le i \le n - 2 \tag{6.65}$$

$$L_{ad_f^{n-1} g} h(x_0) \ne 0 \tag{6.66}$$

These relations lead to the following result, in which we employ the following notation:

$$\mathcal{G}_i = \text{span} \{ g(x) \quad \text{ad}_f g(x) \quad \cdots \quad \text{ad}_f^i g(x) \} \tag{6.67}$$

Proposition 6.195 *The SISO linear control system (6.62) is exactly feedback linearizable around x_0 if and only if*

1. *the distribution $\mathcal{G}_{n-2}$ is involutive on a neighborhood of x_0.*

2. $\text{rank}\, \mathcal{G}_{n-1}(x_0) = n$

Proof: Notice that

$$L_{ad_f^i g} L_f^k h(x) = \langle dL_f^k h(x), \text{ad}_f^i g \rangle$$

so that Lemma (6.179) leads us to conclude, for a system of relative degree r, the matrix

$$\begin{pmatrix} dh(x_0) \\ dL_f h(x_0) \\ \vdots \\ dL_f^{r-1} h(x_0) \end{pmatrix} (g(x_0) \quad \text{ad}_f g(x_0) \quad \cdots \quad \text{ad}_f^{r-1} g(x_0)) =$$

$$\begin{pmatrix} 0 & & \cdots & 0 & \langle dL_f^{r-1} h(x_0), \text{ad}_f^{r-1} g(x_0) \rangle \\ \vdots & & & \bullet & * \\ 0 & \bullet & & & \vdots \\ \langle dL_f^{r-1} h(x_0), g(x_0) \rangle & * & \cdots & & * \end{pmatrix}$$

has rank r. Thus, if $h(x)$ exists, producing relative degree n, we have necessity of condition (2).

If (2) holds then the distribution $\mathcal{G}_{n-2}(x)$ is nonsingular and of dimension $n - 1$ on a neighborhood of x_0. Equations (6.65) can be written in the form

$$dh(x)\, (g(x) \quad \text{ad}_f g(x) \quad \cdots \quad \text{ad}_f^{n-2} g(x)) = 0$$

This implies that the covector field $dh(x)$ is a basis for the codistribution $\mathcal{G}_{n-2}^\perp$ (alternatively, ann $\mathcal{G}_{n-2}$). As a consequence, the Frobenius theorem implies that $\mathcal{G}_{n-2}$ is involutive, establishing the necessity of (1).

Conversely, if (2) holds the distribution $\mathcal{G}_{n-2}$ is nonsingular and $n-1$ dimensional around x_0. If (1) also holds, then the Frobenius theorem implies existence of the function $h(x)$ on a neighborhood of x_0 such that $dh(x)$ solves (6.65). Moreover, $dh(x_0)$ spans the one one dimensional linear subspace $\mathcal{G}_{n-2}(x_0)^\perp$. Thus, (6.66) is also satisfied because otherwise $dh(x_0)$ would annihilate a set of n linear independent vectors, a contradiction.   ∎

When a system is exactly linearizable several methods are available for constructing the coordinate transformation and computing the required feedback control. One approach transforms the system into a normal form in which the linearizing control is obvious. We consider a simple example.

Example 6.196 (Feedback linearization) *The function*

FeedbackLinearizable

implements the specified test. Consider the system:

$$\begin{bmatrix} \dot{x}_1 \\ \dot{x}_2 \\ \dot{x}_3 \end{bmatrix} = \begin{bmatrix} \theta x_1^3 + x_2 \\ x_3 \\ 0 \end{bmatrix} + \begin{bmatrix} 0 \\ 0 \\ 1 \end{bmatrix} u$$

First check for linearizability

In[198]:= *f30 = {θ x1³ + x2, x3, 0};*
g30 = {0, 0, 1};
var30 = {x1, x2, x3};
FeedbackLinearizable[f30, g30, var30]
Out[198]= True

From these computations, we see that this system is linearizable. Thus, we can proceed to obtain the exact feedback linearizing state transformation with the function SIExactFBL, *which implements the feedback linearization algorithm described in [1] Chapter 4.2.*

In[199]:= *Trans = SIExactFBL[f30, g30, var30, True]*
Out[199]= {x1, θ x1³ + x2, 3 θ x1² (θ x1³ + x2) + x3}

To obtain the linearizable system in normal form coordinates, we need to first invert the transformation using

InverseTransformation

and then use

TransformSystem

to obtain the system representation in the new coordinates. These computations are illustrated below.

In [200] :=   InvTrans = InverseTransformation[var30, {z1, z2, z3}, Trans];

In [201] :=   TransformSystem[f30, g30, var30, {z1, z2, z3}, Trans, InvTrans]
Out [201]= {{z2, z3, 3 θ z1 (2 z2² + z1 z3)}, {0, 0, 1}}

In these coordinates it is seen that the control:

$$u = -3\theta z_1(2z_2^2 + z_1 z_3) + v$$

reduces the system to

$$\begin{bmatrix} \dot{z}_1 \\ \dot{z}_2 \\ \dot{z}_3 \end{bmatrix} = \begin{bmatrix} z_2 \\ z_3 \\ 0 \end{bmatrix} + \begin{bmatrix} 0 \\ 0 \\ 1 \end{bmatrix} v$$

The control can be obtained as a function of the original state variables by using the transformation equations.

Now, we turn to the MIMO case. Formally, the MIMO exact feedback linearization problem is defined as follows.

Definition 6.197 *Given a control system (1), it is said to be exactly feedback linearizable if there exists a coordinate transformation $x = \phi(z)$ on a neighborhood X of the origin of R^n and a feedback control $u = \varphi(x) + \Phi(x)v$, also defined on X with $\Phi(x)$ nonsingular on X such that the transformed system is of the form*

$$\dot{z} = Az + Bv$$

with (A, B) controllable.

To determine if a system is feedback linearizable, we can use the conditions established in the following proposition that generalizes Proposition (6.195) to the MIMO case.

Proposition 6.198 *Suppose the matrix $G(0)$ is nonsingular and the distributions*

$$\mathcal{G}_j = \{ad_f^k g_i : 0 \le k \le j, 1 \le i \le m\}, \quad 0 \le j \le n - 1,$$

where g_i are the columns of G, have constant dimension near the origin. Then the system (1) is exactly feedback linearizable (near the origin) if and only if:

1. *the distribution $\mathcal{G}_{n-1}$ has dimension n,*

2. *for each j, $0 \le j \le n - 2$, the distribution $\mathcal{G}_j$ is involutive.*

Notice that $\mathcal{G}_{n-1} = \Delta_{L_0}$ so that local controllability is necessary for exact feedback linearizability.

Example 6.199 (Feedback linearizability)
Another illustration of the function FeedbackLinearizable *is:*

$$\begin{bmatrix} \dot{x}_1 \\ \dot{x}_2 \\ \dot{x}_3 \end{bmatrix} = \begin{bmatrix} 1 + x_1 + x_3 \\ 1 + x_2 \\ -x_3 \end{bmatrix} + \begin{bmatrix} 1 + x_1 + x_3 & 0 \\ 1 + x_2 & 0 \\ 0 & e^{x_3} + x_1 x_2 \end{bmatrix} v$$

In[202]:=   f33 = {1 + x1 + x3, 1 + x2, −x3};
           g33 = {{1 + x1 + x3, 0}, {1 + x2, 0}, {0, exp[x3] + x1 x2}};
           var33 = {x1, x2, x3};

In[203]:=   FeedbackLinearizable[f33, g33, var33]
Out[203]= False

We see that the system is not feedback linearizable. This fact is not remarkable, but the way in which linearizability fails is interesting. First, we test controllability and find that the system is controllable.

In[204]:=   Controllability[f33, g33, var33]
Out[204]= True

Actually, we need a stronger version of controllability, i.e., rank $\Delta_{L_0} = n = 3$. *The calculation is*

In[205]:=   Rank[{g1, g2, Ad[f33, g1, var33], Ad[f33, g2, var33],
             Ad[f33, g1, var33, 2], Ad[f33, g2, var33, 2]}]
Out[205]= 3

The system satisfies this more restrictive controllability condition. Now, let us test for involutivity of the requisite distributions. There are two of them and one fails.

In[206]:=   {g1, g2} = Transpose[g33];
             Involutive[{g1, g2}, var33]
Out[206]= False

In[207]:=   Involutive[{g1, g2, Ad[f33, g1, var33], Ad[f33, g2, var33]}, var33]
Out[207]= True

6.7   Control via Dynamic Inversion

Control design based on input-output linearization breaks down if the decoupling matrix $\rho(x)$ does not have an inverse. Nevertheless, the basic ideas can be extended to a wider class of systems with some modification. The approach we take is from the vantage point of system invertibility. Given a control system such as (6.1), with initial state fixed, we can define both a right and a left inverse. Roughly speaking, a right inverse generates a control u that will produce a given output y, and a left inverse generates the control that produced an observed output.

Definition 6.200 (Invertible) *1. The system (6.1) is invertible at $x_0 \in R^n$ if whenever $u_1(t)$ and $u_2(t)$ are distinct admissible (real, analytic) controls, $y(\cdot, u_1, x_0) \neq y(\cdot, u_2, x_0)$.*

2. The system (6.1) is strongly invertible at $x_0 \in R^n$ if there exists a neighborhood V of x_0 such that for all $x \in V$ the system is ivertible at x.

3. The system (6.1) is strongly invertible if it is strongly invertible at x_0 for all $x_0 \in R^n$.

First, observe that if the system (6.1) is square $(p = m)$ and can be input–output linearized as described above, that is if $\det\{\rho(x)\} \neq 0$, then both right and left inverses exist. Notice that the linearized input–output dynamics (6.43) can be written

$$y^{(r)} = v$$

where r is the vector relative degree and

$$y^{(r)} = (y_1^{(r_1)}, \ldots, y_m^{(r_m)})^T$$

Consequently, in view of (6.40), (6.41) and (6.42), the inverse can be explicitly represented:

$$\begin{aligned}
\dot{\xi} &= F(\xi, z, y_R^{(r)}) \\
\dot{z} &= Az + Ey_R^{(r)} \\
u &= \rho^{-1}(x(\xi, z))\{-\alpha(x(\xi, z)) + y_R^{(r)}\}
\end{aligned} \tag{6.68}$$

The system (6.68) can serve either as a right ($y_R^{(r)}$ is a prescribed reference output) or a left ($y_R^{(r)}$ is an observed output) inverse. As a right inverse, we consider $y_R(t)$ to be prescribed and sufficiently smooth so that all of its derivatives, including the highest order derivatives which drive (6.68) are known, $y_R^{(r)} = \left[y_{R,1}^{(r_1)}, .., y_{R,m}^{(r_m)}\right]$. As a left inverse the required smoothness is automatic if $u(t)$ is piecewise continuous. Note that (6.68) is equivalent to:

$$\begin{aligned}
\dot{x} &= f(x) + G(x)u \\
u &= \rho^{-1}(x)\left\{-\alpha(x) + y_R^{(r)}\right\}
\end{aligned} \quad or \quad \begin{aligned}
\dot{x} &= f(x) + G(x)u \\
y_R^{(r)} &= \alpha(x) + \rho(x)u
\end{aligned} \tag{6.69}$$

Equations (6.68) and (6.69) represent the same system described in different state coordinates.

Equation (6.68) clearly displays the relationship between input–output linearization and inversion. We have seen above that if the decoupling matrix is nonsingular then a system inverse exists. On the other hand, singularity of the decoupling matrix does not imply that an inverse fails to exist. We seek a more general construction for an inverse with the goal of identifying a larger class of control laws. The basic tool for constructing a right inverse is the *structure algorithm* introduced by Hirshorn [7] and Singh [8]. If the system (6.1) has an inverse, then application of the structure algorithm leads to identification of a finite integer β and matrices $H_\beta(x)$, $C_\beta(x)$, $D_\beta(x)$ such that

$$H_\beta(x)Y_\beta(t) = C_\beta(x) + D_\beta(x)u \tag{6.70}$$

where

$$Y_\beta(t) = [y^{(1)^T}, y^{(2)^T}, .., y^{(\beta)^T}]^T \tag{6.71}$$

and $D_\beta(x)$ is an $p \times m$ matrix with rank $\min(m,p)$. Thus, (6.70) may be thought of as a generalization of the second equation of (6.69). Suppose $p = \min(m,p)$. It follows that $D_\beta(x)$ has a right (matrix) inverse $D_\beta^\dagger(x)$. Consequently, the right system inverse is defined by:

$$\begin{aligned} \dot{x} &= f(x) + G(x)u \\ u &= D_\beta^\dagger(x)\{-C_\beta(x) + H_\beta(x)Y_\beta(t)\} \end{aligned} \tag{6.72}$$

In this case, given a reference $y_R(t)$, a control $u(t)$ is obtained that will reproduce it when applied to the system (assuming the correct initial state). While $u(t)$ is not unique (if $p < m$), it does the job. On the other hand if $m = \min(m,p)$, then $D_\beta(x)$ has a left (matrix) inverse $D_\beta^\dagger(x)$ and (6.72) defines a left system inverse. In this case an observed $y(t)$ drives the (left) inverse system which produces the unique control that generated $y(t)$. However, there may be different observed outputs that result in the same control.

The following summarizes the Structure algorithm.

Algorithm 6.201 (Structure Algorithm) *Consider the system (6.1).*

1. *Step 1 Compute*

$$\dot{y} = \frac{\partial h}{\partial x}[f + Gu] =: L_f h(x) + L_G h(x)u \tag{6.73}$$

   and define $r_1 := \operatorname{rank} L_G h(x)^1$. *Permute the output components so that the first* r_1 *rows of* $L_G h$ *are independent. Since the last* $p - r_1$ *rows are linearly dependent on the first* r_1 *rows, combinations of the first rows can be used to zero out the last rows. Let* E_1^1 *and* $E_1^2(x)$ *be the permutation and row zeroing matrices. Then define*

$$z_1 = E_1^2(x)E_1^1 \dot{y} \tag{6.74}$$

$$z_1 = E_1^2(x)E_1^1(L_f h(x) + L_G h(x)u) \tag{6.75}$$

   Now, write $z_1 = (\bar{z}_1^T, \hat{z}_1^T)^T$, *with* $\bar{z}_1 \in R^{r_1}$, $\hat{z}_1 \in R^{p-r_1}$. *From the first* r_1 *rows of (6.75)*

$$\bar{z}_1 = \bar{c}_1(x) + \bar{D}_1(x)u, \quad \operatorname{rank} \bar{D}_1 = r_1 \tag{6.76}$$

---

[1]By the notation $L_G h$, we mean the matrix whose columns are $L_{g_i} h$.

and from the last $p - r_1$ *rows of*

$$\widehat{z}_1 = \widehat{c}_1(x) \tag{6.77}$$

Finally, define System 1 *to be*

$$\dot{x} = f(x) + G(x)u$$

$$z_1 = \begin{bmatrix} \bar{z}_1 \\ \widehat{z}_1 \end{bmatrix} = c_1(x) + D_1(x)u$$

where

$$c_1(x) = \begin{bmatrix} \bar{c}_1(x) \\ \widehat{c}_1(x) \end{bmatrix}, \ D_1(x) = \begin{bmatrix} \bar{D}_1(x) \\ 0 \end{bmatrix}$$

2. *Step 2 Differentiate* $\widehat{z}_1$ *to obtain*

$$\dot{\widehat{z}}_1 = \frac{\partial \widehat{z}_1}{\partial x} [f + Gu]$$

which can be written as

$$\dot{\widehat{z}}_1 = L_f \widehat{c}_1(x) + L_G \widehat{c}_1(x)u$$

Now, consider

$$\begin{bmatrix} \bar{z}_1 \\ \dot{\widehat{z}}_1 \end{bmatrix} = \begin{bmatrix} \bar{c}_1(x) \\ L_f \widehat{c}_1(x) \end{bmatrix} + D(x)u, \ D(x) := \begin{bmatrix} \bar{D}_1(x) \\ L_G \widehat{c}_1(x) \end{bmatrix}$$

Let $r_2 = \operatorname{rank} D$. *Then permute the rows of* D *to make the first* r_2 *rows independent and the zero out the last rows. Let* E_2^1 *and* $E_2^2(x)$ *be the permutation and row zeroing matrices. Define*

$$z_2 = E_2^2(x) E_2^1 \begin{bmatrix} \bar{z}_1 \\ \dot{\widehat{z}}_1 \end{bmatrix}$$

and divide z_2 *into* $\bar{z}_2 \in R^{r_2}$ *and* $\widehat{z}_2 \in R^{p - r_2}$:

$$\bar{z}_2 = \bar{c}_2(x) + \bar{D}_2(x)u$$

$$\widehat{z}_2 = \widehat{c}_2(x)$$

Finally, define System 2 *to be*

$$\dot{x} = f(x) + G(x)u$$

$$z_2 = \begin{bmatrix} \bar{z}_2 \\ \widehat{z}_2 \end{bmatrix} = c_2(x) + D_2(x)u$$

$$c_2(x) = \begin{bmatrix} \bar{c}_2(x) \\ \widehat{c}_2(x) \end{bmatrix}, \ D_2(x) = \begin{bmatrix} \bar{D}_2(x) \\ 0 \end{bmatrix}$$

3. *Step $k + 1$ Suppose that in steps 1 through k, the integers $r_1, \ldots, r_k$ and the functions $\bar{z}_1 \in R^{r_1}, \ldots, \bar{z}_k \in R^{r_k}, \hat{z}_k \in Rp - r_k$ have been defined so that we have System k:*

$$\dot{x} = f(x) + G(x)u$$

$$z_k = \begin{bmatrix} \bar{z}_k \\ \hat{z}_k \end{bmatrix} = c_k(x) + D_k(x)u$$

with

$$c_k(x) = \begin{bmatrix} \bar{c}_k(x) \\ \hat{c}_k(x) \end{bmatrix}, \ D_k(x) = \begin{bmatrix} \bar{D}_k(x) \\ 0 \end{bmatrix}$$

Then differentiate $\hat{z}_k$

$$\dot{\hat{z}}_k = \frac{\partial z_k}{\partial x}[f + Gu] \tag{6.78}$$

which can be rewritten as

$$\dot{\hat{z}}_k = L_f \hat{c}_k(x) + L_G \hat{c}_k(x)u \tag{6.79}$$

Now, consider

$$\begin{bmatrix} \bar{z}_k \\ \dot{\hat{z}}_k \end{bmatrix} = \begin{bmatrix} \bar{c}_k(x) \\ L_f \hat{c}_k(x) \end{bmatrix} + D(x)u, \ D(x) := \begin{bmatrix} \bar{D}_k(x) \\ L_G \hat{c}_k(x) \end{bmatrix}$$

Let $r_{k+1} := \text{rank } D$. Permute the rows of D to make the first r_{k+1} rows independent. Use combinations of these rows to zero out the remaining (dependent) rows. Denote the permutation matrix and row zeroing matrices E_{k+1}^1 and $E_{k+1}^2(x)$, respectively. Then define

$$z_{k+1} = E_{k+1}^2(x)E_{k+1}^1 \begin{bmatrix} \bar{z}_k \\ \dot{\hat{z}}_k \end{bmatrix} \tag{6.80}$$

Finally, define System $k + 1$

$$\dot{x} = f(x) + G(x)u$$

$$z_{k+1} = \begin{bmatrix} \bar{z}_{k+1} \\ \hat{z}_{k+1} \end{bmatrix} = c_{k+1}(x) + D_{k+1}(x)u$$

with

$$c_{k+1}(x) = \begin{bmatrix} \bar{c}_{k+1}(x) \\ \hat{c}_{k+1}(x) \end{bmatrix}, \ D_{k+1}(x) = \begin{bmatrix} \bar{D}_{k+1}(x) \\ 0 \end{bmatrix}$$

4. *Stop By construction the integers r_i satisfy $r_1 \leq r_2 \leq \ldots \leq r_k$. Moreover, there is a smallest positive integer k^* such that $r_{k^*} \leq \min(m, p)$ is maximal. If the procedure terminates in finite steps, it does so at step k^* with $r_{k^*} = \min(m, p)$ and an inverse can be constructed (a right inverse if $p = \min(m, p)$ and a left inverse if $m = \min(m, p)$). The relative order β is k^* if the procedure terminates in finite steps, otherwise $\beta = \infty$. The number β identifies the highest order derivative required to drive the inverse. Thus, it can not be greater than the number of states, n. Thus, the procedure should not proceed beyond n steps.*

In *ProPac*, the structure algorithm is implemented in the function

StructureAlgorithm.

Example 6.202 (Inverse system, output restrictions) *This first example is considered in both Hirschorn [7] and Singh [9].*

$In[208] :=$ $x = \{x1, x2, x3\};$
$f = \{0, x3, 0\}; G = \{\{x1, 0\}, \{-x3, 0\}, \{0, x1\}\};$
$h = \{x1, x2\};$
$\{DD, CC, HH, ZZ\} = StructureAlgorithm[f, G, h, x, t];$

$In[209] :=$ $u = Simplify[RightInverse[DD].(-CC + ZZ)];$
$(f + G.u)//MatrixForm$

$Out[209] =$ $$\begin{pmatrix} y1'[t] \\ x3 - \dfrac{x3\ y1'[t]}{x1} \\ \dfrac{-x3\ y1'[t]^2 + x1\ (x3\ y1''[t] + x1\ y2''[t])}{x1\ (x1 - y1'[t])} \end{pmatrix}$$

It is to be anticipated that a complete discussion of system inverses would include a characterization of the system input and output spaces. In [7], Hirschorn gives sufficient conditions for the existence of an inverse and shows that they do not apply in the case of this example. Modifying the arguments in [7], Singh in [9] derives sufficient conditions for existence of a left system inverse that apply with restrictions imposed on the system output space. These conditions, applied to this example, establish the existence of a left inverse provided $y_1^{(1)} \neq x_1$. Clearly, if our inverse is to be meaningful, this condition must be satisfied.

Example 6.203 (Inverse system) *The following example is from Neijmeijer and Van der Schaft [2].*

$In[210] :=$ $x = \{x1, x2, x3, x4\};$
$f = \{0, x3, x4, 0\}; G = \{\{1, 0\}, \{x4, 0\}, \{0, 0\}, \{0, 1\}\};$
$h = \{x1, x2\};$
$\{DD, CC, HH, ZZ\} = StructureAlgorithm[f, G, h, x, t];$

$In[211] :=$ $u = RightInverse[DD].(-CC + ZZ);$
$(f + G.u)//MatrixForm$

$Out[211] =$ $$\begin{pmatrix} y1'[t] \\ x3 + x4\ y1'[t] \\ x4 \\ \dfrac{-x4 - x4\ y1''[t] + y2''[t]}{y1'[t]} \end{pmatrix}$$

6.7.1 Tracking Control

There are a number of ways in which the results of the structure algorithm can be used to construct feedback controllers. It is an important fact that not all of the elements in Y_β actually survive multiplication by $H_\beta(x)$. Let us denote by n_i and N_i the lowest and highest order derivative of y_i appearing in $H_\beta(x)Y_\beta$. Then, we can write

$$H_\beta(x)Y_\beta = \widetilde{H}\widetilde{y} \tag{6.81}$$

where

$$\widetilde{y} = [y_1^{n_1}, ..y_1^{N_1}, .., y_l^{n_l} .., y_l^{N_l}]^T \tag{6.82}$$

so that (6.72) can be written as:

$$\begin{aligned} \dot{x} &= f(x) + G(x)u \\ u &= D_\beta^\dagger(x)\left\{-C_\beta(x) + \widetilde{H}(x)\widetilde{y}(t)\right\} \end{aligned} \tag{6.83}$$

One approach to tracking control is based on the concept of using the inverse system to compute a feedforward control and then add a perturbation controller based on the tracking error (see Figure (6.3)). For instance if $y_R(t)$ is the reference trajectory, we could implement the control

$$u = D_\beta^\dagger(x)\left\{-C_\beta(x) + \widetilde{H}(x)\widetilde{y}_R(t) + v(t)\right\} \tag{6.84}$$

where $v(t)$ is the perturbation controller.

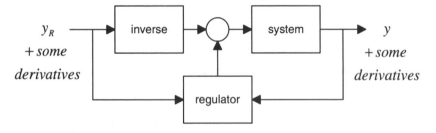

Figure 6.3: A tracking control concept based on the system right inverse.

One choice for the perturbation controller is [8]:

$$v_i = \gamma_{i0}\int(y_{Ri} - y_i)dt + \sum_{j=0}^{n_i-1} p_{ij}(y_{Ri}^{(j)} - y_i^{(j)}), \quad i = 1, .., l \tag{6.85}$$

Stability of the closed loop requires that the following polynomials are Hurwitz:

$$\sum_{j=0}^{n_i+1} \gamma_{ij} \cdot s^j = 0, \quad i = 1, .., l \tag{6.86}$$

with $\gamma_{ij} = p_{i,j-1}, j = 1, .., n_i + 1$.

A variant of this controller is given by Singh [9]. Here (6.84) is replaced by

$$u = D_\beta^\dagger(x)\{-C_\beta(x) + M(x)\widetilde{y}_0 + N(x)v(t)\} \tag{6.87}$$

where

$$H_\beta(x)Y_\beta = M(x)\widetilde{y}_0 + N(x)y^{(n)} \tag{6.88}$$

$$\widetilde{y}_0 = [y_1^{(n_1+1)}, \ldots, y_1^{(N_1)}, \ldots, y_l^{(n_l+1)}, \ldots, y_l^{(N_l)}]^T \quad \text{and} \quad y(n) = [y_1^{n_1}, \ldots, y_l^{(n_l)}]^T \tag{6.89}$$

and $N(x)$ is an invertible matrix. It is easily verified, by substituting (6.87) and (6.88) into (6.70), that this is a decoupling controller that reduces the input–output equations to

$$y_i^{(n_i)} = v_i, \quad i = 1, \ldots, l \tag{6.90}$$

The perturbation controller is

$$v_i = \gamma_{i,0} \int (y_{Ri} - y_i)dt + \sum_{j=0}^{n_i-1} p_{ij}(y_{Ri}^{(j)} - y_i^{(j)}) + y_{Ri}^{(n_j)}. \tag{6.91}$$

Substituting the perturbation control yields the closed loop error dynamics

$$\varepsilon_i^{(n_i+1)} + \gamma_{i,n_i}\varepsilon_i^{(n_i)} + \gamma_{i,n_i-1}\varepsilon_i^{(n_i-1)} + \ldots + \gamma_{i,0} = 0$$

where $\varepsilon_i := y_i - y_{Ri}$. Implementation of this controller requires measurement of the state x and output derivatives up to $y_i^{(N_i)}$, and knowledge of the reference trajectory and its derivatives up to $y_{Ri}^{(n_i)}$. While output derivatives up to order $y_i^{(n_i-1)}$ are indeed necessary to implement a perfect tracking controller, the higher order derivatives are not. The need to measure or compute the higher order derivatives of the output, specifically $\widetilde{y}_0$ can be a serious problem, because they may be noncausal in the sense that computing them may require computing derivatives of the input u. But this can be remedied as follows in the next section.

6.7.2 Dynamic Decoupling Control

Let us rewrite (6.83) in the form

$$u = D_\beta^\dagger(x)\left\{-C_\beta(x) + \widetilde{H}_1(x)\widetilde{y}_1(t) + \widetilde{H}_2(x)y^{(N)}\right\} \tag{6.92}$$

$$\widetilde{y}_1 = [y_1^{n_1}, ..y_1^{N_1-1}, .., y_l^{n_l}..,y_l^{N_l-1}]^T \quad \text{and} \quad y^{(N)} = [y_1^{(N_1)}, .., y_l^{(N_l)}]^T \tag{6.93}$$

so that we have pulled out the highest order derivative of y. Now, let $\eta_i \in R^{N_i-n_i}, i = 1, \ldots, l$, set $v = y^{(N)}$ and consider the dynamical systems

$$\dot{\eta}_i = A_i\eta_i + B_i\nu_i, \quad i = 1, \ldots, l \tag{6.94}$$

with (A_i, B_i) in Brunovsky canonical form. Notice that $y_i^{(N_i - j)} = \eta_i^j$, $j = 1, \ldots, N_i - n_i$. The control (6.87) can be written

$$u = D_\beta^\dagger(x)\{-C_\beta(x) + \tilde{H}_1(x)\eta + \tilde{H}_2(x)v\}. \tag{6.95}$$

The dynamic compensator defined by (6.94) and (6.95) is a decoupling, input–output linearizing controller. Its application results in the input–output dynamics:

$$y_i^{(N_i)} = v_i, \quad i = 1, \ldots, l \tag{6.96}$$

Thus, we can apply the controller

$$v_i = \gamma_{i,0} \int (y_{Ri} - y_i)dt + \sum_{j=0}^{N_i-1} p_{ij}(y_{Ri}^{(j)} - y_i^{(j)}) + y_{Ri}^{(N_i)}.$$

to obtain the closed loop error dynamics

$$\varepsilon_i^{(N_i+1)} + \gamma_{i,N_i}\varepsilon_i^{(N_i)} + \gamma_{i,N_i-1}\varepsilon_i^{(N_i-1)} + \ldots + \gamma_{i,0} = 0$$

where $\varepsilon_i := y_i - y_{Ri}$.

Implementation of this controller requires measurement of the state x, and either computation or measurement of the output derivatives $y_i, y_i^{(1)}, \ldots, y_i^{(n_i-l)}$, for $i = 1, \ldots, l$. Such an implementation is illustrated in Figure (6.4).

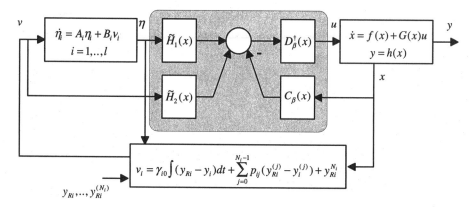

Figure 6.4: This figure illustrates the implementation of a tracking controller using dynamic decoupling.

6.8   Dynamic Extension

Another way of dealing with the singularity of the decoupling matrix is a process known as dynamic extension [1, 11]. Dynamic extension entails augmenting the dynamical equations with integrators added at the input channels. The algorithm of Descusse and Moog [11] converges in a finite number of steps if the original system is strongly invertible. It has been implemented in *ProPac* as the function DynamicExtension.

To motivate the procedure, consider the simple linear system shown in Figure (6.5):

$$\dot{x}_1 = x_2 + u_1$$
$$\dot{x}_2 = u_2$$
$$\dot{x}_3 = 2x_2 + u_1$$

$$y_1 = x_1$$
$$y_2 = x_3$$

Let us try to decouple the system by following the input–output linearization

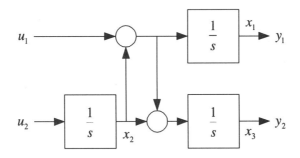

Figure 6.5: This simple linear system illustrates the delayed appearance of the second control input.

procedure of Section 6.5. To do this, successively diffenerentiate y_1 and y_2 until a control appears,

$$\dot{y}_1 = x_2 + u_1$$

$$\dot{y}_2 = 2x_2 + u_1$$

Thus, we see that

$$\rho = \begin{bmatrix} 1 & 0 \\ 1 & 0 \end{bmatrix}$$

It is singular and the system does not have a well-defined relative degree vector. But the second control has not had an influence because of the premature appearance of u_1 simultaneously in both outputs. Let us delay the appearance of u_1 by placing an integrator before the first input. Thus, the new state equations are

$$\dot{x}_1 = x_2 + u_1$$
$$\dot{x}_2 = u_2$$
$$\dot{x}_3 = 2x_2 + u_1$$
$$\dot{u}_1 = v$$

Now attempt to decouple the augmented system. Successive differentiation leads to

$$\ddot{y}_1 = u_2 + v$$
$$\ddot{y}_2 = 2u_2 + v$$

So

$$\rho = \begin{bmatrix} 1 & 1 \\ 2 & 1 \end{bmatrix}$$

The augmented system decoupling matrix is nonsingular and the relative degree is well defined $\mathbf{r} = \{2, 2\}$. A systematic procedure for attempting to achieve well-defined relative degree by integrator augmentation of the input channels is given in the following Algorithm.

Algorithm 6.204 (Dynamic Extension) *Consider the system (6.1) with $p = m$ (a square system).*

1. *Compute the matrix $\rho(x)$. If rank $\rho = m$ Stop!*

2. *if rank $\rho = s < m$, perform elementary column operations to make the first s columns independent and the last $m - s$ columns zero. Let $E(x)$ denote the square, nonsingular matrix that does this: $\rho_1(x) = \rho(x)E(x)$.*

3. *Suppose there are q columns (say, $i_1, \ldots, i_q$) each having two or more elements that are not identically zero around x_0. If $q = 0$, Stop! The process fails.*

4. *If $q \neq 0$, define the index set $\alpha = \{i_1, \ldots, i_q\}$ and let $\bar{\alpha}$ denote its complment. Put an integrator in series with the q corresponding controls to obtain a new augmented system*

$$\begin{bmatrix} \dot{x} \\ \dot{u}_\alpha \end{bmatrix} = \begin{bmatrix} f(x) + \sum_{i \in \alpha} g_i(x)u_i \\ 0 \end{bmatrix} + \begin{bmatrix} \sum_{i \in \bar{\alpha}} g_i(x)u_i \\ v_\alpha \end{bmatrix}$$

5. *Go to step 1 and repeat the process with the new augmented system.*

Example 6.205 (Dynamic inversion revisited) *Let us reconsider Examples (6.202) and (6.203). First, Example (6.202). Recall that D[f,{x,n}] is the Mathematica function for the nth partial derivative with respect to x.*

$In[212] :=$ {*fnew, Gnew, hnew, xnew*} = *DynamicExtension*[f, G, h, x];

$In[213] :=$ {$\rho, \alpha, r0, u$} = *IOLinearize*[*fnew, Gnew, hnew, xnew*];
$u = (u/.\{v1->D[y1[t], \{t, r0[[1]]\}], v2->D[y2[t], \{t, r0[[2]]\}]\});$
Simplify[(*fnew + Gnew.u*)]//*MatrixForm*

`Computing Decoupling Matrix`

`Computing linearizing/decoupling control`

$$
Out[213] = \begin{pmatrix} x1\ z1 \\ x3 - x3\ z1 \\ \dfrac{x1\ x3\ z1^2 - x3\ y1''[t] - x1\ y2''[t]}{x1\ (-1 + z1)} \\ -z1^2 + \dfrac{y1''[t]}{x1} \end{pmatrix}
$$

Now, let us turn to the system from Example (6.203). Dynamic extension can also be used to construct the inverse.

$In[214] :=$ *IOLinearize*[f, G, h, x]
$\{fnew, Gnew, hnew, xnew\} = DynamicExtension[f, G, h, x];$

Computing Decoupling Matrix

Decoupling matrix is singular!

Dynamic extension completed

$In[215] :=$ $\{\rho, \alpha, r0, u\} = IOLinearize[fnew, Gnew, hnew, xnew];$
$u = (u/.\{v1- > D[y1[t], \{t, r0[[1]]\}], v2- > D[y2[t], \{t, r0[[2]]\}]\});$
$(fnew + Gnew.u)//MatrixForm$

Computing Decoupling Matrix

Computing linearizing/decoupling control

$$
Out[215] = \begin{pmatrix} z1 \\ x3 + x4\ z1 \\ x4 \\ -\dfrac{x4\ y1''[t]}{z1} + \dfrac{-x4 + y2''[t]}{z1} \\ y1''[t] \end{pmatrix}
$$

Example 6.206 (Dynamic extension) *Here is another example for which the decoupling matrix is singular and thus feedback linearization can not be employed directly:*

$$
\begin{bmatrix} \dot{x}_1 \\ \dot{x}_2 \\ \dot{x}_3 \end{bmatrix} = \begin{bmatrix} \cos(x_3) & 0 \\ \sin(x_3) & 0 \\ 0 & 1 \end{bmatrix} \begin{bmatrix} u_1 \\ u_2 \end{bmatrix}
$$

$$
\begin{bmatrix} y_1 \\ y_2 \end{bmatrix} = \begin{bmatrix} x_1 \\ x_2 \end{bmatrix}
$$

$In[216] :=$ $f12 := \{0, 0, 0\};$
$g12 := Transpose[\{\{\cos[x3], \sin[x3], 0\}, \{0, 0, 1\}\}];$
$h12 := \{x1, x2\}; var12 := \{x1, x2, x3\}$
$\{fnew, Gnew, hnew, xnew\} = DynamicExtension[f12, g12, h12, var12]$

Dynamic extension completed

$Out[216] = \{\{z1\ Cos[x3], z1\ Sin[x3], 0, 0\}, \{\{0, 0\}, \{0, 0\}, \{0, 1\}, \{1, 0\}\},$
$\{x1, x2\}, \{x1, x2, x3, z1\}\}$

Thus, we find that the extended state vector is (x_1, x_2, x_3, z_1) and f , G and h are modified to:

$$f = \begin{bmatrix} z_1 \cos(x_3) \\ z_1 \sin(x_3) \\ 0 \\ 0 \end{bmatrix}, \quad G = \begin{bmatrix} 0 & 0 \\ 0 & 0 \\ 0 & 1 \\ 1 & 0 \end{bmatrix}, \quad h = \begin{bmatrix} x_1 \\ x_2 \end{bmatrix}$$

The extended system has a nonsingular decoupling matrix.

6.9　Nonlinear Observers

Observers are essential for implementing model-based control systems. We will discuss only one approach to observer design for nonlinear systems.

6.9.1　Basics of Nonlinear Observers

Consider the nonlinear system

$$\dot{x} = f(x, u), \quad y = h(x) \tag{6.97}$$

and suppose the system has an equilibrium point at (x^*, u^*), i.e. $f(x^*, u^*) = 0$. A *local observer* for (6.97) is another dynamical system, driven by inputs $y(t)$ and $u(t)$, that produces an estimate $\hat{x}(t)$ of $x(t)$ such that the error $e(t) = x(t) - \hat{x}(t)$ converges to zero as time tends to infinity:

$$\|x(t) - \hat{x}(t)\| \to 0, \quad \text{as } t \to \infty \tag{6.98}$$

provided $x(t)$ remains sufficiently close to x^*.

One approach to observer design mimics the structure of full state observers for linear systems. Namely, the measurement error (or residual) is used to drive a replica of the system so that the observer equations are

$$\dot{\hat{x}} = f(\hat{x}, u) + \kappa(\hat{x}, y - h(\hat{x})) \tag{6.99}$$

The error dynamics can be computed

$$\dot{e} = \dot{x}(t) - \dot{\hat{x}}(t) = f(x, u) - f(x - e, u) + \kappa(x - e, h(x) - h(x - e)) \tag{6.100}$$

Let us consider this as a differential equation that defines e, the error response to an exogenous input $x(t)$ (of course, it must be a solution of (6.97)). Notice that $e = 0$ does indeed correspond to an equilibrium point ($\dot{e} = 0$) of (6.100) for arbitrary $x(t)$ provided $\kappa(x, 0) = 0$. Hence, the system (6.99) is an observer if $\kappa(\cdot, \cdot)$ can be chosen so that this equilibrium point is asymptotically stable.

6.9.2 Exponential Detectability

A system is said to be exponentially detectable at (x^*, u^*) if there exists a function $\gamma(\xi, y)$ defined on a neighborhood of $(x^*, y^* = h(x^*)) \in R^{n+q}$ that satisfies

(i) $\gamma(x^*, y^*) = 0$,

(ii) $\gamma(\xi, h(\xi)) = f(\xi, u^*)$,

(iii) the equilibrium point $\xi = x^*$ of $\dot{\xi} = \gamma(\xi, y^*)$ is exponentially stable.

Exponential detectability implies that the system

$$\dot{\widehat{x}} = f(\widehat{x}, u) - f(\widehat{x}, u^*) + \gamma(\widehat{x}, y) \tag{6.101}$$

is a local observer. To see this, compute the error dynamics

$$\dot{e} = f(x, u) - f(x - e, u) + f(x - e, u^*) - \gamma(x - e, h(x)) \tag{6.102}$$

and notice that in view of (ii), $e = 0$ is an equilibrium point for all exogenous inputs $x(t), u(t)$. Moreover, exponential stability is assured by (iii)– as can be verified by linearizing (6.102) with respect to e at the equilibrium point $e = 0$.

Conditions (ii) and (iii) generalize the linear case in a natural way. It is easy to see that $\gamma(\xi, y) := A\xi - L(y - C\xi)$ satisfies (ii), and (iii) is satisfied as well provided L is chosen such that $A + LC$ is asymptotically stable. Moreover, the nonlinear observer (6.99) corresponds to the choice $\gamma(\xi, y) := f(\xi, u^*) + \kappa(\xi, y - h(\xi))$ provided $\kappa(\cdot, \cdot)$ can be chosen to provide exponential stability, i.e., to satisfy (iii).

The function ExponentialObserver implements the above construction. Its calling syntax is

```
{fhat,xhat,eigs}=ExponentialObserver[f,h,x,u,y,x0,u0,delta]
```

Thus, given: the system as defined by equations (6.97), an equilibrium point (x_0, u_0) and a specified decay rate δ the function returns an observer of the form

$$\dot{\widehat{x}} = \widehat{f}(\widehat{x}, u, y)$$

The eigenvalues of the linearized observer at the equilibrium point (x_0, u_0) are also returned.

Example 6.207 (An exponential observer) *Here is a simple example of an exponential observer.*

```
In[217]:= f = {x1^2 + cos[x2] * x3 - x2, x2 * cos[x1], sin[x3] + cos[x1] + u1};
          h = {x1 + x1 * x2 + x3, x2 + x3^3};
```

In[218]:= ExponentialObserver[f, h, {x1, x2, x3}, {u1}, {y1, y2}, {0, 0, 0}, {0}, 4]

Out[218]= {{x1hat^2 − x2hat + 27 (−x1hat − x1hat x2hat − x3hat + y1)+

     5 (−x2hat − x3hat^3 + y2) + x3hat Cos[x2hat],

     −10 (−x2hat − x3hat^3 + y2) + x2hat Cos[x1hat],

     u1 − 45 (−x1hat − x1hat x2hat − x3hat + y1)−

     5 (−x2hat − x3hat^3 + y2) + Cos[x1hat] + Sin[x3hat]},

     {x1hat, x2hat, x3hat}, {−5, −5, −4}}

Notice that the observer states are assigned names that are the original state names extended with 'hat.' Alternatively, the designer can specify a symbol upon which state names will be based.

6.10   Problems

6.10.1   Controllability

Problem 6.208 *Check controllability for the system:*

$$\dot{x}_1 = u_1 x_3 + u_2$$
$$\dot{x}_2 = u_1 x_1$$
$$\dot{x}_3 = u_1 x_2$$

Problem 6.209 *Consider the control systems given below. Compute and describe the maximal integral manifolds of the controllability distributions Δ_C and Δ_{C_0}.*

(a) the linear system

$$\dot{x} = \begin{bmatrix} 1 & 0 & 0 \\ 0 & -1 & 0 \\ 0 & 0 & 1 \end{bmatrix} x + \begin{bmatrix} 1 \\ 0 \\ 0 \end{bmatrix} u$$

(b) the bilinear system

$$\dot{x} = \begin{bmatrix} 0 & 1 & 0 \\ -1 & 0 & 0 \\ 0 & 0 & 0 \end{bmatrix} x + u \begin{bmatrix} 0 & 0 & 1 \\ 0 & 0 & 0 \\ -1 & 0 & 0 \end{bmatrix} x$$

Problem 6.210 *Investigate the controllability properties of the nonlinear system*

$$\dot{x} = \begin{bmatrix} x_2 + x_2^2 + x_3^2 \\ x_3 + \sin(x_1 - x_3) \\ x_3^2 \end{bmatrix} + \begin{bmatrix} 1 \\ 0 \\ 1 \end{bmatrix} u$$

Problem 6.211 *Consider the (angular) velocity control of a sattelite in space using gas jets. The dynamical equations are:*

$$J\dot\omega = \omega \times (J\omega) + \tau$$

where ω is the angular velocity, τ is the control torque vector, and $J = \mathrm{Diag}\{J_x, J_y, J_z\}$ is the inertia matrix, all in principal axis body coordinates.

(a) *Show that the system is not (locally) controllable if any one actuator is used.*

(b) *Suppose $J_x > J_y > J_z$. Show that the system is controllable if any two of the actuators are used.*

(c) *Suppose $J_x = J_y > J_z$. Show that the system is controllable if τ_1 and τ_3 are used or if τ_2 and τ_3, but not controllable if τ_1 and τ_2 are used.*

(d) *Suppose $J_x = J_y = J_z$. Show that the system is not controllable unless all three actuators are used.*

Problem 6.212 *A simplied model for the configuration of a vehicle moving in a plane is*

$$\begin{bmatrix} \dot x \\ \dot y \\ \dot\theta \end{bmatrix} = \begin{bmatrix} \sin\theta \\ \cos\theta \\ 0 \end{bmatrix} u_1 + \begin{bmatrix} 0 \\ 0 \\ 1 \end{bmatrix} u_2$$

Here x, y denote the planar location of the vehicle and θ denotes its orientation. The model is purely kinematic. It is assumed that the forward velocity u_1 and angular velocity u_2 are control inputs. Show that the system is controllable.

Problem 6.213 *A purely kinematic model for the rolling penny is*

$$\frac{d}{dt}\begin{bmatrix} \phi \\ \psi \\ x \\ y \end{bmatrix} = \begin{bmatrix} 1 \\ 0 \\ 0 \\ 0 \end{bmatrix} u_1 + \begin{bmatrix} 0 \\ 1 \\ \cos\phi \\ \sin\phi \end{bmatrix} u_2$$

Is it controllable?

Problem 6.214 (Sleigh, continued) *Consider the sleigh of Problem (5.155) in Chapter 5 (consider the case $\alpha = 0$).*

(a) *Show that the sleigh is locally controllable.*

(b) *Show that the origin is an equilibrium point with controls $, F = 0$, $T = 0$ but that it can not be asymptotically stabilized by any smooth feeback control. Hint: recall Lemma (2.16) of Chapter 2.*

6.10.2   Feedback Linearization

Problem 6.215 *Consider a linear system characterized by the SISO transfer function*

$$G(s) = K \frac{s^m + a_{m-1}s^{m-1} + \cdots + a_0}{s^n + b_{n-1}s^{n-1} + \cdots + b_0}, \ n \geq m$$

Determine a state space realization for this system. Show that the definition of relative degree given in Definition (6.177) is consistent with the traditional concept of relative degree in SISO linear systems. Compute the zero dynamics.

Problem 6.216 *Consider the simplified vehicle model*

(a) Vehicle dynamics:

$$m\ddot{x} = F - \rho\dot{x}^2\text{sgn}(\dot{x}) - d_m$$

(b) Engine:

$$\dot{\xi} = \frac{\xi}{\tau(\dot{x})} + \frac{u}{m\tau(\dot{x})}, \quad F = m\xi$$

m, ρ, d_m *are constants, and u is the throttle input. Write these equations in state space form and examine the linearizability of the system from input u to output x. Determine the relative degree and zero dynamics of the system.*

Problem 6.217 *Find a feedback linearizing and stabilizing control such that the closed loop poles are -1,-2,-3, for the system:*

$$\dot{x}_1 = \sin x_2$$
$$\dot{x}_2 = \sin x_3$$
$$\dot{x}_3 = u$$

Taylor linearize the system and find a linear perturbation control that places the poles at -1,-2,-3. Compare the two controllers.

Problem 6.218 *Consider the bilinear system*

$$\dot{x} = \begin{bmatrix} 0 & 0 & 3 \\ 0 & 0 & 6 \\ 0 & 0 & -2 \end{bmatrix} x + \begin{bmatrix} 1 & 2 & 4 \\ 2 & 2 & 0 \\ 0 & 0 & 3 \end{bmatrix} x\,u$$

(a) Determine if this system is locally controllable around the origin.

(b) Consider the Taylor linearization of this system and determine if it is controllable.

(c) Determine if the system is exactly feedback linearizable around the origin.

(d) Design an asymptotically stabilizing state feedback controller.

Problem 6.219 *(from [2]) Consider the system*

$$\dot{x} = f(x) + G(x)u$$
$$y = h(x)$$

with $f(x_0) = 0$. Denote its linearization around $x = x_0$, $u = 0$ by $\dot{\chi} = A\chi + Bu$ with $A = \partial f(x_0)/\partial x$ and $B = G(x_0)$. Suppose the system is exactly feedback linearizable around x_0. Show that the system can be transformed via state transformation and feedback to the linear system $\dot{z} = Az + Bv$ with A, B as given above.

Problem 6.220 *(from [2]) Consider the Hamiltonian control system*

$$\dot{q}_i = \frac{\partial H(q,p)}{\partial p_i}, \quad \dot{p}_i = -\frac{\partial H(q,p)}{\partial q_i} + u_i$$

where $i = 1, \ldots, n$ and the Hamiltonian is

$$H(q,p) = \tfrac{1}{2}p^T G(q)p + V(q)$$

with $G(q)$ a positive definite matrix for each q and $\partial V(q_0)/\partial q = 0$. Check feedback linearizability about the point $(q_0, 0)$.

Problem 6.221 (Overhead crane, continued) *Consider the overhead crane of Problems (4.131) and (5.153).*

(a) *Design a feedback linearizing control that steers the payload to a specified location, $x = x_d$, $z = z_d$, in the plane with the arm pointing straight down, $\phi = 0$. Use all three control inputs.*

(b) *Compute the zero dynamics.*

(c) *Specify numerical values for the parameters and simulate the closed loop behavior.*

Problem 6.222 *([12], Example 6.14) Consider the system*

$$\dot{x}_1 = x_1^2 x_2$$
$$\dot{x}_2 = 3x_2 + u$$

(a) *Taylor linearize the system at the origin $u = 0, x_1 = 0, x_2 = 0$ and show that the linear system is not controllable. Examine the controllability of the nonlinear system and explain your findings.*

(b) *Test to see if the system is exactly feedback linearizable. If so, compute the normal form transformation and feedback linearizing control. What's wrong with this picture?*

(c) *Consider the output $y = -2x_1 - x_2$. Find the input–output linearizing control. Obtain the zero dynamics and evaluate their stability. Can the system be asymptotically stabilized?*

Problem 6.223 (Stabilizing Nonminimum Phase Systems) *Consider the following system*

$$\dot{x}_1 = -x_1^3 x_2^2 + (2 + x_1^2)u$$
$$\dot{x}_2 = x_3$$
$$\dot{x}_3 = -x_3^3 + x_2^5 - x_1 x_2$$

(a) *Transform the system to normal form and determine the feedback linearizing control law. Show that the feedback linearized system is*

$$\dot{z}_1 = v$$
$$\dot{z}_2 = z_3$$
$$\dot{z}_3 = -z_1 z_2 + z_2^5 - z_3^3$$

where v is the new control.

(b) *Determine if the zero dynamics have a stable origin.*

(c) *Now, stabilize the system using a two-step (backstepping) procedure:*

(1) *Consider z_1 to be a psuedo-control and use it to stabilize the zero dynamics. Hint: Consider the Lyapunov function (why this function?)*

$$V_0 = \tfrac{1}{2}z_3^2 + \tfrac{1}{6}z_2^6$$

for the system

$$\dot{z}_2 = z_3$$
$$\dot{z}_3 = -\mu z_2 + z_2^5 - z_3^3$$

and pick $\mu = \mu(z_2, z_3)$ to insure stability.

(2) *Choose v to stablize the full system using the Lyapunov function*

$$V = V_0 + \tfrac{1}{2}\left(z_1 - \mu(z_2, z_3)\right)^2$$

(d) *What is the final control, $u(x_1, x_2, x_3)$?*

Problem 6.224 (Issues with Decoupling) *Consider the system*

$$\dot{x}_1 = x_1^3 + x_1 x_2 + (1 + x_2)u$$
$$\dot{x}_2 = -x_2 + (1 + 2x_2)^2 x_1$$

$$y = x_1$$

(a) Put the system in normal form and show that the zero dynamics are globally exponentially stable.

(b) Use a feedback linearizing control law to stablize the linearizable part so that $y \to 0$. Show, by simulation or other means, that if the initial value $z_1(0)$ (the new first state) is large enough, then the trajectory is unbounded. Thus, even though the linearized part and the zero dynamics are both globally exponentially stable, the closed loop system is not globally stable.

Problem 6.225 (Tracking Contol) *Consider a SISO system*

$$\dot{x} = f(x) + g(x)u, \quad y = h(x)$$

Assume the system has well-defined relative degree r, so that the feedback linearized input-output dynamics are:

$$y^{(r)} = v$$

Suppose $y_R(t)$ is a smooth (continuous derivatives up to $y_R^{(r)}$ will suffice) reference trajectory.

(a) Design a tracking controller based on stabilizing the error dynamics.

(b) Consider the following flexible joint robot:

$$I\ddot{q}_1 + mgl \sin q_1 + k(q_1 - q_2) = 0$$

$$J\ddot{q}_2 - k(q_1 - q_2) = u$$

Take $I = 1$, $mgl = 1$, $J = 1$, $k = 10$. Design a tracking controller so that $q_1(t)$ tracks $y_R(t) = 1 - e^{-t/5}$.

(c) Add friction to the joint

$$I\ddot{q}_1 + mgl \sin q_1 + k(q_1 - q_2) + c(\dot{q}_1 - \dot{q}_2) = 0$$

$$J\ddot{q}_2 - k(q_1 - q_2) - c(\dot{q}_1 - \dot{q}_2) = u$$

Take $c = 0.05$ and repeat the above design.

(d) Apply both controllers to the system with friction and discuss your results.

6.11 References

1. Isidori, A., Nonlinear Control Systems. 3 ed. 1995, London: Springer-Verlag.

2. Nijmeijer, H. and H.J. van der Schaft, Nonlinear Dynamical Control Systems. 1990, New York: SpringerVerlag.

3. Hermann, R. and A.J. Krener, Nonlinear Controllability and Observability. IEEE Transactions on Automatic Control, 1977. 22(5): p. 728-740.

4. Vidyasagar, M., Nonlinear Systems Analysis. 2 ed. 1993, Englewood Cliffs: Prentice Hall.

5. Von Westenholz, C., Differential Forms in Mathematical Physics. 1981, New York: North-Holland Publishing Co.

6. Brockett, R.W. Feedback Invariants for Nonlinear Systems. Proceedings IFAC World Congress. 1978. Helsinki.

7. Hirschorn, R.M., Invertibility of Multivariable Nonlinear Control Systems. IEEE Transactions on Automatic Control, 1979. AC-24(6): p. 855-865.

8. Singh, S.N., Decoupling of Invertible Nonlinear Systems with State Feedback and Precompensation. IEEE Transactions on Automatic Control, 1980. AC-25(6): p. 1237-1239.

9. Singh, S.N., A Modified Algorithm for Invertibility in Nonlinear Systems. IEEE Transactions on Automatic Control, 1981. AC-26(2): p. 595-598.

10. Singh, S.N., Asymptotic Reproducibility in Nonlinear Systems and Attitude Control of a Gyrostat. IEEE Transactions on Aerospace and Electronics, 1984. 20(2): p. 94.

11. Descusse, J. and C.H. Moog, Decoupling with Dynamic Compensation for Strong Invertible Affine Nonlinear Systems. International Journal of Control, 1985. 42(6): p. 13871398.

12. Slotine, J-J. E., and W. Li, Applied Nonlinear Control. 1991, Englewood Cliffs: Prentice Hall.

Chapter 7

Robust and Adaptive Control Systems

7.1 Introduction

Developments in nonlinear geometric control theory have had a substantial impact on the theory and design of robust and adaptive controls for nonlinear systems. We will describe some of the more established approaches and the associated computations. Our primary interest, in this chapter and the next, is the application of feedback linearizaton methods to uncertain systems.

Any model-based control design method is vulnerable to model errors. This is obviously a concern with feedback linearization techniques since exact cancellation is a basic ingredient of the method. On the other hand, models used for control system design are never infinitely precise. Thus, it is necessary to investigate the impact of model uncertainty on closed loop system performance and to devise methods for insuring adequate performance when a controller is applied to systems that deviate from the design model.

Model uncertainty is generally of two types. *Unmodeled dynamics* refers to dynamics that are neglected because they act on a time scale (typically) much faster than the time scale of interest. *Functional uncertainty* means uncertainty in the functions f, G, h that define the affine differential equation model. The effect of unmodeled dynamics is often analyzed using singular perturbation or averaging methods. While unmodeled dynamics are extremely important, our focus will be on functional uncertainty. Normally, functional uncertainty is characterized by perturbing a nominal (certain) part by an appropriately bounded, but otherwise unspecified function. In some cases the uncertainty is characterized in terms of an uncertain parameter.

Feedback linearization methods typically begin with the transformation of the design model to a normal form. Thus, we begin, in Section 2, with an investigation of the consequences of applying a state transformation derived on the basis of a nominal model to a perturbation of it. When the uncertainties satisfy certain structural conditions, the transformed system assumes a triangular form that can facilitate analysis. The design of robust stabilizers, i.e., controllers that guarantee closed loop stability for all admissible perturbations of a nominal plant, for systems with matched uncertainties is considered in Section 3 using the method of *Lyapunov redesign*. In Section 4 we consider the design of robust stabilizing controllers for systems with a class of nonmatched uncertainties. The design we describe employs a *backstepping* method. Backstepping will reappear with some variation in our discussion of adaptive control and variable structure control.

We then turn to the case where the uncertainty can be characterized in terms of uncertain parameters. Thus, we describe the design of parameter-adaptive controls. Section 5 introduces a basic adaptive regulator. In general, this controller requires measurement of the system states and some of the transformed states. The need to measure transformed states can be avoided by using the backstepping approach discussed in Section 6. Section 7 describes an adaptive tracking controller based on dynamic inversion. Computational tools are described and illustrated for each method.

7.2 Perturbations of Feedback Linearizable Systems

We will examine perturbations of systems of known relative degree. There are several important applications of such an analysis. For example, a perturbation may arise as an uncertainty applied to a nominal system. Or it may simply be convenient, for a variety of reasons, to divide a system into a nominal system plus a perturbation in order to isolate certain terms. The main point is that under certain constraints on the perturbation, the perturbed system can be transformed into a triangular or near triangular form that can be exploited for purposes of control system design.

7.2.1 SISO Case

We will consider the 'perturbed' SISO system

$$\begin{aligned}
\dot{x} &= f(x) + \varphi(x) + [g(x) + \gamma(x)]\, u \\
y &= h(x)
\end{aligned} \tag{7.1}$$

where $x \in R^n$, $u \in R$, $y \in R$ and $\varphi(x)$, $\gamma(x)$ represent a perturbation applied to the nominal system (f, g, h). Previously, we discussed the reduction of the nominal system to normal form via a transformation of coordinates. Under certain conditions, the nominal transformation when applied to the perturbed system still produces a useful form of the system equations. The following definitions apply constraints on the structure of the perturbation $\varphi(x)$, along the lines of [1]. Recall

$$G_i = \text{span} \left\{ g, \ldots, \text{ad}_f^i g \right\}, \quad 0 \le i \le n - 1$$

Definition 7.226 *Suppose the system (7.1) is of relative degree r. We say that the perturbation, $\varphi(x)$ satisfies:*

1. *the* triangularity condition *if* $\text{ad}_\varphi G_i \in G_{i+1}$, $0 \le i \le r - 3$

2. *the* strict triangularity condition *if* $\text{ad}_\varphi G_i \in G_i$, $0 \le i \le r - 2$

3. *the* extended matching condition *if* $\varphi \in G_1$

4. *the* matching condition *if* $\varphi \in G_0$

Notice that these conditions are listed from weakest to strongest, i.e.,

$$matching \Rightarrow extended\ matching \Rightarrow strict\ triangularity \Rightarrow triangularity$$

Consider the following result which perturbs only the drift term.

Proposition 7.227 *Assume*

1. *the nominal system (f, g, h) has relative degree r at the $x_0 \in R^n$.*

2. *the perturbed system $(f + \varphi, g, h)$ satisfies the* strict triangularity *assumption on a neighborhood of x_0: $\text{ad}_\varphi G_i \subset G_i$, $0 \le i \le r - 2$.*

There exists a local transformation on a neighborhood of x_0 that reduces the perturbed system to

$$\dot{\xi} = \widehat{F}(\xi, z), \ \xi \in R^{n-r}$$
$$\dot{z}_1 = z_2 + \phi_1(\xi, z_1)$$
$$\dot{z}_2 = z_3 + \phi_2(\xi, z_1, z_2)$$
$$\vdots$$
$$\dot{z}_r = \alpha(x(\xi, z)) + \phi_r(\xi, z_1, \ldots, z_r) + \rho(x(\xi, z))u$$

Proof: The nominal system has relative degree r. Then there exists a transformation $x \mapsto (\xi, z)$, $\xi \in R^{n-r}$, $z_i \in R$ with $z_i(x) = L_f^{i-1} h(x)$, $1 \le i \le r$, that takes the nominal system to

$$\dot{\xi} = F(\xi, z)$$
$$\dot{z}_1 = z_2$$
$$\vdots$$
$$\dot{z}_r = L_f^r h(x(\xi, z)) + L_g L_f^{r-1} h(x(\xi, z))u$$

Now, we apply this transformation to the system (7.1). First compute

$$
\dot{\xi} = \left.\frac{\partial \xi}{\partial x}\dot{x}\right|_{x \mapsto (\xi, z)} = L_f \xi(x) + L_\varphi \xi(x) + L_g \xi(x)u|_{x \mapsto (\xi, z)}
$$
$$
= F(\xi, z) + L_\varphi \xi(x)|_{x \mapsto (\xi, z)}
$$
$$
= \widehat{F}(\xi, z)
$$

Now compute

$$\dot{z}_1 = \dot{y} = L_f h(x) + L_\varphi h(x) + L_g h(x)u = z_2 + L_\varphi h(x)$$
$$\dot{z}_2 = L_f^2 h(x) + L_\varphi L_f h(x) + L_g L_f h(x)u = z_3 + L_\varphi L_f h(x)$$
$$\vdots$$
$$\dot{z}_r = L_f^r h(x) + L_\varphi L_f^{r-1} h(x) + L_g L_f^{r-1} h(x)u$$

Define the function $\phi(\xi, z)$

$$
\begin{bmatrix}
L_\varphi h(x(\xi, z)) \\
L_\varphi L_f h(x(\xi, z)) \\
\vdots \\
L_\varphi L_f^{r-1} h(x(\xi, z))
\end{bmatrix} = \phi(\xi, z)
$$

so that we can write

$$\dot{z}_1 = z_2 + \phi_1(\xi, z))$$
$$\dot{z}_2 = z_3 + \phi_2(\xi, z)$$
$$\vdots$$
$$\dot{z}_r = L_f^r h(x) + \phi_r(\xi, z) + L_g L_f^{r-1} h(x)u$$

So, we see that under transformation $\varphi(x) \mapsto \phi(\xi, z)$.

It is necessary to establish the triangular dependence of ϕ on $z_1, \ldots, z_r$. We do the required calculations in the transformed $- (\xi, z) -$ coordinates. Under transformation f and g become

$$
f(x) \mapsto \widehat{f}(\zeta) =
\begin{bmatrix}
F(\xi, z) \\
z_2 \\
\vdots \\
z_r \\
L_f^r h(x(\xi, z))
\end{bmatrix}, \quad
g(x) \mapsto \widehat{g}(\zeta) =
\begin{bmatrix}
0 \\
\vdots \\
\vdots \\
0 \\
L_g L_f^{r-1} h(x(\xi, z))
\end{bmatrix}
$$

Thus, we see that

$$\mathrm{ad}_f g = \left[\frac{\partial \hat{g}}{\partial \zeta}\right]\hat{f} - \left[\frac{\partial \hat{f}}{\partial \zeta}\right]\hat{g} = \begin{bmatrix} 0 \\ \vdots \\ 0 \\ 0 \\ L_g L_f^r h \end{bmatrix} - \begin{bmatrix} 0 \\ \vdots \\ 0 \\ L_g L_f^{r-1} h \\ L_g L_f^r h \end{bmatrix} = -\begin{bmatrix} 0 \\ \vdots \\ 0 \\ L_g L_f^{r-1} h \\ 0 \end{bmatrix}$$

Similarly, we compute

$$\mathrm{ad}_f^i g = \begin{bmatrix} 0 \\ \vdots \\ 0 \\ (-1)^i L_g L_f^{r-1} h \\ 0 \\ \vdots \\ 0 \end{bmatrix} \leftarrow (n-i), \quad 1 \le i \le r-1$$

Thus, we have

$$\mathcal{G}_i = \mathrm{span}\left\{ \begin{bmatrix} 0 \\ \vdots \\ \vdots \\ \vdots \\ \vdots \\ 0 \\ L_g L_f^{r-1} h \end{bmatrix} \begin{bmatrix} 0 \\ \vdots \\ \vdots \\ 0 \\ -L_g L_f^{r-1} h \\ 0 \end{bmatrix} \cdots \begin{bmatrix} 0 \\ \vdots \\ 0 \\ (-1)^i L_g L_f^{r-1} h \\ 0 \\ \vdots \\ 0 \end{bmatrix} \right\}$$

$$\underbrace{\qquad\qquad\qquad\qquad\qquad\qquad\qquad\qquad}_{i \text{ terms}}$$

Now, consider

$$\mathrm{ad}_\varphi g = \frac{\partial \varphi}{\partial \zeta} g - \frac{\partial g}{\partial \zeta}\varphi = \frac{\partial \varphi}{\partial z_n} L_g L_f^{r-1} h - \begin{bmatrix} 0 \\ \vdots \\ 0 \\ \frac{\partial L_g L_f^{r-1} h}{\partial \zeta}\varphi \end{bmatrix} \in \mathrm{span}\left\{ \begin{bmatrix} 0 \\ \vdots \\ 0 \\ L_g L_f^{r-1} h \end{bmatrix} \right\}$$

This implies

$$\frac{\partial \varphi_1}{\partial z_r} = 0, \ldots, \frac{\partial \varphi_{n-1}}{\partial z_r} = 0$$

Next consider

$$\text{ad}_\varphi \text{ad}_f g = \frac{\partial \varphi}{\partial \zeta} \text{ad}_f g - \frac{\partial \text{ad}_f g}{\partial \zeta} \varphi$$

$$= \frac{\partial \varphi}{\partial z_{r-1}} L_g L_f^{r-1} h - \begin{bmatrix} 0 \\ \vdots \\ 0 \\ \frac{\partial L_g L_f^{r-1} h}{\partial \zeta} \varphi \\ 0 \end{bmatrix} \in \text{span} \left\{ \begin{bmatrix} 0 \\ \vdots \\ \vdots \\ 0 \\ L_g L_f^{r-1} h \end{bmatrix} \begin{bmatrix} 0 \\ \vdots \\ 0 \\ L_g L_f^{r-1} h \\ 0 \end{bmatrix} \right\}$$

This implies

$$\frac{\partial \varphi_1}{\partial z_{r-1}} = 0, \ldots, \frac{\partial \varphi_{n-2}}{\partial z_{r-1}} = 0$$

Continuing in this way, we find

$$\frac{\partial \varphi_1}{\partial z_{r-i}} = 0, \ldots, \frac{\partial \varphi_{n-i-1}}{\partial z_{r-i}} = 0, \quad 0 \le i \le r-2$$

providing the desired result. ∎

The above result can be easily generalized to allow perturbations in the control gain of the form $\gamma \in \mathcal{G}_0$. Although restrictive, this class of perturbations is nontheless useful in applications. Notice that for any scalar function or vector field $w(x)$, if $L_g w(x) = 0$ then if the *matching condition*, $\gamma \in \mathcal{G}_0$, is true we have $L_\gamma w(x) = 0$. Thus, in the calculations in the above proof, the only change is in the last of the transformed state equations. In fact, we have for the transformed system:

$$\dot{\xi} = \widehat{F}(\xi, z), \; \xi \in R^{n-r}$$
$$\dot{z}_1 = z_2 + \phi_1(\xi, z_1)$$
$$\dot{z}_2 = z_3 + \phi_2(\xi, z_1, z_2)$$
$$\vdots$$
$$\dot{z}_r = \alpha(x(\xi, z)) + \phi_r(\xi, z_1, \ldots, z_r) + [\rho(x(\xi, x)) + \widehat{\rho}(x(\xi, x))] u$$

where $\widehat{\rho}(x(\xi, x)) = L_\gamma L_f^{r-1} h(x(\xi, x))$.

Similar results obtain for the other conditions of Definition (7.226). The following is a summary:

Suppose the nominal part of the control system (7.1) has local relative degree r. Then the perturbation conditions of Definition (7.226) assure local transformation to triangular forms as follows:

1. the triangularity condition implies

$$\dot{\xi} = \widehat{F}(\xi, z), \; \xi \in R^{n-r}$$
$$\dot{z}_i = z_{i+1} + \phi_i(\xi, z_1, \ldots, z_{i+1}), \; 1 \le i \le r-1 \qquad (7.2)$$
$$\dot{z}_r = \alpha(x(\xi, z)) + \phi_r(\xi, z_1, \ldots, z_r) + \rho(x(\xi, z))u$$

2. the strict triangularity condition implies

$$
\begin{aligned}
\dot{\xi} &= \widehat{F}(\xi, z), \ \xi \in R^{n-r} \\
\dot{z}_i &= z_{i+1} + \phi_i(\xi, z_1, \ldots, z_i), \ 1 \le i \le r-1 \\
\dot{z}_r &= \alpha(x(\xi, z)) + \phi_r(\xi, z_1, \ldots, z_r) + \rho(x(\xi, z))u
\end{aligned}
\tag{7.3}
$$

3. the extended matching condition implies

$$
\begin{aligned}
\dot{\xi} &= \widehat{F}(\xi, z), \ \xi \in R^{n-r} \\
\dot{z}_i &= z_{i+1}, \qquad 1 \le i \le r-2 \\
\dot{z}_{r-1} &= z_r + \phi_{r-1}(\xi, z_1, \ldots, z_r) \\
\dot{z}_r &= \alpha(x(\xi, z)) + \phi_r(\xi, z_1, \ldots, z_r) + \rho(x(\xi, z))u
\end{aligned}
\tag{7.4}
$$

4. the matching condition implies

$$
\begin{aligned}
\dot{\xi} &= F(\xi, z), \ \xi \in R^{n-r} \\
\dot{z}_i &= z_{i+1}, \qquad 1 \le i \le r-1 \\
\dot{z}_r &= \alpha(x(\xi, z)) + \phi_r(\xi, z_1, \ldots, z_r) + \rho(x(\xi, z))u
\end{aligned}
\tag{7.5}
$$

Example 7.228 (Strict Triangularity) *Consider the following system:*

In[219]:= $f = \{x2, x3 - x1 - x2, x4, x1 - x3 - (1/2)x4\};$
$g = \{0, 0, 0, 1\};$
$h = x1;$

In[220]:= $DF = \{0, -DF1[x1, x2], 0, -DF2[x3, x4]\};$

First, let us show that the uncertainty satisfies the strict triangularity condition. To do this, we compute the relative degree,

In[221]:= *VectorRelativeOrder*$[f, g, \{h\}, \{x1, x2, x3, x4\}]$
Out[221]= $\{4\}$

and the required distributions $\mathcal{G}_0$, $\mathcal{G}_1$ *and* $\mathcal{G}_2$:

In[222]:= $G0 = Span[\{g\}]$
$G1 = Span[\{g, Ad[f, g, \{x1, x2, x3, x4\}, 1]\}]$
$G2 = Span[Join[G1, \{Ad[f, g, \{x1, x2, x3, x4\}, 2]\}]]$
Out[222]= $\{\{0, 0, 0, 1\}\}$
Out[222]= $\{\{0, 0, 1, 0\}, \{0, 0, 0, 1\}\}$
Out[222]= $\{\{0, 1, 0, 0\}, \{0, 0, 1, 0\}, \{0, 0, 0, 1\}\}$

Now, test the uncertainty

In[223]:= $Map[Ad[DF, \#, \{x1, x2, x3, x4\}, 1]\&, G0]$
Out[223]= $\{\{0, 0, 0, DF2^{(0,1)}[x3, x4]\}\}$

In[224]:= $Map[Ad[DF, \#, \{x1, x2, x3, x4\}, 1]\&, G1]$
Out[224]= $\{\{0, 0, 0, DF2^{(1,0)}[x3, x4]\}, \{0, 0, 0, DF2^{(0,1)}[x3, x4]\}\}$

In[225]:= $Map[Ad[DF, \#, \{x1, x2, x3, x4\}, 1]\&, G2]$

Out[225]= $\{\{0, \mathrm{DF1}^{(0,1)}[\mathrm{x1}, \mathrm{x2}], 0, 0\},$
$\{0, 0, 0, \mathrm{DF2}^{(1,0)}[\mathrm{x3}, \mathrm{x4}]\}, \{0, 0, 0, \mathrm{DF2}^{(0,1)}[\mathrm{x3}, \mathrm{x4}]\}\}$

and notice that the conditions are indeed satisifed. The transformation that places the nominal system is obtained:

In[226]:= $\{T1, T2\} = SISONormalFormTrans[f, g, x1, \{x1, x2, x3, x4\}]$
Out[226]= $\{\{\mathrm{x1}, \mathrm{x2}, -\mathrm{x1} - \mathrm{x2} + \mathrm{x3}, \mathrm{x1} - \mathrm{x3} + \mathrm{x4}\}, \{\}\}$

Then, its inverse.

In[227]:= $InvTrans = InverseTransformation[\{x1, x2, x3, x4\}, \{z1, z2, z3, z4\}, T1];$
"InverseTransformation : " $\{\mathrm{x1}, \mathrm{x2}, \mathrm{x3}, \mathrm{x4}\}$" = "$\{\mathrm{z1}, \mathrm{z2}, \mathrm{z1} + \mathrm{z2} + \mathrm{z3}, \mathrm{z2} + \mathrm{z3} + \mathrm{z4}\}$

Application of the transformation to the nominal system confirms the reduction to normal form.

In[228]:= $\{fnew, gnew, hnew\} = TransformSystem[f, g, h, \{x1, x2, x3, x4\},$
$\{z1, z2, z3, z4\}, T1, InvTrans]$
Out[228]= $\left\{\left\{\mathrm{z2}, \mathrm{z3}, \mathrm{z4}, -\mathrm{z3} - \dfrac{3}{2} \ (\mathrm{z2} + \mathrm{z3} + \mathrm{z4})\right\}, \{0, 0, 0, 1\}, \mathrm{z1}\right\}$

Now, apply the transformation to the perturbed system, to obtain:

In[229]:= $\{ff, gg, hh\} = TransformSystem[f + DF, g, h, \{x1, x2, x3, x4\},$
$\{z1, z2, z3, z4\}, T1, InvTrans];$

In[230]:= $ff // MatrixForm$

Out[230]=
$$\begin{pmatrix} \mathrm{z2} \\ \mathrm{z3} - DF1[\mathrm{z1}, \mathrm{z2}] \\ \mathrm{z4} + DF1[\mathrm{z1}, \mathrm{z2}] \\ \dfrac{1}{2} \ (-3 \ \mathrm{z2} - 5 \ \mathrm{z3} - 3 \ \mathrm{z4} - 2 \ DF2[\mathrm{z1} + \mathrm{z2} + \mathrm{z3}, \mathrm{z2} + \mathrm{z3} + \mathrm{z4}]) \end{pmatrix}$$

The system is indeed in the strict triangular form anticipated by Proposition (7.227).

The perturbation is readily modified so that the strict triangularity assumption fails.

In[231]:= $DF = \{0, -DF1[x1, x2, x3], 0, -DF2[x4]\};$

In[232]:= $Map[Ad[DF, \#, \{x1, x2, x3, x4\}, 1]\&, G0]$
Out[232]= $\{\{0, 0, 0, \mathrm{DF2}'[\mathrm{x4}]\}\}$

In[233]:= $Map[Ad[DF, \#, \{x1, x2, x3, x4\}, 1]\&, G1]$
Out[233]= $\{\{0, \mathrm{DF1}^{(0,0,1)}[\mathrm{x1}, \mathrm{x2}, \mathrm{x3}], 0, 0\}, \{0, 0, 0, \mathrm{DF2}'[\mathrm{x4}]\}\}$

In[234]:= $\{ff, gg, hh\} = TransformSystem[f + DF, g, h, \{x1, x2, x3, x4\},$
$\{z1, z2, z3, z4\}, T1, InvTrans];$

In[235]:= $ff // MatrixForm$

Out[235]=
$$\begin{pmatrix} \mathrm{z2} \\ \mathrm{z3} - DF1[\mathrm{z1}, \mathrm{z2}, \mathrm{z1} + \mathrm{z2} + \mathrm{z3}] \\ \mathrm{z4} + DF1[\mathrm{z1}, \mathrm{z2}, \mathrm{z1} + \mathrm{z2} + \mathrm{z3}] \\ \dfrac{1}{2} \ (-3 \ \mathrm{z2} - 5 \ \mathrm{z3} - 3 \ \mathrm{z4} - 2 \ DF2[\mathrm{z2} + \mathrm{z3} + \mathrm{z4}]) \end{pmatrix}$$

In this case we do not achieve strict triangular reduction of the perturbed system because $\mathrm{ad}_\varphi \mathcal{G}_1 \notin \mathcal{G}_1$. However, we do have a (non-strict) triangular form since $\mathrm{ad}_\varphi \mathcal{G}_1 \in \mathcal{G}_2$ (as well as $\mathrm{ad}_\varphi \mathcal{G}_0 \in \mathcal{G}_1$).

7.2.2 MIMO Case

We now turn to the multi-input multi-output system with uncertainty in the drift term

$$\begin{aligned}\dot{x} &= f(x) + \varphi(x) + G(x)u \\ y &= h(x)\end{aligned} \tag{7.6}$$

where $x \in R^n$, $u \in R^m$, $y \in R^m$. Define the distributions

$$\mathcal{G}_i = \mathrm{span}\left\{\mathrm{ad}_f^j g_k \,|\, 0 \le j \le i, 1 \le k \le m\right\}, \quad 0 \le i \le n-2$$

Then Definition (7.226) can be adapted to the MIMO case.

Definition 7.229 *Suppose the system (7.1) is of (vector) relative degree* $\mathbf{r} = \{r_1, \ldots, r_m\}$, *with* $r = r_1 + \cdots + r_2$. *We say that the perturbation satisfies:*

1. *the* triangularity condition *if* $\mathrm{ad}_\varphi \mathcal{G}_i \in \mathcal{G}_{i+1}$, $0 \le i \le r-3$

2. *the* strict triangularity condition *if* $\mathrm{ad}_\varphi \mathcal{G}_i \in \mathcal{G}_i$, $0 \le i \le r-2$

3. *the* extended matching condition *if* $\varphi \in \mathcal{G}_1$

4. *the* matching condition *if* $\varphi \in \mathcal{G}_0$

The following result generalizes Proposition (7.227).

Proposition 7.230 *Suppose the nominal part of the control system (7.6) has vector relative degree* $\{r_1, \ldots, r_m\}$ *at* x_0 *with* $r_1 \ge r_2 \ge \ldots \ge r_m$ *and* $r = r_1 + \ldots + r_m$. *Moreover assume that the strict triangularity condition applies. Then there exists a local transformation of coordinates such that the perturbed equations take the form:*

$$\begin{aligned}\dot{\xi} &= \widehat{F}(\xi, z, u), \quad \xi \in R^{r-1} \\ \dot{z}_j^i &= z_{j+1}^i + \phi_j^i(\xi, z_{r_1-r_i+1}^1, \ldots, z_{r_1-r_i+j}^1, \ldots, z_{r_m-r_i+1}^m, \ldots, z_{r_m-r_i+j}^m), \\ & \qquad 1 \le j \le r_i - 1 \\ \dot{z}_{r_i}^i &= \alpha_i(x(\xi, z)) + \rho_i.(x(\xi, z))u + \phi_j^i(\xi, z^1, \ldots, z^m) \\ & \qquad 1 \le i \le m\end{aligned}$$

on a neighborhood of x_0.

Proof: The proof proceeds precisely as in Proposition (7.227), although the calculations are considerably more tedious. First, as in Proposition (7.227), we compute

$$\dot{\xi} = \frac{\partial \xi}{\partial x}\dot{x} = F(\xi, z, u) + L_\varphi \xi(x)|_{x \mapsto (\xi, z)} = \widehat{F}(\xi, z, u)$$

Now, compute

$$\dot{z}_1^i = z_2^i + L_\varphi h_i(x)$$
$$\vdots$$
$$\dot{z}_{r_i}^i = L_f^{r_i} h_i(x) + L_\varphi L_f^{r_i - 1} h_i(x) + \sum_{j=1}^{m} L_g L_f^{r_i - 1} h_i(x) u_j$$

and define the functions

$$\phi^i(\xi, z) = \begin{bmatrix} L_\varphi h_i(x(\xi, z)) \\ L_\varphi L_f h_i(x(\xi, z)) \\ \vdots \\ L_\varphi L_f^{r_i - 1} h_i(x(\xi, z)) \end{bmatrix}$$

for $i = 1, \ldots, m$. In (ξ, z) coordinates

$$g_k = \begin{bmatrix} 0 \\ L_{g_k} L_f^{r_1 - 1} h_1 \\ \vdots \\ 0 \\ L_{g_k} L_f^{r_m - 1} h_m \end{bmatrix} \qquad \begin{matrix} \leftarrow \text{ row } r_1 \\ \vdots \\ \vdots \\ \leftarrow \text{ row } n = r_1 + \cdots + r_m \end{matrix}$$

and

$$\mathrm{ad}_f^j g_k = \begin{bmatrix} 0 \\ L_{g_k} L_f^{r_1 - 1} h_1 \\ \vdots \\ 0 \\ L_{g_k} L_f^{r_m - 1} h_m \\ 0 \end{bmatrix} \qquad \begin{matrix} \leftarrow \text{ row } r_1 - j \\ \vdots \\ \vdots \\ \leftarrow \text{ row } n - j = r_1 + \cdots + r_m - j \end{matrix} \quad , \quad 1 \le j \le r_1 - 1$$

Notice that

$$\mathcal{G}_1 = \mathrm{span} \left\{ \begin{bmatrix} \vdots \\ 0 \\ 1 \\ 0 \\ \vdots \end{bmatrix}, \begin{bmatrix} \vdots \\ 0 \\ 0 \\ 1 \\ 0 \end{bmatrix}, \ldots, \begin{bmatrix} \vdots \\ 0 \\ \vdots \\ 0 \\ 1 \end{bmatrix} \right\} \qquad \begin{matrix} \leftarrow \text{ row } r_1 \\ \\ \leftarrow \text{ row } r_1 + r_2 \\ \vdots \\ \leftarrow \text{ row } r_1 + \cdots + r_m = r \end{matrix}$$

$$
\mathcal{G}_2 = \text{span} \left\{ \begin{bmatrix} \vdots \\ 0 \\ 1 \\ 0 \\ \vdots \\ 0 \\ \vdots \end{bmatrix}, \begin{bmatrix} \vdots \\ 0 \\ 0 \\ 1 \\ \vdots \\ 0 \\ \vdots \\ 1 \end{bmatrix}, \dots, \begin{bmatrix} \vdots \\ 0 \\ 0 \\ \vdots \\ 0 \\ 1 \end{bmatrix}, \begin{bmatrix} 0 \\ 1 \\ 0 \\ \vdots \\ 1 \\ 0 \\ \vdots \end{bmatrix}, \begin{bmatrix} \vdots \\ 0 \\ 1 \\ 0 \\ \vdots \\ 0 \\ 1 \\ 0 \end{bmatrix}, \dots, \begin{bmatrix} \vdots \\ 0 \\ \vdots \\ 0 \\ 1 \\ 0 \end{bmatrix} \right\} \begin{array}{l} \leftarrow \text{row } r_1 - 1 \\ \\ \leftarrow \text{row } r_1 + r_2 - 1 \\ \vdots \\ \leftarrow \text{row } r_1 + \cdots + r_m - 1 \end{array}
$$

and so on.

Now, as in Proposition (7.227), we apply the triangularity asumption to obtain:

$$
\frac{\partial \phi_1^i}{\partial z_{r_j-k}^j} = 0, \dots, \frac{\partial \phi_{r_j-k-1}^i}{\partial z_{r_j-k}^j} = 0, \quad 0 \le k \le r_i - 2, \quad 1 \le i,j \le m
$$

These relations establish the conclusion of the theorem.   ∎

As in the SISO case uncertainty in the control input matrix of the matched type is easily accommodated. Consider a perturbation of the form $G(x) \rightarrow G(x) + \Gamma(x)$, $\Gamma(x) = [\gamma_1(x) \cdots \gamma_m(x)]$ and $\gamma_i \in \mathcal{G}_0$ for each $i = 1, \dots m$. Then the transformed equations are unchanged other than

$$
\rho_{ij} = L_{g_j} L_f^{r_i-1} h_i + L_{\gamma_j} L_f^{r_i-1} h_i
$$

7.3   Lyapunov Redesign for Matched Uncertainty

Very often a control system is designed on the basis of a nominal model. If the uncertainty is confined to a suitably characterized admissible class, it may be possible to augment the nominal control with a robustifying component that insures asymptotic stability for all admissible uncertainties. When this is accomplished using Lyapunov methods the technique is referred to as *Lyapunov redesign* (see, for example, [2], Chapter 5).

Suppose that the multi-input system (7.6) has well defined relative degree with $r = n$ (i.e., the exact state linearizable case) and the uncertainty satisfies the matching condition. Then, it is reducible, by state transformation, to the form

$$
\dot{z} = Az + E[\alpha(z) + \Delta(z, u, t) + \rho(z)u] \tag{7.7}
$$

A nominal feedback control designed on the basis of the feedback linearization approach is

$$
u^*(z) = \rho^{-1}(z) \{-\alpha(z) + Kz\} \tag{7.8}
$$

where K is chosen such that $(A + EK)$ is Hurwitz. Thus, the nominal closed loop system is described by the equation

$$
\dot{z} = (A + EK)z \tag{7.9}
$$

Moreover, it may be associated with a Lyapunov function $V(z) = z^T P z$, where P satisfies the Lyapunov equation

$$P(A + EK) + (A + EK)^T P = -Q - I, \quad Q = Q^T > 0 \qquad (7.10)$$

and $\dot{V} = -z^T Q z - \|z\|^2 < 0$ along trajectories of the closed loop nominal system. In the sequal it will become apparent why it is convenient to require the right hand side of (7.10) to be more negative than $-I$.

The nominal control u^* does provide some protection against plant unertainty. Indeed, along trajectories of (7.7) with $u = u^*$, we have

$$\dot{V} = -z^T Q z - \|z\|^2 + 2z^T P E \Delta \qquad (7.11)$$

If the uncertainty has a bound $\|\Delta\| \leq \gamma \|z\|$, with constant $\gamma \geq 0$, then

$$\dot{V} \leq - (\lambda_{\min}(Q) + 1) \|z\|^2 + 2\gamma \|PE\| \|z\|^2$$

Thus, stability is assured provided $\gamma < (\lambda_{\min}(Q) + 1) / 2 \|PE\|$.

Now, consider the system with uncertainty and apply a control $u = u^* + \rho^{-1}\mu$, that includes a 'robustifying' component, μ intended to compensate for the uncertainty. Assume that the uncertainty satisfies the condition $\Delta(0, t) = 0$, $\forall t$, and the bounding condition

$$\left\| \Delta(z, u^* + \rho^{-1}\mu, t) \right\| \leq \sigma(z) \|z\| + k \|\mu\|, \quad 0 \leq k < 1 \qquad (7.12)$$

for some known, smooth bounding function $\sigma(z) > 0$. We wish to choose μ so that the closed loop is asymptotically stable for any admissible uncertainty. The actual system closed loop equation is

$$\dot{z} = (A + EK) z + E \left(\mu + \Delta(z, u^*(z) + \rho^{-1}\mu, t)\right) \qquad (7.13)$$

The derivative of $V(x)$ along trajectories of (7.13) is

$$\dot{V} = -z^T Q z - \|z\|^2 + 2z^T P E (\mu + \Delta) \qquad (7.14)$$

Notice that the first two terms arise from the nominal system and the next is due to the uncertainty, Δ, and the control, μ, that is intended to compensate for it. For the time being, write $w^T = z^T P E$ and observe that the design objective is achieved if μ can be chosen such that $- \|z\|^2 + w^T (\mu + \Delta) \leq 0$.

In view of (7.12)

$$\begin{aligned} w^T \mu + w^T \Delta \;&\leq w^T \mu + \|w\| \|\Delta\| \\ &\leq w^T \mu + \|w\| [\sigma(z) \|z\| + k \|\mu\|] \end{aligned} \qquad (7.15)$$

For now, let us proceed in a fashion that leads to a smooth control. Set $\mu = -w\kappa$, where $\kappa(z) > 0$ is a scalar valued function not yet defined. Then

$$\begin{aligned} - \|z\|^2 + w^T \mu + w^T \Delta \;&\leq - \|z\|^2 - \|w\|^2 \kappa + \|w\| [\sigma(z) \|z\| + k \|w\| \|\kappa\|] \\ &\leq - \|w\|^2 \kappa (1 - k) + \|w\| \sigma(z) \|z\| - \|z\|^2 \end{aligned}$$

Now choose

$$\kappa = \tfrac{1}{4(1-k)}\sigma^2$$

so that

$$
\begin{aligned}
-\|z\|^2 + w^T\mu + w^T\Delta &\le -\|w\|^2 \tfrac{1}{4}\sigma^2(z) + \|w\|\,\sigma(z)\,\|z\|\,1\,\|z\|^2 \\
&= -\left(\tfrac{1}{2}\|w\|\,\sigma(z) - \|z\|\right)^2
\end{aligned}
$$

Consequently, a robust control is achieved with $\kappa > \tfrac{1}{4(1-k)}\sigma^2$. In particular, suppose $\sigma_0 > 0$ is a constant and take

$$\mu = -\left(\sigma_0 + \frac{\sigma^2(z)}{4(1-k)}\right) w = -\left(\sigma_0 + \frac{\sigma^2(z)}{4(1-k)}\right) E^T P z \tag{7.16}$$

These calculations establish the following proposition.

Proposition 7.231 *Consider the uncertain system (7.7) and assume it has an isolated equilibrium point at the origin. Suppose that a nominal control u^*, given by (7.8), is associated with the Lyapunov function $V(z) = z^T P z$, where P satisfies (7.10). Then the control $u = u^* + \rho^{-1}\mu$, where*

$$\mu = -\left(\sigma_0 + \frac{\sigma^2(z)}{4(1-k)}\right) E^T P z, \quad \sigma_0 > 0 \tag{7.17}$$

globally stabilizes the origin of (7.7) for all uncertainties that satisfy (7.12).

Proof: Taking $V = z^T P z$, direct calculation as above leads to

$$\dot{V} = -z^T Q z - \sigma_0 \left\|E^T P z\right\|^2 - \left(\tfrac{1}{2}\left\|E^T P z\right\|\sigma(z) - \|z\|\right)^2$$

Thus, $\dot{V} < 0$ everywhere except at the origin $z = 0$. ∎

Remark 7.232 *As an alternative to the smooth control given above, we could proceed, as in [2], to design a discontinuous control. For example, choose*

$$\mu = -\frac{\eta(z)}{1-k}\frac{w}{\|w\|}, \quad \eta(z) \ge \sigma(z)\,\|z\| \tag{7.18}$$

A simple computation verifies that this control achieves $\dot{V} < 0$,

$$
\begin{aligned}
w^T\mu + w^T\Delta &\le -\tfrac{\eta(z)}{1-k}\|w\| + \sigma(z)\,\|z\|\,\|w\| + \tfrac{k\eta(z)}{1-k}\|w\| \\
&\le -\eta(z) + \sigma(z)\,\|z\|
\end{aligned}
\tag{7.19}
$$

However, there are subtleties with discontinuous controls and we will consider them fully in the next chapter.

Remark 7.233 *The second term in the uncertainty bounding condition (7.12) can be interpreted in the following way. Suppose that both the nominal system*

$$\dot{x} = f(x) + G(x)u$$

and the actual (perturbed) system

$$\dot{x} = (f(x) + \varphi(x)) + (G(x) + \mu(x))\, u$$

are both exactly feedback linearizable. Then, they are respectively reducible to the normal forms

$$\dot{\widehat{z}} = A\widehat{z} + E\left[\widehat{\alpha}(\widehat{z}) + \widehat{\rho}(\widehat{z})u\right]$$

and

$$\dot{z} = Az + E[\alpha(z) + \rho(z)u]$$

The feedback linearizing control for the nominal system is $u = \widehat{\rho}^{-1}\left[-\widehat{\alpha} + v\right]$. *Using this control, the perturbed system can be expressed*

$$\begin{aligned}
\dot{z} &= Az + E\left[\alpha + \rho\left(\widehat{\rho}^{-1}\left[-\widehat{\alpha} + v\right]\right)\right] \\
&= Az + E\left[v + \left(\alpha - \rho\widehat{\rho}^{-1}\widehat{\alpha} + \rho\widehat{\rho}^{-1}v - v\right)\right]
\end{aligned}$$

In the nominal case, $\widehat{\rho} = \rho$, $\widehat{\alpha} = \alpha$, *the equation reduces to*

$$\dot{z} = Az + Ev$$

As a result, we identify

$$\Delta(z,v) = \alpha - \rho\widehat{\rho}^{-1}\widehat{\alpha} + \left[\rho\widehat{\rho}^{-1} - I\right]v$$

Now choose $v = v^* + \mu$, $v^* = Kz$ *so that*

$$\Delta(z, v^* + \mu) = \alpha - \rho\widehat{\rho}^{-1}\widehat{\alpha} + \left[\rho\widehat{\rho}^{-1} - I\right]Kz + \left[\rho\widehat{\rho}^{-1} - I\right]\mu$$

Thus,

$$\left\|\Delta(z, v^* + \mu)\right\| \le \left\|\alpha - \rho\widehat{\rho}^{-1}\widehat{\alpha} + \left[\rho\widehat{\rho}^{-1} - I\right]Kz\right\| + \left\|\left[\rho\widehat{\rho}^{-1} - I\right]\right\| \left\|\mu\right\|$$

The requirement $0 \le k < 1$ *implies*

$$0 \le \left\|\left[\rho\widehat{\rho}^{-1} - I\right]\right\| < 1$$

Consequently, there is a specific limit on the tolerable variation of the control gain matrix.

Example 7.234 (Linearization with Matched Uncertainty) *This simple example illustrates the effectiveness of robust feedback linearization. Consider the system:*

$$\begin{bmatrix} \dot{x}_1 \\ \dot{x}_2 \end{bmatrix} = \begin{bmatrix} x_2 \\ -0.1x_2 + x_1^3/2 \end{bmatrix} + \begin{bmatrix} 0 \\ 1 \end{bmatrix} \{u + \underbrace{\kappa x_1^3 + au}_{uncertainty}\}$$

The parameters $\kappa \in [0,1]$ and $a \in [-0.1, 0.1]$ are uncertain. We will design a stabilizing feedback control for the nominal system using feedback linearization. Then we will make it robust via Lyapunov redesign and evaluate performance of both controllers when the system is subject to a perturbation.

First, enter the system definition, and design the nominal system.

```
In[236]:= f0 = {x2, -0.1 x2 + x1 + x1^3/2};
          g0 = {0, 1};
          f1 = {0, kap x1^3} ;
          g1 = {0, a};
          x = {x1, x2};
```

```
In[237]:= alpha0 = -0.1 x2 + x1 + x1^3/2;
          alpha1 = alpha0 + kap x1^3;
          rho0 = 1;
          rho1 = rho0 + a;
          K = {-1, -2};
          ustar = (1/rho0) (-alpha0 + K.x);
```

Now, turn to the redesign process.

```
In[238]:= A = {{0, 1}, {-1, -2}}; Q = 2 IdentityMatrix[2];
          P = LyapunovEquation[A, Q];
```

A bound for the uncertainty needs to be established in order to define the robustifying control component, μ, and assemble the control, u.

```
In[239]:= Del1 = Chop[Simplify[alpha1 - (rho1/rho0) alpha0 + (rho1/rho0 - 1) K.x]]
```

$$Del2 = (rho1/rho0 - 1);$$

$$Out[239]= \text{kap } x1^3 + a \left(-2 \text{ x1} - \frac{x1^3}{2} - 1.9 \text{ x2} \right)$$

```
In[240]:= AA = Chop[Coefficient[Del1, {x1, x2, x1^3}]];
          sig2 =
            Simplify[(AA[[1]] + AA[[3]] x1^2)^2 + AA[[2]]^2]/.{a -> -0.1, kap -> 1};
          μ = -(1 + (sig2/4)(1/(1 - 0.1)) ){0, 1}.P.x;
          u = Simplify[ustar + μ];
```

$$Out[240]= -3.02114 \text{ x1} - 0.616667 \text{ x1}^3 - 0.30625 \text{ x1}^5 - 2.92114 \text{ x2} - 0.116667 \text{ x1}^2 \text{ x2} - 0.30625 \text{ x1}^4 \text{ x2}$$

Now, the equations can be assembled and computations performed. First, the redesigned control is applied to the nominal system.

```
In[241]:= ReplacementRules = Inner[Rule, {x1, x2}, {x1[t], x2[t]}, List];
          Eqns = Chop[MakeODEs[{x1, x2}, f0 + f1 + (g0 + g1) u, t]]/.{a -> 0, kap -> 0};
          InitialConds = {x1[0] == 1.5, x2[0] == 0};
          VSsols = NDSolve[Join[Eqns, InitialConds], {x1[t], x2[t]}, {t, 0, 10},
            AccuracyGoal -> 2, PrecisionGoal- > 1, MaxStepSize- > 10/60000,
            MaxSteps -> 60000];
```

In[242] := *Plot*[*Evaluate*[{*x1*[*t*]} /. *VSsols*],
           {*t*, 0, 10}, *PlotRange*– > *All*, *AxesLabel* → {*t*, *x1*}];
         Plot[*Evaluate*[{*x2*[*t*]} /. *VSsols*],
           {*t*, 0, 10}, *PlotRange*– > *All*, *AxesLabel* → {*t*, *x2*}];

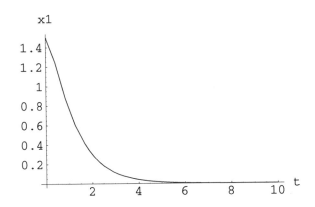

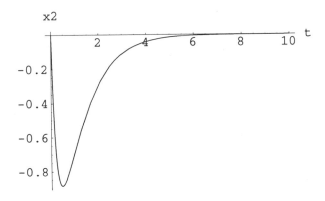

Now apply the control to a perturbed system.

In[243] := *Eqns* =
         Chop[*MakeODEs*[{*x1*, *x2*}, *f0* + *f1* + (*g0* + *g1*) *u*, *t*]]/.{*a* → −0.1, *kap* → 1};
         InitialConds = {*x1*[0] == 1.5, *x2*[0] == 0};
         VSsols = *NDSolve*[*Join*[*Eqns*, *InitialConds*], {*x1*[*t*], *x2*[*t*]}, {*t*, 0, 10},
           AccuracyGoal → 2, *PrecisionGoal*– > 1, *MaxStepSize*– > 10/60000,
           MaxSteps → 60000];

In[244]:= *Plot[Evaluate[{x1[t]} /. VSsols],*
    {t, 0, 10}, PlotRange− > All, AxesLabel → {t, x1}];
    Plot[Evaluate[{x2[t]} /. VSsols],
    {t, 0, 10}, PlotRange− > All, AxesLabel → {t, x2}];

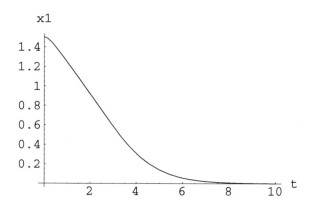

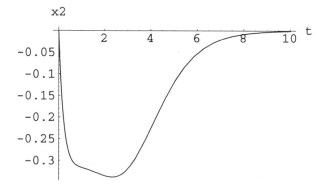

Notice that there is performance degradation but the system remains stable. For comparison, let us apply the nominal control to the perturbed system.

In[245]:= *Eqns =*
    Chop[MakeODEs[{x1, x2}, f0 + f1 + (g0 + g1) ustar, t]]/.{a → −0.1, kap → 1};
    InitialConds = {x1[0] == 1.5, x2[0] == 0};
    VSsols = NDSolve[Join[Eqns, InitialConds], {x1[t], x2[t]}, {t, 0, 1.55},
      AccuracyGoal → 2, PrecisionGoal− > 1, MaxStepSize− > 1.55/60000,
      MaxSteps → 60000];

In[246]:=   *Plot[Evaluate[{x1[t]} /. VSsols],*
              {t, 0, 1.5}, PlotRange− > All, AxesLabel → {t, x1}];
              Plot[Evaluate[{x2[t]} /. VSsols],
              {t, 0, 1.5}, PlotRange− > All, AxesLabel → {t, x2}];

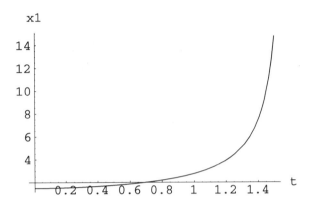

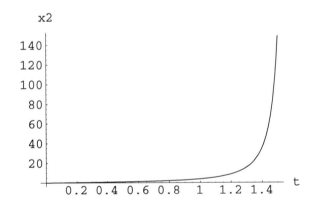

Clearly, the system is unstable − the feedback linearizing control without redesign can not cope with the perturbation.

7.4   Robust Stabilization via Backstepping

When a system involves nonmatched uncertainty, a backstepping procedure might be appropriate. We will describe such an approach to the design of a robust stabilizing controller for SISO systems that satisfy the triangularity conditions. Backstepping will be revisited below for the design of adaptive and variable structure controllers.

Suppose that nominal part (f, g, h) of the system (7.1) has relative degree $r = n$ (alternatively, (f, g) is exactly feedback linearizable) around the origin, and that the strict triangularity condition applies. Moreover, suppose that $\varphi(0, t) = 0$ and $|\varphi(x, t)| \leq \hat{\sigma}(x) \|x\|$ $\forall t$. It follows from Proposition (7.227) that we may as well begin with the triangular form

$$
\begin{aligned}
\dot{x}_i &= x_{i+1} + \Delta_i(x_1, \ldots, x_i, t), \quad 1 \leq i \leq n-1 \\
\dot{x}_n &= \alpha(x) + \rho(x)u + \Delta_n(x, t)
\end{aligned}
\tag{7.20}
$$

with $\det \rho(x) \neq 0$ at least around the origin, $\Delta(0, t) = 0$ and $|\Delta_i(x, t)| \leq \sigma_i(x) \|x\|$, for some smooth bounding function $\sigma_i(x) \geq 0$. An uncertainty vector $\Delta(x, t)$ that satisifies these two conditions along with the triangular structure exhibited in (7.20) will be called *admissible*. The *robust stabilization problem* is to design a state feedback control such that the origin $x = 0$ is asymptotically stable for every admissible uncertainty.

Now, we design the control sequentially. At each of n steps we design a 'psuedo-control' v_k. At the kth step we consider the system (with $v_0 = 0$)

$$
\begin{aligned}
\dot{x}_i &= x_{i+1} + \Delta_i(x_1, \ldots, x_i, t), \quad 1 \leq i \leq k-1 \\
\dot{x}_k &= v_k + \Delta_k(x_1, \ldots, x_k, t) \\
y_k &= x_k - v_{k-1}(x_1, \ldots, x_{k-1})
\end{aligned}
\tag{7.21}
$$

and choose a control v_k to stabilize the input-output behavior (drive $y_k \to 0$ for all initial conditions and all admissible uncertainties). Finally, the actual control u is defined by $u = \rho^{-1}(-\alpha + v_n)$. The process proceeds as follows.

1. $k = 1$ At the first step we have

$$
\begin{aligned}
\dot{x}_1 &= v_1 + \Delta_1(x_1, t) \\
y_1 &= x_1
\end{aligned}
\tag{7.22}
$$

Notice that we can write Equation (7.22)

$$
\dot{y}_1 = v_1 + y_1 \bar{\Delta}_1(y_1, t) := f_1(y_1)
\tag{7.23}
$$

Now, consider the function

$$
V_1 = \tfrac{1}{2} y_1^2
$$

and compute it's derivative along trajectories of Equation (7.23).

$$
L_{f_1} V_1 = y_1 (v_1 + \Delta_1(y_1, t))
$$

Choose $v_1 = -k_1 y_1 - y_1 \kappa_1(y_1)$ which yields

$$
L_{f_1} V_1 = -k_1 y_1^2 - \kappa_1(y_1) y_1^2 + y_1 \bar{\Delta}_1
$$

Now take $\kappa_1(y_1) > \frac{1}{4}\sigma_1^2(x_1)$ for $x_1 \neq 0$. This insures that

$$
\begin{aligned}
L_{f_1}V_1 &\leq -(k_1 - 1)\,y_1^2 - \left(y_1^2 \tfrac{1}{4}\sigma_1^2 + y_1^2\right)\\
&\leq -(k_1 - 1)\,y_1^2 - \left(y_1^2 \tfrac{1}{4}\sigma_1^2 - \sigma_1 y_1^2 + y_1^2\right)\\
&\leq -(k_1 - 1)\,y_1^2 - \left(\tfrac{1}{2}\sigma_1\,|y_1| - |y_1|\right)^2\\
&\leq -(k_1 - 1)\,y_1^2
\end{aligned}
$$

is negative definite (provided $k_1 > 1$) so that the origin $x_1 = 0$ is asymptotically stable.

2. k=2 Now consider the system

$$
\begin{aligned}
\dot{x}_1 &= x_2 + \Delta_1(x_1, t)\\
\dot{x}_2 &= v_2 + \Delta_2(x_1, x_2, t)\\
y_2 &= x_2 - v_1(x_1)
\end{aligned}
\tag{7.24}
$$

Consider the coordinate change $(x_1, x_2) \mapsto (y_1, y_2)$ and compute

$$
\dot{y}_2 = v_2 + \Delta_2(x_1, x_2, t) - \frac{\partial v_1}{\partial x_1}(x_2 + \Delta_1(x_1, t)) := v_2 + \bar{\Delta}_2(y_1, y_2, t)
$$

The differential equations (7.24) in the new coordinates are

$$
\begin{aligned}
\dot{y}_1 &= v_1(y_1) + y_2 + \Delta_1(y_1, t)\\
\dot{y}_2 &= v_2 + \bar{\Delta}_2(y_1, y_2, t)
\end{aligned}
\quad := f_2(y_1, y_2)
\tag{7.25}
$$

Notice that the derivative of V_1 along trajectories of Equation (7.25) is

$$
L_{f_2}V_1 = y_1(v_1 + y_2 + \bar{\Delta}_1) = L_{f_1}V_1 + y_1 y_2
$$

Define the function

$$
V_2 = V_1 + \tfrac{1}{2}y_2^2
$$

and compute

$$
\begin{aligned}
L_{f_2}V_2 &= L_{f_1}V_1 + y_1 y_2 + y_2(v_2 + \bar{\Delta}_2)\\
&= L_{f_1}V_1 + y_2\left(y_1 + v_2 + \bar{\Delta}_2\right)
\end{aligned}
$$

Now, set $v_2 = -y_1 - k_2 y_2 - y_2\kappa_2(y_1, y_2)$ so that

$$
\begin{aligned}
L_{f_2}V_2 &= L_{f_1}V_1 - k_2 y_2^2 - y_2^2\kappa_2(y_1, y_2) + y_2\bar{\Delta}_2\\
&\leq -(k_1 - 1)\,y_1^2 - k_2 y_2^2 - y_2^2\kappa_2(y_1, y_2) + |y_2|\,\|Y_2\|\,\bar{\sigma}_2(y_1, y_2)\\
&\leq -(k_1 - 2)\,y_1^2 - (k_2 - 1)\,y_2^2 - y_2^2\kappa_2(y_1, y_2)\\
&\quad + |y_2|\,\|Y_2\|\,\bar{\sigma}_2(y_1, y_2) - \|Y_2\|^2\\
&\leq -(k_1 - 2)\,y_1^2 - (k_2 - 1)\,y_2^2 - \left(\tfrac{1}{2}\,|y_2|\,\bar{\sigma}_2(Y_2) - \|Y_2\|\right)^2\\
&\leq -(k_1 - 2)\,y_1^2 - (k_2 - 1)\,y_2^2
\end{aligned}
\tag{7.26}
$$

3. $k = 3 \ldots n$ We continue in the same fashion. Suppose we have completed i steps $(i = 1, \ldots, n-1)$. So, we have already defined the new states $y_1, \ldots, y_i$ and psuedo controls $v_1, \ldots, v_i$ and the functions

$$V_j = V_{j-1} + \tfrac{1}{2}y_j^2, \quad 1 \le j \le i$$

Now, we wish to compute v_{i+1}. Define $y_{i+1} = x_{i+1} - v_i$, and organize the equations

$$\left. \begin{aligned} \dot{y}_j &= v_j(Y_j) + y_{j+1} + \bar{\Delta}(Y_j, t), \quad 1 \le j \le i \\ \dot{y}_{i+1} &= v_{i+1} + \bar{\Delta}(Y_{i+1}, t) \end{aligned} \right\} := f_{i+1} \qquad (7.27)$$

As above, we have

$$L_{f_{i=1}}V_i = L_{f_i}V_i + y_i y_{i+1}$$

so that

$$L_{f_{i+1}}V_{i+1} = L_{f_i}V_i - y_{i+1}\left(y_i + v_{i+1} + \bar{\Delta}_{i+1}(Y_{i+1}, t)\right)$$

Choose $v_{i+1} = -y_i - k_{i+1}y_{i+1} - y_{i+1}\kappa_{i+1}$ and $\kappa_{i+1}(Y_{i+1}) > \tfrac{1}{4}\bar{\sigma}_{i+1}^2(Y_{i+1})$ to obtain

$$
\begin{aligned}
L_{f_{i+1}}V_{i+1} \;\le\; & -\sum_{j=1}^{i+1} \left(k_j - (i+2-j)\right) y_j^2 - y_{i+1}^2 \kappa_{i+1} \\
& + |y_{i+1}| \, \|Y_{i+1}\| \, \bar{\sigma}_{i+1}(Y_{i+1}) - \|Y_{i+1}\|^2 \\
\le\; & -\sum_{j=1}^{i+1} \left(k_j - (i+2-j)\right) y_j^2 - \left(\tfrac{1}{2}|y_{i+1}|\bar{\sigma}_{i+1}(Y_{i+1}) - \|Y_{i+1}\|\right)^2 \\
\le\; & -\sum_{j=1}^{i+1} \left(k_j - (i+2-j)\right) y_j^2
\end{aligned}
$$

$$(7.28)$$

These calculations establish the following proposition.

Proposition 7.235 (Smooth Robust Stabilization) *Consider the system (7.1) and suppose*

1. *the nominal system (f, g) is exactly feedback linearizable,*

2. *the strict trangularity condition is true,*

3. *the uncertainty satisfies the conditions: $\varphi(0, t) = 0$ and $|\varphi(x, t)| \le \hat{\sigma}(x)\,\|x\|$ $\forall t$.*

Then there exists a smooth state feedback controller such that the origin, $x = 0$, is asymptotically stable for all admissible uncertainties $\varphi(x, t)$.

Proof: Apply the nominal system normal form transformation to the actual uncertain system and follow the construction of Equations (7.20) through (7.28) to obtain

$$L_{f_n}V_n \le \sum_{j=1}^{n} (k_j - n - 1 + j)y_j^2$$

Notice that we need to choose $k_j > n + 1 - j$, $j = 1, \ldots, n$. ∎

7.5   Adaptive Control of Linearizable Systems

The essential idea is easy to develop. Consider a parameter dependent system reduced to local regular form:

$$\dot{\xi} = F(\xi, z, \vartheta) \tag{7.29}$$

$$\dot{z} = Az + E[\alpha(x, \vartheta) + \rho(x, \vartheta)u] \tag{7.30}$$

$$y = Cz \tag{7.31}$$

Here A, E, C are of the special Brunovsky form as indicated in Section 6.5 and independent of all system parameters. Now, suppose that the control u is different from the ideal decoupling control, $u^* = \rho^{-1}\{-\alpha + v\}$, because it is based on current estimates of the uncertain parameters:

$$u = \rho(x, \widehat{\vartheta})^{-1}\{-\alpha(x, \widehat{\vartheta}) + v\} \tag{7.32}$$

equivalently,

$$\alpha(x, \widehat{\vartheta}) + \rho(x, \widehat{\vartheta})u = v$$

Then we can compute

$$\dot{z} = Az + E[\alpha(x, \vartheta) + \rho(x, \vartheta)u] \tag{7.33}$$

$$\dot{z} = Az + E[v + \Delta] \tag{7.34}$$

where

$$\Delta = [\alpha(x, \vartheta) + \rho(x, \vartheta)u] - [\alpha(x, \widehat{\vartheta}) + \rho(x, \widehat{\vartheta})u] \tag{7.35}$$

Assumption 1: $\Delta(\xi, z, \widehat{\vartheta}, \vartheta, u)$ is linear in the parameter estimation error, i.e.,

$$\Delta(\xi, z, \widehat{\vartheta}, \vartheta, u) = \Psi(\xi, z, \widehat{\vartheta}, u)(\vartheta - \widehat{\vartheta}) \tag{7.36}$$

Thus, we have the

$$\dot{\xi} = F(\xi, z, \vartheta) \tag{7.37}$$

$$\dot{z} = A_c z + E\Psi(\vartheta - \widehat{\vartheta}) \tag{7.38}$$

We are in a position to employ the standard Lyapunov argument to derive an update law for the parameter estimate.

Proposition 7.236 *Asymptotic output stabilization, $y \to 0$, is achieved with the parameter estimator*

$$\dot{\widehat{\vartheta}} = Q\Psi^T(\xi, z, \widehat{v}, u)E^T Pz$$

where P is a symmetric, positive definite solution of

$$(A + EK)^T P + P(A + EK) = -I$$

and Q is any symmetric positive definite matrix.

Proof: Choose a candidate Lyapunov function

$$V = z^T P z + (\vartheta - \widehat{\vartheta})^T Q^{-1}(\vartheta - \widehat{\vartheta})$$

Differentiate with respect to time to obtain

$$
\begin{aligned}
\dot{V} &= 2z^T P \dot{z} - 2\dot{\widehat{\vartheta}}^T Q^{-1}(\vartheta - \widehat{\vartheta}) \\
&= 2z^T P(A_c z + E\Psi(\vartheta - \widehat{\vartheta})) - 2\dot{\widehat{\vartheta}}^T Q^{-1}(\vartheta - \widehat{\vartheta}) \\
&= 2z^T P A_c z + 2(z^T P E\Psi - \dot{\widehat{\vartheta}}^T Q^{-1})(\vartheta - \widehat{\vartheta}) \\
&= z^T(P A_c + A_c^T P)z + 2(z^T P E\Psi - \dot{\widehat{\vartheta}}^T Q^{-1})(\vartheta - \widehat{\vartheta})
\end{aligned}
$$

The assumptions reduce this to

$$\dot{V} = -z^T z$$

∎

There are many variants of this basic construction. One model reference adaptive control configuration is illustrated in Figure (7.1). The key point is that the input–output linearizing and decoupling control absorbs all of the parameter dependencies so that only this part of the control law has to be adjusted.

Remark 7.237 *The regressor* $\Psi(\xi, z, \widehat{\vartheta}, u)$ *is particularly easy to compute if* $\alpha(x, \vartheta)$ *and* $\rho(x, \vartheta)u$ *are linear in the uncertain parameters:*

$$\alpha(x, \vartheta) = \alpha_0(x) + \widetilde{\alpha}(x)\vartheta$$

$$\rho(x, \vartheta)u = \rho_0(x, u) + \widetilde{\rho}(x, u)\vartheta$$

Then

$$\Delta = \{\widetilde{\alpha}(x) + \widetilde{\rho}(x, u)\}(\vartheta - \widehat{\vartheta}),$$

and

$$\Psi = \{\widetilde{\alpha}(x) + \widetilde{\rho}(x, u)\}$$

In the implementation of the controller illustrated in Figure (7.1), it is necessary to measure or estimate both x and z. Notice that z can not be computed from the normal coordinate relations $z(x, \vartheta)$ because they now depend on the unknown parameter ϑ.

Example 7.238 (Adaptive Regulator) *We will illustrate adaptive regulation using the following example adapted from Isidori [3]:*

$$
\begin{bmatrix} \dot{x}_1 \\ \dot{x}_2 \\ \dot{x}_3 \end{bmatrix} = \begin{bmatrix} 0 \\ x_1 + x_2^2 \\ \theta x_1 - x_2 \end{bmatrix} + \begin{bmatrix} e^{x_2} \\ e^{x_2} \\ 0 \end{bmatrix} u
$$

$$y = x_3$$

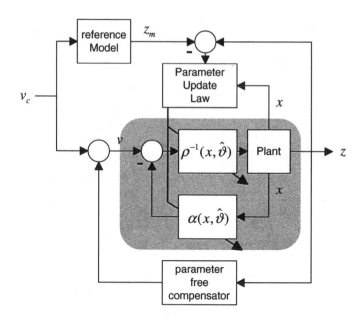

Figure 7.1: A model reference, adaptive tracking control configuration based on decoupling control.

We need to specify the desired closed loop pole locations. It is necessary to do this in groups according to the vector relative degree. So we first compute the vector relative degree and then use the function AdaptiveRegulator. *Here are the computations.*

$In[247] :=$ $var32 := \{x1, x2, x3\};$
$f32 := \{0, \ x1 \ + \ x2\hat{\ }2, \ \theta * x1 \ - \ x2\};$
$g32 := \{\exp[x2], \exp[x2], 0\};$
$h32 := \{x3\};$
$ro = VectorRelativeOrder[f32, g32, h32, var32]$
$Out[247] = \{2\}$

$In[248] :=$ $Poles = \{\{-2, -2\}\};$
$\{Parameters, ParameterEstimates, UpdateLaw, Control\} =$
$AdaptiveRegulator[f32, g32, h32, var32, t, \{\theta\}, \{\}, Poles]$

Computing Decoupling Matrix

Computing linearizing/decoupling control

Finished Linearizing Control

NewParameters $= \{\theta\}$

Finished Stabilizing Control

Finished Stabilizing Control

Finished regressor computation

Finished Lyapunov Equation

Finished parameter update law

$Out[248]=$ $\left\{\{0\}, \{\text{thetahat1}\}, \left\{ \dfrac{125.\ \text{AdaptGain1}\ e^{x2}\ z1\ (x1 + x2^2 - 4\ z1 - 4\ z2)}{-e^{x2} + e^{x2}\ \text{thetahat1}} + \right.\right.$

$\qquad\qquad \left.\dfrac{156.25\ \text{AdaptGain1}\ e^{x2}\ (x1 + x2^2 - 4\ z1 - 4\ z2)\ z2}{-e^{x2} + e^{x2}\ \text{thetahat1}} \right\},$

$\qquad \left.\left\{ \dfrac{x1 + x2^2 - 4\ z1 - 4\ z2}{-e^{x2} + e^{x2}\ \text{thetahat1}} \right\}\right\}$

While this approach to adaptive control has limitations, it does have its place in applications. See, for example, [4, 5].

We can provide a modest but useful generalization of the above result. Once again, consider a parameter dependent system reduced to local regular form:

$$\begin{aligned}
\dot{\xi} &= F(\xi, z, \vartheta) \\
\dot{z} &= Az + E\left[\alpha(x, \vartheta) + \rho(x, \vartheta)\,\phi(u, \vartheta)\right] \\
y &= Cz
\end{aligned} \qquad (7.39)$$

Here matrices A, E, C are of the special Brunovsky form and independent of all system parameters; $\alpha : R^{n \times p} \to R^m$, $\rho : R^{n \times p} \to R^{m \times m}$ and $\phi : R^{n \times p} \to R^m$ are piecewise smooth in x and continuous in the parameter ϑ; for each admissible ϑ, and the map ϕ has a piecewise smooth inverse $\phi^{-1}(\cdot, \vartheta)$. The inclusion of the map ϕ allows us to treat systems with certain types of control saturation, backlash and similar input nonlinearities (see [6]). If the parameter ϑ is known, it is possible to implement the ideal decoupling control law

$$u^* = \phi^{-1}\left(\rho^{-1}(x, \vartheta)\left(-\alpha(x, \vartheta) + v\right), \vartheta\right)$$

to obtain $\dot{z} = Az + Ev$. If the zero dynamics $\dot{\xi} = F(\xi, 0, \vartheta)$ are stable the control renders the loop stable. On the other hand, if the parameter ϑ is uncertain, then we can implement a control based on estimates of the parameter:

$$u = \phi^{-1}\left(\rho^{-1}\left(\widehat{x}, \widehat{\vartheta}\right)\left(-\alpha\left(\widehat{x}, \widehat{\vartheta}\right) + v\right), \widehat{\vartheta}\right) \qquad (7.40)$$

equivalently, u satisfies

$$\alpha\left(\widehat{x}, \widehat{\vartheta}\right) + \rho\left(\widehat{x}, \widehat{\vartheta}\right)\phi(u, \widehat{\vartheta}) = v$$

In the present case we have

$$\dot{z} = Az + E\left[v + \Delta\right]$$

where

$$\Delta = \left[\alpha(x, \vartheta) + \rho(x, \vartheta)\phi(u, \vartheta)\right] - \left[\alpha(\widehat{x}, \widehat{\vartheta}) + \rho(\widehat{x}, \widehat{\vartheta})\phi(u, \widehat{\vartheta})\right] \qquad (7.41)$$

with $x = x\left(\xi, z, \vartheta\right)$ and $\widehat{x} = x\left(\xi, z, \widehat{\vartheta}\right)$. The following basic assumption replaces Assumption 1.

Assumption 2: The control error $\Delta\left(\xi, z, \vartheta, \widehat{\vartheta}, u\right)$ has the form

$$\Delta\left(\xi, z, \vartheta, \widehat{\vartheta}, u\right) = \Psi\left(\xi, z, \widehat{\vartheta}, u\right)\left(\vartheta - \widehat{\vartheta}\right) + \varphi_0\left(\xi, z, \vartheta, \widehat{\vartheta}, u\right)$$

where φ_0 is a bounded, piecewise smooth function.

Assume, further, that upper and lower bounds, $\vartheta_{i_{\min}}, \vartheta_{i_{\max}}$, respectively, are known for each uncertain parameter ϑ_i. In conjunction with the control (7.40), we implement a parameter update rule:

$$\dot{\widehat{\vartheta}} = \Omega\Psi^T\left(\xi, z, \widehat{\vartheta}, u\right)E^T Pz - \Omega\sigma\left(\widehat{\vartheta}\right) \tag{7.42}$$

where $\Omega > 0$ is a (matrix) design parameter, P satisfies the Lyapunov equation

$$(A + EK)^T P + P(A + EK) = -Q, \quad Q > 0 \tag{7.43}$$

and the function $\sigma\left(\widehat{\vartheta}\right)$ is defined by

$$\sigma(\widehat{\vartheta}) = \left[\sigma_1(\widehat{\vartheta}), \ldots, \sigma_p(\widehat{\vartheta})\right]^T \tag{7.44}$$

$$\sigma_i\left(\widehat{\vartheta}\right) := \begin{cases} \kappa_i & \widehat{\vartheta}_i > \vartheta_{i_{\max}} \\ 0 & \vartheta_{i_{\min}} \leq \widehat{\vartheta}_i \leq \vartheta_{i_{\max}} \\ -\kappa_i & \widehat{\vartheta}_i < \vartheta_{i_{\min}} \end{cases} \quad \kappa_i > 0, \quad i = 1, \cdots, p \tag{7.45}$$

Remark 7.239 (Implementation) *In order to implement the control (7.40) and (7.42), we require direct measurement or estimates of the system state in either the original coordinates x or the normal form coordinates (ξ, z). In many practical cases (ξ, z) are the natural coordinates for measurement. In fact, it is not uncommon with electromechanical systems for the components of z to be a subset of the original states x. In general if the measurements are (ξ, z), then when computing the regressor it is necessary to use the state transformation $\widehat{x} = x\left(\xi, z, \widehat{\vartheta}\right)$ in ρ and α. On the other hand, if x is the natural measurement and z needs to be computed from the parameter dependent state transformation, then it is necessary to proceed quite differently, i.e. via backstepping as described below.*

The closed loop system that obtains when the control (7.40) with update law (7.42) is applied to the system (7.39) enjoys three basic properties:

1. the parameter trajectory $\widehat{\vartheta}(t), t > 0$ is bounded,

2. the partial state trajectory $z(t), t > 0$, is bounded and enters a neighborhood of the origin whose size is proportional to the bound on φ_0,

3. under mild additional conditions $z(t), y(t) \rightarrow 0$ as $t \rightarrow \infty$.

Proposition 7.240 (Bounded states and parameter estimates) *Consider the closed loop system composed of the plant (7.39) and control (7.40)-(7.45). Suppose that Assumption 2 is satisfied. Then the partial state trajectory $z(t), t > 0$ and the parameter estimate $\widehat{\vartheta}(t), t > 0$ are bounded. Furthermore, the state trajectory $z(t)$ eventually enters the disk*

$$D = \left\{ z \in R^r \;\middle|\; \left\| Q^{1/2} z - \mathbf{r} \right\|^2 \leq \|\mathbf{r}\|^2 \right\}$$

where $\mathbf{r} = \left(Q^{-1/2} P E \varphi_0 \right)_{\max}$.

Proof: : Choose a candidate Lyapunov function

$$V = z^T P z + (\vartheta - \widehat{\vartheta})^T \Omega^{-1} (\vartheta - \widehat{\vartheta}) \tag{7.46}$$

and compute

$$\dot{V} = -\left\| Q^{1/2} z - Q^{-1/2} P E \varphi_0 \right\|^2 + \left\| Q^{-1/2} P E \varphi_0 \right\|^2 + 2 \left(z^T P E \Psi - \dot{\widehat{\vartheta}}^T \Omega^{-1} \right) (\vartheta - \widehat{\vartheta}) \tag{7.47}$$

$$\dot{V} = -\left\| Q^{1/2} z - Q^{-1/2} P E \varphi_0 \right\|^2 + \left\| Q^{-1/2} P E \varphi_0 \right\|^2 - \sigma(\widehat{\vartheta})(\vartheta - \widehat{\vartheta}) \tag{7.48}$$

which is clearly negative provided

$$\left\| Q^{1/2} z - Q^{-1/2} P E \varphi_0 \right\|^2 > \left\| Q^{-1/2} P E \varphi_0 \right\|^2 \tag{7.49}$$

For each fixed φ_0 this condition defines a circular disk in R^r (in the coordinates $Q^{1/2} z$) of radius $\left\| Q^{-1/2} P E \varphi_0 \right\|^2$ and centered at $Q^{-1/2} P E \varphi_0$. Since $\varphi 0$ is bounded there exists a largest disk (of maximum radius) that contains all others. This is the disk D. D lifts to a cylinder in the state space (R^{r+p}). $\dot{V} < 0$ outside of this cylinder. The minimum value of V ($V = 0$) occurs on its boundary. Hence all trajectories must reach a neighborhood of the cylinder in finite time. Since $z(t)$ is continuous for $t > 0$, it is bounded. In view of (7.48) we have

$$\dot{V} \leq \left\| Q^{-1/2} P E \varphi_0 \right\|^2_{\max} - \sigma(\widehat{\vartheta})(\vartheta - \widehat{\vartheta}) \tag{7.50}$$

Thus, $\dot{V} < 0$ outside of the rectangular domain

$$\left[\vartheta_{i_{\min}} - \frac{\left\| Q^{-1/2} P E \varphi_0 \right\|^2_{\max}}{\kappa_i}, \vartheta_{i_{\max}} + \frac{\left\| Q^{-1/2} P E \varphi_0 \right\|^2_{\max}}{\kappa_i} \right], \quad i = 1, \ldots, p$$

So that $\widehat{\vartheta}(t), t > 0$ is bounded. ∎

Notice that we would like to insure that $\widehat{\vartheta}(t)$ begins and remains in the parameter cube

$$C_{param} = \left\{ \vartheta \in R^p \mid \vartheta_{i_{\min}} \le \widehat{\vartheta} \le \vartheta_{i_{\max}}, i = 1, \ldots, p \right\}$$

Since the estimate can always be initialized within the cube, it is desired to insure that trajectories beginning in the cube can not leave it. This is often done by simple projection. However, by choosing each κ_i sufficiently large we can guarantee that the parameter will remain within C_{param} so long as the state trajectory remains within any prespecified bound.

Define the sets

$$S_{ext} = \left\{ (z, \vartheta) \in R^{r+p} \left\| \left\| Q^{1/2}z - Q^{-1/2}PE\varphi_0 \right\|^2 > \left\| Q^{-1/2}PE\varphi_0 \right\|^2 \right. \right\}$$

$$S_{int} = \left\{ (z, \vartheta) \in R^{r+p} \left\| \left\| Q^{1/2}z - Q^{-1/2}PE\varphi_0 \right\|^2 < \left\| Q^{-1/2}PE\varphi_0 \right\|^2 \right. \right\}$$

Notice that S_{ext} and S_{int} share a common boundary that we denote $\partial S :=$ $\partial S_{ext} := \partial S_{int}$. ∂S includes points that satisfy

$$\left\| Q^{1/2}z - Q^{-1/2}PE\varphi_0 \right\|^2 = \left\| Q^{-1/2}PE\varphi_0 \right\|^2$$

as well as points at which φ_0 is undefined. Observe that all points with $z = 0$ belong to ∂S. With this notation we can state the second key result:

Proposition 7.241 (Output convergence) *Suppose Assumption 2 is satisfied and, in addition: (i) the only invariant set of the closed loop system contained in ∂S corresponds to $z = 0$, (ii) all trajectories beginning in S_{int} with initial estimates in C_{param} remain in C_{param} while they are in S_{int}. Then all trajectories of the closed loop system beginning in C_{param} satisfy $z(t) \to 0$ (hence $y(t) \to 0$) as $t \to \infty$.*

Proof: Once again consider the Lyapunov function

$$V = z^T P z + (\vartheta - \widehat{\vartheta})^T \Omega^{-1}(\vartheta - \widehat{\vartheta})$$

Along closed loop trajectories we have

$$\dot{V} = - \left\| Q^{1/2}z - Q^{-1/2}PE\varphi_0 \right\|^2 + \left\| Q^{-1/2}PE\varphi_0 \right\|^2 - \sigma(\widehat{\vartheta})(\vartheta - \widehat{\vartheta})$$

Thus, $\dot{V} < 0$ on S_{ext} and $\dot{V} > 0$ along all trajectories in S_{int} that begin in C_{param}. Furthermore, $\inf_{(z,\vartheta) \in S_{ext}} V(z, \vartheta)$ occurs on ∂S so that trajectories beginning in S_{ext} eventually reach ∂S. Similarly, for each fixed $\widehat{\vartheta}$, say $\widehat{\vartheta} = \widehat{\vartheta}^*$, $\sup_{(z,\widehat{\vartheta}^*) \in S_{int}} V(z, \widehat{\vartheta}^*)$ occurs on ∂S. Consequently, trajectories beginning in S_{int} with initial estimates in C_{param} also reach ∂S. By assumption, the only invariant set in ∂S corresponds to $z = 0$, so all trajectories beginning in C_{param} tend to $z = 0$. ∎

Example 7.242 (Example (7.234) Revisited) *Let us reconsider the uncertain system of example (7.234). Previously we designed a controller using the Lyapunov redesign approach. For comparison, we now design an adaptive controller. First, the data:*

$In[249] :=$ $f0 = \{x2, -0.1\ x2\ + x1 + x1\hat{\ }3/2\}; g0 = \{0, 1\};$
$\qquad f1 = \{0,\ kap\ x1\hat{\ }3\}\ ; (*\ 0 < kap < 1*)$
$\qquad g1 = \{0, a\}; (*\ -0.1 < a < 0.1*)$
$\qquad x = \{x1, x2\};$
$\qquad f = f0 + f1; g = g0 + g1; h = \{x1\};$

Now, design the adaptive regulator.

$In[250] :=$ $Poles = \{\{-1, -2\}\};$
$\qquad \{Parameters, ParameterEstimates, UpdateLaw, Control\} =$
$\qquad\qquad AdaptiveRegulator[f, g, h, x, t, \{a, kap\}, \{0.00002, 0.0005\}, Poles];$

Computing Decoupling Matrix

Computing linearizing/decoupling control

Finished Linearizing Control

NewParameters $= \{a, kap\}$

Finished Stabilizing Control

Finished regressor computation

Finished Lyapunov Equation

Finished parameter update law

We replace occurrences of transformed z variables with the measured x variables (in this example, the transformation is independent of the uncertain parameters) and add the parameter range limit switch to the update law.

$In[251] :=$ $u = Control/.\{z1 \to x1, z2 \to x2\}$

$\qquad ULaw = (UpdateLaw/.\{z1 \to x1, z2 \to x2\}) -$
$\qquad\qquad Sig[\{thetahat1, thetahat2\}, \{10, 1\}, \{-0.1, 0\}, \{0.1, 1\}];$

For a baseline, simulate with perfectly known parameters.

$In[252] :=$ $Eqns = Chop[MakeODEs[\{x1, x2, thetahat1, thetahat2\},$
$\qquad\qquad Join[f + g\ u[[1]], ULaw], t]]/.\{a \to 0, kap \to 0\};$
$\qquad InitialConds =$
$\qquad\qquad \{x1[0] == 1.5, x2[0] == 0, thetahat1[0] == 0, thetahat2[0] == 0\};$
$\qquad VSsols = NDSolve[Join[Eqns, InitialConds], \{x1[t], x2[t],$
$\qquad\qquad thetahat1[t], thetahat2[t]\}, \{t, 0, 10\}, AccuracyGoal \to 2,$
$\qquad\qquad PrecisionGoal- > 1, MaxStepSize- > 10/60000, MaxSteps \to 60000];$

$In[253] :=$ $Plot[Evaluate[\{x1[t]\}\ /.\ VSsols],$
$\qquad\qquad \{t, 0, 10\}, PlotRange- > All, AxesLabel \to \{t, x1\}];$
$\qquad Plot[Evaluate[\{x2[t]\}\ /.\ VSsols],$
$\qquad\qquad \{t, 0,\ 10\}, PlotRange- > All, AxesLabel \to \{t, x2\}];$

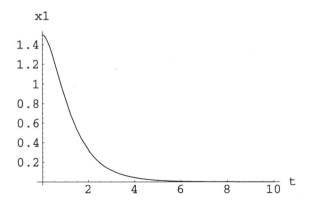

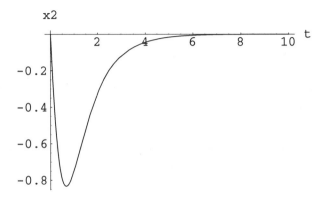

Now, suppose that the parameters are unknown.

In[254]:= $Eqns = Chop[MakeODEs[\{x1, x2, thetahat1, thetahat2\},$
        $Join[f + g \ u[[1]], \ ULaw], \ t]]/.\{a \rightarrow -0.1, kap \rightarrow 1\};$
     $InitialConds =$
        $\{x1[0] == 1.5, x2[0] == 0, thetahat1[0] == 0, thetahat2[0] == 0\};$
     $VSsols = NDSolve[Join[Eqns, InitialConds], \{x1[t], x2[t],$
        $thetahat1[t], thetahat2[t]\}, \{t, 0, 10\}, AccuracyGoal \rightarrow 2,$
        $PrecisionGoal- > 1, MaxStepSize- > 10/60000, MaxSteps \rightarrow 60000];$

In[255]:= $Plot[Evaluate[\{x1[t]\} \ /. \ VSsols],$
        $\{t, 0, 10\}, PlotRange- > All, AxesLabel \rightarrow \{t, x1\}];$
     $Plot[Evaluate[\{x2[t]\} \ /. \ VSsols],$
        $\{t, 0, \ 10\}, PlotRange- > All, AxesLabel \rightarrow \{t, x2\}];$

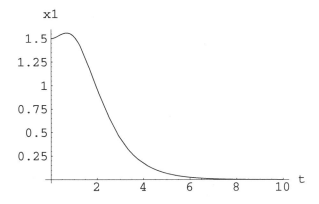

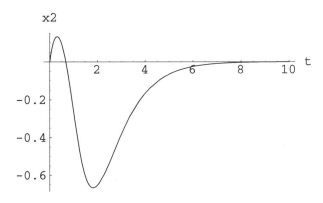

These results should be compared with trajectories in example (7.234).

The following figures illustrate the parameter estimates. Note that there is no guarantee that the estimates will converge to the true parameter values. It is only assured that the loop is stable and that the states $z(t)$ converge to zero.

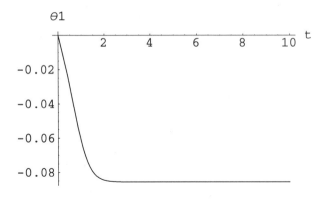

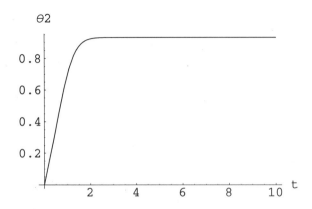

7.6   Adaptive Control via Backstepping

The adaptive control controllers discussed above require access to the full state, x, and the transformed partial state, z. Also, there is a constraint imposed on the uncertainty structure imposed by equation (7.36) - Assumption 1 - that cannot be validated *a priori*. Another approach to adaptive control design has been described in [7] by Kanellakopolis, Kokotovic and Morse that requires access only to the system state, x, and characterizes the uncertainty constraint directly in terms of the way in which the parameters appear in the differential equations.

It is assumed that the single-input, feedback linearizable system is of the form

$$\dot{\zeta} = f(\zeta, \theta) + g(\zeta, \theta)u \qquad (7.51)$$

where $\zeta \in R^n$ is the state, $u \in R$ is the control, and $\theta \in R^p$ is the uncertain parameter vector. Moreover, f, and g are linear in the parameters:

$$f(\zeta,\theta) = f_0(\zeta) + \sum_{i=1}^{p} \theta_i f_i(\zeta), \quad g(\zeta,\theta) = g_0(\zeta) + \sum_{i=1}^{p} \theta_i g_i(\zeta) \qquad (7.52)$$

and $f_i(\zeta), g_i(\zeta), 0 \le i \le p$ are smooth vector fields in a neighborhood of the origin $\zeta = 0$ with $f_i(0) = 0, 0 \le i \le p$ and $g_0(0) \ne 0$. A fundamental assumption is that there exists a parameter-independent diffeomorphism $x = \phi(\zeta)$ that transforms the system into *parametric-pure-feedback form*:

$$\dot{x}_i = x_{i+1} + \theta^T \gamma_i(x_1,\ldots,x_{i+1}), \quad 1 \le i \le n-1$$
$$\dot{x}_n = \gamma_0(x) + \theta^T \gamma_n(x) + [\beta_0(x) + \theta^T \beta(x)]u \qquad (7.53)$$

with

$$\gamma_i(0) = 0, \quad 1 \le i \le n \quad \text{and} \quad \beta_0(0) \ne 0 \qquad (7.54)$$

Necessary and sufficient conditions for such a transformation are given in the following proposition from [7].

Proposition 7.243 *A diffeomorphism $x = \phi(\zeta)$ with $\phi(0) = 0$, transforming (7.51) and (7.52) into (7.53) and (7.54) exists in a neighborhood $B_x \subset U$ of the origin if and only if the following conditions are satisfied in U.*

i) Feedback Linearization Condition: The Distributions

$$\mathcal{G}_i = span\{g_0, ad_{f_0}g_0, \ldots, ad_{f_0}^i\}$$

are involutive and of constant rank $i + 1$.

ii) Parametric-Pure-Feedback Condition:

$$g_i \in \mathcal{G}_0$$
$$[X, f_i] \in \mathcal{G}_{j+1}, \forall X \in \mathcal{G}_j, 0 \le j \le n-3, 1 \le i \le p$$

Conditions i) and ii) can be restated in more compact form:

i) $\mathcal{G}_{n-2}$ is involutive, and $\mathcal{G}_{n-1}$ has constant rank n.

ii) $[ad_{f_0}^j g_0, f_i] \in \mathcal{G}_{j+1}, 0 \le j \le n-3, 1 \le i \le p$

A special case of the parametric-pure-feedback form is the so-called *parametric-strict-feedback* form. The system (7.51) and (7.52) is of the parametric-strict-feedback type if it is diffeomorphically equivalent to:

$$\dot{x}_i = x_{i+1} + \theta^T \gamma_i(x_1,\ldots,x_i), \quad 1 \le i \le n-1$$
$$\dot{x}_n = \gamma_0(x) + \theta^T \gamma_n(x) + \beta_0(x)u \qquad (7.55)$$

Necessary and sufficient conditions for the existence of the required diffeomorphism are given by the following proposition [7].

Proposition 7.244 *Suppose there exists a global diffeomorphism, $x = \phi(\zeta)$ with $\phi(0) = 0$, that transforms the (nominal) system*

$$\dot{\zeta} = f_0(\zeta) + g_0(\zeta)u$$

into

$$\dot{x}_i = x_{i+1}, \ 1 \leq i \leq n - 1$$

$$\dot{x}_n = \gamma_0(x) + \beta_0(x)u$$

with $\gamma_0(0) = 0$ and $\beta_0(x) \neq 0, \forall x \in R^n$. Then the system (7.51) and (7.52) is globally diffeomorphically equivalent to (7.55) if and only if the following conditions hold:

(i) $g_i \equiv 0$

(ii) $[X, f_i] \in \mathcal{G}_{j+1}, \forall X \in \mathcal{G}_j, 0 \leq j \leq n - 2, 1 \leq i \leq p$

The backstepping procedure for adaptive control design is given [7, 8]. We will summarize the constructions for the simpler case of parametric-strict-feedback form in order to explain the basic ideas. More details can be obtained from [7, 8] and their references. Suppose the system has been reduced to parametric-strict-feedback form, (7.55). Then the backstepping procedure sequentially generates:

(i) a (parameter-dependent) state transformation to new coordinates z, $z = z(x, \widehat{\theta})$,

(ii) a feedback control law $u = u(x, \widehat{\theta})$,

(iii) a parameter update law $\dot{\widehat{\theta}} = \tau(x, \widehat{\theta})$

When the state transformation and feedback control are applied, the closed loop equations in the z-coordinates have the form

$$\dot{z} = \left[\text{diag}(-c_1, \ldots, -c_n) + \Phi(z, \widehat{\theta})\right] z + \Psi(z, \widehat{\theta})\widetilde{\theta}$$

where $\Phi(z, \widehat{\theta})$ is an antisymmetric matrix, i.e., $\Phi^T(z, \widehat{\theta}) = -\Phi(z, \widehat{\theta})$, and $\widetilde{\theta}$ is the parameter estimation error, $\widetilde{\theta} = \theta - \widehat{\theta}$. Stability can be established via standard Lyapunov arguments. Choose a candidate Lyapunov function in the form

$$V(z, \widehat{\theta}) + \frac{1}{2}z^T z = \widetilde{\theta}^T Q \widetilde{\theta}, \quad Q^T = Q > 0$$

Differentiate and use the closed loop equations to obtain

$$\dot{V} = z^T \left[\text{diag}(-c_1, \ldots, -c_n) + \Phi(z, \widehat{\theta})\right] z + \left(z^T \Psi + \dot{\widehat{\theta}}^T Q\right) \widetilde{\theta}$$

Because of asymmetry, $z^T \Phi(z, \widehat{\theta})z = 0$, so we can choose the update law

$$\dot{\widehat{\theta}} = -Q^{-1}\Psi^T z$$

to obtain

$$\dot{V} = z^T \text{diag}(-c_1, \dots, -c_n)z$$

Provided $c_i > 0, i = 1, \dots, n$, this establishes the uniform stability of the equilibrium point $z = 0, \widehat{\theta} = \theta$, which, corresponds to $x = 0$. Moreover, from the LaSalle invariance theorem we can obtain

$$\lim_{t \to \infty} z(t) = 0, \lim_{t \to \infty} \dot{z}(t) = 0, \lim_{t \to \infty} \dot{\widehat{\theta}} = 0$$

The update law is implemented in the form

$$\dot{\widehat{\theta}} = \tau(x, \widehat{\theta}) = -Q^{-1}\Psi^T(z(x, \widehat{\theta}), \widehat{\theta})z(x, \widehat{\theta})$$

Remark 7.245 *1. Computations for the parametric-pure-form of the equations are somewhat more complicated but lead to similar results.*

2. The controller consists of a feedback law

$$u = \frac{1}{\beta_0(x)} \left[-\gamma_0(x) + \alpha_n(z_1, \dots, z_n, \widehat{\theta}) \right]$$

and parameter estimator equations

$$\dot{\widehat{\theta}} = \tau_n(z_1, \dots, z_n, \widehat{\theta})$$

where α_n, τ_n are the last of recursively computed sequences as defined in the above references. In actual computation, they are obtained, successively, directly as functions of x (rather than z) which is the way in which the controller is to be implemented. This is easier, and avoids the need to invert the state transformation equations $z = z(x)$.

ProPac implements three functions that assist in the design of backstepping adaptive controllers: `AdaptiveBackstepRegulator`, `PSFFCond`, and `PSFFSolve`. The following example provides an illustration of their use.

Example 7.246 (Backstepping Adaptive Regulator) *Consider the following single input example. First we put the system in parametric strict feedback form using* `PSFFSolve`. *Then, we design an adaptive regulator. We only display the control and update law.*

```
In[256]:=  var30 = {x1, x2};
           f30 = {θ x1³ + sin[x2], x2};
           g30 = {0, 1};
```

In[257] := *PSFFSolve[f30, g30, var30, {θ}]*

LinearizingOutputSolutions : {x1}

PSFFTransformationz = T({x1, x2})

z1 = x1

z2 = Sin[x2]

Out[257] = {{x1, Sin[x2]}, {θ z1³ + z2, x2 Cos[x2]}, {0, Cos[x2]}}

In[258] := {*control, update, zcoords*} = *AdaptiveBackstepRegulator[f30, g30,*
                var30, {θ}, {{AdGain}}, {c1, c2, c3}];

In[259] := *control*

Out[259] = $-x1 - c1\ c2\ x1 - c1$ thetahat1 $x1^3 - c2$ thetahat1 $x1^3 -$
       3 thetahat1² $x1^5$ − AdGain $x1^7$ − AdGain c1² $x1^7 -$
       4 AdGain c1 thetahat1 $x1^9 -$
       3 AdGain thetahat1² $x1^{11} - x2 - c1\ x2 - c2\ x2 - 3$ thetahat1 $x1^2\ x2 -$
       AdGain c1 $x1^6\ x2 - 3$ AdGain thetahat1 $x1^8\ x2$

In[260] := *update*

Out[260] = {AdGain $x1^3\ (x1 + (c1 + 3$ thetahat1 $x1^2)\ (c1\ x1 +$ thetahat1 $x1^3 + x2))$}

7.7   Adaptive Tracking via Dynamic Inversion

ProPac contains the function `AdaptiveTracking` that produces an adaptive version of the tracking controller defined by (6.87), (6.88) and (6.91) [9]. In this case, the system is assumed to depend on an uncertain parameter vector ϑ. Then the control (6.87) and (6.88) also depends explicitly on ϑ. This control is implemented with an estimate of the parameter and a parameter update law in the form:

$$u = D_\beta^\dagger(x, \vartheta) \left\{ -C_\beta(x, \vartheta) + M(x, \vartheta)\widetilde{y}_0 + N(x, \vartheta)y^{(n)} + v(t) \right\} \tag{7.56}$$

$$\dot{\widehat{\vartheta}} = -\Omega^{-1}W^T R\varepsilon \tag{7.57}$$

where

$$\varepsilon := [e_1, .., e_1^{n_1-1}, .., e_l, .., e_l^{n_l-1}]^T \tag{7.58}$$

Computing the Regressor

Recall that application of the exact control (6.87) reduces the input-output dynamics to (6.90). When the 'inexact' control is applied the input-out dynamics can always be expressed in the form

$$y^{(n)} = v + \Delta(x, \vartheta) \tag{7.59}$$

Our goal is to compute $\Delta(x, \vartheta)$. Notice that combining (6.70) and (6.88) and making the parameter dependence explicit, we have

$$N(x, \vartheta)y^{(n)} + M(x, \vartheta)\widetilde{y}_0 = C_\beta(x, \vartheta) + D_\beta(x, \vartheta)u \tag{7.60}$$

In view of (7.58), the control satisfies

$$N(x, \widehat{\vartheta})v + M(x, \widehat{\vartheta})\widetilde{y}_0 = C_\beta(x, \widehat{\vartheta}) + D_\beta(x, \widehat{\vartheta})u \tag{7.61}$$

Now we make the following assumption:

Assumption 3: The matrices $C_\beta(x, \vartheta), D_\beta(x, \vartheta)$ and $C_\beta(x, \vartheta)$ are linear in the uncertain parameters.

This allows us to write

$$\begin{array}{rcl}
C_\beta(x, \vartheta) & = & C_{\beta_0}(x) + \widetilde{C}_\beta(x)\vartheta \\
D_\beta(x, \vartheta)u & = & D_{\beta_0}(x, u) + \widetilde{D}_\beta(x, u)\vartheta \\
M(x, \vartheta)\widetilde{y}_0 & = & M_0(x, \widetilde{y}_0) + \widetilde{M}(x, \widetilde{y}_0)\vartheta \\
N(x, \vartheta)y^{(n)} & = & N_0(x, y^{(n)}) + \widetilde{N}(x, y^{(n)})\vartheta
\end{array} \tag{7.62}$$

Subtracting (7.60) from (7.61) and using (7.62) yields

$$N(x, \widehat{\vartheta})(v - y^{(n)}) = \\
\widetilde{C}_\beta(x)(\widehat{\vartheta} - \vartheta) + \widetilde{D}_\beta(x, u)(\widehat{\vartheta} - \vartheta) - \widetilde{M}(x, \widetilde{y}_0)(\widehat{\vartheta} - \vartheta) - \widetilde{N}(x, y^{(n)})(\widehat{\vartheta} - \vartheta)$$

so that

$$\begin{array}{rcl}
\Delta & = & -N^{-1}(x, \widehat{\vartheta})\left\{\widetilde{C}_\beta(x) + \widetilde{D}_\beta(x, u) - \widetilde{M}(x, \widetilde{y}_0) - \widetilde{N}(x, y^{(n)})\right\}(\widehat{\vartheta} - \vartheta) \\
& = & -W(\widehat{\vartheta} - \vartheta)
\end{array} \tag{7.63}$$

The function `AdaptiveTracking` performs two key operations in assembling the regressor W. First it sorts through the matrices C_β, D_β, M and N to identify groups of physical parameters that can be combined to form new parameters that fit the linearity assumption. One of the outputs of `AdaptiveTracking` is a list of these parameter transformation rules. Then the matrices are expanded as in (7.62) in terms of the new parameters so that W can be assembled as in (7.63).

Example 7.247 (Adaptive Tracking Controller) *The following example is the same as Example 7.238 (see also [9]). After defining the system equations we define a reference signal, then specify the desired closed loop pole locations and, finally, compute the control. Because of the length of the output, we give only the control and update law.*

```
In[261]:=  RefSig =
              Table[ToExpression[yd <> ToString[i1] <> [t]], {i1, 1, Length[h32]}];
           Poles = {{-2, -2}};
           {Parameters, ParameterEstimates, UpdateLaw,
           Control, DecoupMatrix, DerivativeOrders} = AdaptiveTracking[
              f32, Transpose[{g32}], h32, var32, t, {θ}, RefSig, {}, Poles];
           Control
```

Symbol	Definition	Parameter Value
ω	motor speed	
i_a	armature current	
i_f	field current	
e_a	armature applied voltage	
e_f	field applied voltage	
J	motor inertia	$0.1 \ kg - m^2$
B	motor damping	$0.01 \ Kg - m^2 - sec^{-1}$
K	electromechanical transduction	$0.30 \ nm/a^2$
T_L	load torque	0 to $15 \ nm$
L_a	armature inductance	$1.0 \ H$
L_f	field inductance	$10.0 \ H$
R_a	armature resistance	$10.0 \ \Omega$
R_f	field resistance	$50.0 \ \Omega$

Table 7.1: **DC motor nomenclature**

$Out[261]=$ $\left\{ \dfrac{x1 + x2^2 + 4 \ (-x3 + yd1[t]) + 4 \ (-y1'[t] + yd1'[t]) + yd1''[t]}{-e^{x2} + e^{x2} \ thetahat1} \right\}$

$In[262]:=$ $UpdateLaw$

$Out[262]=$ $\left\{ -\dfrac{1}{-e^{x2} + e^{x2} \ thetahat1}(125. \ AdaptGain1 \ e^{x2} \ (-x3 + yd1[t]) \ (x1 +\right.$

$x2^2 + 4 \ (-x3 + yd1[t]) + 4 \ (-y1'[t] + yd1'[t]) + yd1''[t])) -$

$\dfrac{1}{-e^{x2} + e^{x2} \ thetahat1}(156.25 \ AdaptGain1 \ e^{x2} \ (-y1'[t]+yd1'[t]) \ (x1+x2^2+4 \ (-x3+yd1[t])+$

$\left. 4 \ (-y1'[t] + yd1'[t]) + yd1''[t])) \right\}$

7.8   Problems

Problem 7.248 (DC drive motor) *A separately excited dc motor is decscribed by the differential equations*

$$J\frac{d\omega}{dt} = -B\omega + Ki_f i_a - T_L$$

$$L_a\frac{di_a}{dt} = -R_a i_a - Ki_f \omega + e_a$$

$$L_f\frac{di_f}{dt} = -R_f i_f + e_f$$

where the variables and parameters are defined in Table (7.1) Consider the control inputs to be the two applied voltages, e_a, e_f. The goal is to regulate speed to a desired value ω_0 and to minimize electrical losses $(R_a i_a^2 + R_f i_f^2)$. Thus, we formulate two outputs

$$y_1 = \omega - \omega_0$$

$$y_2 = R_a i_a - R_f i_f$$

(a) Design a feedback linearizing adaptive controller taking the load torque as an uncertain parameter. Assume that ω_0 is a specified constant.

(b) Assume that the load torque, T_L can be measured or accurately estimated, but that the motor friction coefficient is uncertain with $B \in (0, 0.05)$. Design a feedback linearizing adaptive controller. Via simulation compare the adaptive and nonadaptive $(B = 0.01)$ performance.

Problem 7.249 (Load with backlash and friction) *An inertial load with backlash and friction is illustrated in Figure (7.2). The drive motor angle θ_m is cosidered as the control input to a drive shaft/gear with backlash modeled using the dead zone function:*

$$D(\theta) = \begin{cases} \theta - \varepsilon & \theta \geq \varepsilon \\ 0 & |\theta| < \varepsilon \\ \theta + \varepsilon & \theta \leq -\varepsilon \end{cases}$$

The shaft has stiffness K and the load has inertia J and friction $f(\omega) = b\sin\omega$. $d(t)$ is an external diturbance. Thus, the equations of motion are:

$$\dot{\theta} = \omega$$

$$J\dot{\omega} = -f(\omega) + K D(u - \theta) + d(t)$$

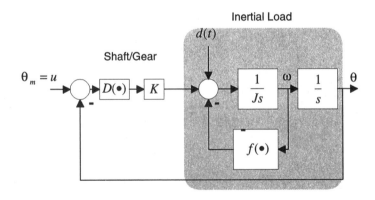

Figure 7.2: A simplified inertial load with backlash and friction.

(a) Assume $\varepsilon \in [0, 0.5]$ and $b \in [0, 1]$ are uncertain parameters within the given bounds and $d(t) = 0$. Design an adaptive feedback linearizing control. Hint: Take the feedback linearizing and stabilizing control to be of the form

$$\hat{u} = \hat{D}^{-1}\left(-(k_1 - 1)\theta - k_2\omega + \hat{f}(\omega)\right)$$

(b) *Take $J = 1$ and $K = 1$ and compute the closed loop system response.*

(c) *Can Lyapunov redesign as described in this chapter be used for this problem?*

(d) *Suppose the disturbance $d(t) = \kappa w(t)$, where $\kappa \in [0, 1]$ is an uncertain parameter with known bounds and $w(t) = 1 + 4\sin(2\pi t)$. Repeat (a) and (b).*

7.9   References

1. Marino, R. and P. Tomei, Nonlinear Control Design: Geometric, Adaptive and Robust. 1995, Upper Saddle River: Prentice-Hall.

2. Khalil, H.K., Nonlinear Systems. 1992, New York: MacMillan.

3. Isidori, A., Nonlinear Control Systems. 3 ed. 1995, London: Springer-Verlag.

4. Bennett, W.H., et al., Nonlinear and Adaptive control of Flexible Space Structures. Transactions ASME, Journal of Dynamic Systems, Measurement and Control, 1993. 115(1): p. 8694.

5. Bennett, W.H., H.G. Kwatny, and M.J. Baek, Nonlinear Dynamics and Control of Articulated Flexible Spacecraft: Application to SSF/MRMS. AIAA Journal on Guidance, Control and Dynamics, 1994. 17(1): p. 3847.

6. Tao, G. and P.V. Kokotovic, Adaptive Control of Systems with Actuator and Sensor Nonlinearities. 1996, New York: John Wiley and Sons, Inc.

7. Kanellakapoulos, I., P.V. Kokotovic, and A.S. Morse, Systematic design of Adaptive Controllers for Feedback Linearizable Systems. IEEE Transactions on Automatic Control, 1991. AC36(11): p. 12411253.

8. Krstic, M., I. Kanellakopoulos, and P.V. Kokotovic, Adaptive Nonlinear Control Without Overparameterization. Systems and Control Letters, 1992. 19: p. 177-185.

9. Ghanadan, R. and G.L. Blankenship, Adaptive Output Tracking of Invertible MIMO Nonlinear Systems, 1992, University of Maryland.

Chapter 8

Variable Structure Control

8.1 Introduction

Variable structure control systems are switching controllers that exhibit certain desirable robustness properties. Consider a nonlinear dynamical system of the form

$$\dot{x} = f(x, u) \tag{8.1}$$

where $x \in R^n, u \in R^m$ and f is a smooth function of x and u. We will focus on switching control systems in which the control functions u_i are discontinuous across smooth surfaces $s_i(x) = 0$, i.e

$$u_i(x) = \begin{cases} u_i^+(x), & s_i(x) > 0 \\ u_i^-(x), & s_i(x) < 0 \end{cases} \quad i = 1, \ldots, m \tag{8.2}$$

and the control functions u_i^+, u_i^- are smooth functions of x.

The design of switching control systems of the type (8.1), (8.2) often focuses on the deliberate introduction of sliding modes [1]. If there exists an open submanifold, M_s, of any intersection of discontinuity surfaces, $s_i(x) = 0$ for $i = 1, \ldots, p \le m$, such that $s_i \dot{s}_i < 0$ in the neighborhood of almost every point in M_s, then it must be true that a trajectory once entering M_s remains in it until a boundary of M_s is reached. M_s is called a *sliding manifold* and the motion in M_s is called a *sliding mode*. Since the control is not defined on the discontinuity surfaces, the sliding dynamics are not characterized by equations (8.1) and (8.2). However, sliding mode dynamics may often be determined by imposing the constraint $s(x) = 0$ on the motion defined by the differential equation (8.1). Under appropriate circumstances this is sufficient to define an 'effective' control u_{eq}, called the *equivalent control*, which obtains for motion constrained to lie in M_s. If this control is smooth and unique, then the sliding behavior is well defined.

Variable structure control system design entails specification of the switching functions $s_i(x)$ and the control functions $u_i^+(x)$ and $u_i^-(x)$. As we will see, the basis for design follows from the observations that the sliding mode dynamics depend on the geometry of M_s, that is, on the switching functions $s_i(x)$, and that sliding can be induced on a desired manifold M_s by designing the control functions $u_i^{\pm}(x)$ to guarantee that M_s is attracting. Thus, control system design is a two step process: 1) design of the 'sliding mode' dynamics by the choice of switching surfaces, and 2) design of the 'reaching' dynamics by the specification of the control functions.

8.2   Basic Properties of Discontinuous Systems

Equation (8.1) when combined with control (8.2) is a special case of the general class of discontinuous dynamical systems

$$\dot{x} = F(x, t) \tag{8.3}$$

where for each fixed t, $F(x, t)$ is smooth ($C^k, k > 0$) on R^n except on m codimension-one surfaces (codimension-one regular submanifolds of R^n) defined by $s_i(x) = 0$, $i = 1, \ldots, m$, on which $F(x, t)$ is not defined. Ordinarily, a solution to (8.3) is a curve $x(t) \subset R^n$ that has the property that $dx/dt = F(x(t), t)$ for each $t \in R$. Such a test, however, would be impossible to apply if the prospective solution contains points on the discontinuity surfaces. Since the set of points for which $F(x, t)$ is not defined has measure zero in R^n, one might simply require that the integral curve property be satisfied only where $F(x, t)$ is defined. This is clearly inadequate because segments of trajectories that lie in a discontinuity surface would be entirely arbitrary. Filippov [2] proposed a satisfactory definition of solutions to (8.3):

Definition 8.250 *A curve $x(t) \subset R^n$, $t \in [t_0, t_1], t_1 > t_0$, is said to be a solution of (8.3) on $[t_0, t_1]$ if it is absolutely continuous on $[t_0, t_1]$ and for each $t \in [t_0, t_1]$*

$$\frac{dx(t)}{dt} \in \widetilde{F}(x(t), t) := \bigcap_{\delta > 0} \operatorname{conv} F(S(\delta, x(t)) - \Lambda(\delta, x(t)), t) \tag{8.4}$$

where $S(\delta, x)$ is the open sphere centered at x and of radius δ, $\Lambda(\delta, x)$ is the subset of measure zero in $S(\delta, x)$ for which F is not defined, and $\operatorname{conv} F(U)$ denotes the convex closure of the set of vectors $\{F(U)\}$.

Remark 8.251 (Remark on notation) *If x is a point in R^n then $S(\delta, x) := \{y \in R^n \,|\, \|y - x\| < \delta\}\}$. If U is a set contained in R^n then*

$$S(\delta, U) := \bigcup_{x \in U} S(\delta, x)$$

We call $S(\delta, U)$ a δ-vicinity of U.

If x does not lie on a discontinuity surface, then the set $\widetilde{F}(x,t) = \{F(x,t)\}$, so that the original differential equation must be satisfied at regular points. However, this definition does help characterize solutions that lie in discontinuity surfaces. Suppose a solution $x(t) \subset R^n$, $t \in [t_0, t_1]$, $t_1 > t_0$, lies entirely in the intersection of some set of p discontinuity surfaces that is a regular embedded submanifold of R^n of dimension $n - p$, which we designate M_s. For each $t^* \in [t_0, t_1]$, $\dot{x}(t^*)$ must belong to the set $\widetilde{F}(x(t^*), t^*)$. In addition, $\dot{x}(t^*)$ must lie in the tangent space to M_s at $x(t^*)$, i.e., $\dot{x}(t^*) \in T_{x(t^*)}M_s$. In many important cases, these two conditions uniquely define solutions that contain segments that lie in M_s.

When we of speak of solutions or, equivalently, trajectories of discontinuous systems we shall mean solutions in the sense of Filippov. One important consequence of the definition is an extension of Lyapunov's direct stability analysis to discontinuous systems [3].

Lemma 8.252 *Suppose that $V : R^n \to R$ is a C^1 function. Then:*

1. *the time derivative of $V(x)$ along trajectories of (8.3) satisfies the set inclusion*

$$\dot{V}(x(t)) \in \left\{ \frac{\partial V}{\partial x}\xi \,\Big|\, \xi \in \widetilde{F}(x(t), t) \right\} \tag{8.5}$$

2. *if $\dot{V} \leq -\rho < 0$ ($\geq \rho > 0$) at all points in an open set $P \subset R^n$ except on a set $\Lambda \subset P$ of measure zero where $F(x,t)$ is not defined, then $\dot{V} \leq -\bar{\rho} < 0$ ($\geq \bar{\rho} > 0$), $\bar{\rho} < \rho$, at all points of P.*

3. *if $\dot{V} \leq -\rho\,\|s(x)\|$, $\rho > 0$, at all points in an open $P \subset R^n$ except on a set $\Lambda \subset P$ of measure zero where $F(x,t)$ is not defined, then $\dot{V} \leq -\rho\,\|s(x)\|$ at all points of P.*

Proof: The first conclusion (8.5) follows directly from the Filippov definition of a trajectory.

To prove the second, first note that at regular points the inclusion reduces to the usual $\dot{V}(x) = [\partial V/\partial x]\,F(x,t)$. Consider the negative definite case. The assumption of definiteness implies that $[\partial V/\partial x]\,F(x,t) \leq -\rho$ at all regular points $x \in P$. Now take any $x^* \in \Lambda$. We need only show that $[\partial V(x^*)/\partial x]\xi \leq -\bar{\rho}$ for each $\xi \in \widetilde{F}(x^*, t)$. Consider a sphere $S(\varepsilon, x^*)$ where $\varepsilon > 0$ is chosen arbitrarily small and so that the sphere is contained in P. By assumption, $[\partial V/\partial x]\,F(x,t) \leq -\rho$ at all regular points in $S(\varepsilon, x^*)$. Since V is C^1, we can choose ε sufficiently small so that $[\partial V(x^*)/\partial x]\xi \leq -\bar{\rho} < 0$ for any specified $\bar{\rho} < \rho$ and all regular x in $S(\varepsilon, x^*)$. By its definition, $\widetilde{F}(x^*, t) \subset F(S(\varepsilon, x^*) - \Lambda, t)$. The conclusion follows.

To prove the third conclusion, consider a point x^* in P. By assumption, the condition $\dot{V} \leq -\rho\,\|s(x)\|$ holds at all regular points. Suppose that x^* is not regular, then the condition holds at almost all points in a sufficiently small neighborhood $S(\varepsilon, x^*)$ of x^*. Now, the smoothness of V 9and s implies the approximations $\partial V(x)/\partial x = \partial V(x^*)/\partial x + O(\varepsilon)$ and $\|s(x)\| = \|s(x^*)\| + O(\varepsilon)$ for all $x \in S(\varepsilon, x^*)$. Thus we have at regular $x \in S(\varepsilon, x^*)$

$$\frac{\partial V(x^*)}{\partial x} F(x, t) \leq -\rho \|s(x^*)\| + O(\varepsilon)$$

Once again, since $\widetilde{F}(x^*, t) \subset F(S(\varepsilon, x^*) - \Lambda, t)$, it follows that

$$\frac{\partial V(x^*)}{\partial x} \xi \leq -\rho \|s(x^*)\| + O(\varepsilon), \forall \xi \in \widetilde{F}(x^*, t)$$

The conclusion follows in the limit $\varepsilon \to 0$.　■

In applications, $\dot{V}$ is often relatively easy to determine at all points in a given domain other than those on the surfaces of discontinuity. The significance of the lemma is that it makes it unnecessary to actually compute $\widetilde{F}(x^*, t)$ in order to determine value of $\dot{V}$ at those points.

Definition 8.253 *Suppose $M_s = \{x \in R^n \mid s(x) = 0\}$ is a regular embedded manifold in R^n and let D_s be an open, connected subset of M_s. D_s is a sliding domain if*

1. *for any $\varepsilon > 0$, there is a $\delta > 0$ such that trajectories of (8.3) which begin in a δ-vicinity of D_s remain in an ε-vicinity of D_s until reaching an ε-vicinity of the boundary of D_s, ∂D_s.*

2. *D_s must not contain any entire trajectories of the 2^m continuous systems defined in the open regions adjacent to M_s and partitioned by the set $M := \bigcup_{i=1,\ldots,m} M_{s_i}$.*

This definition is due to Utkin [1]. By including (2), it is assured that it is the switching mechanism that produces the sliding mode and the possibility of the existence of certain "pathological" sliding domains is excluded.

The definition implies that D_s is invariant with respect to trajectories in the sense of the following rather obvious proposition.

Proposition 8.254 *If D_s is a sliding domain then trajectories of (8.3) which begin in D_s remain in D_s until reaching its boundary, ∂D_s.*

Proof: Since D_s belongs to any δ-vicinity of itself, the definition of a sliding domain implies that trajectories which begin in D_s must remain in every arbitrarily small ε-vicinity of D_s. Hence trajectories beginning in D_s must remain therein until reaching its boundary.　■

Sufficient conditions for the existence of a sliding domain are relatively easy to formulate. One approach is as follows. Define a C^1 scalar function $V : D \subset R^n \to R$ with the following properties

$$V(x) := \begin{cases} = 0 & if \ s(x) = 0 \\ > 0 & otherwise \end{cases} \tag{8.6}$$

Recall that $\dot{V}$ is uniquely defined everywhere but on $M := \bigcup_{i=1,\dots,m} M_{s_i}$ and on M it is still constrained by the set inclusion of Lemma (8.252). Now the following result can be stated.

Proposition 8.255 *Let V be given by (8.6). Suppose that*

1. *D_s is an open, connected subset of M_s*

2. *D is an open connected subset of R^n which contains D_s*

3. *$\dot{V} \leq -\rho \|s(x)\| < 0$ on $D - M$*

Then D_s is a sliding domain.

Proof: Under the stated assumptions, a trajectory cannot leave D_s at any point $x_0 \in D_s$. This is easily proved by contradiction. Suppose a trajectory $x(t)$ does depart D_s from a point $x_0 \in D_s$ at time t_0. Such a departure implies that there is a time $t_1 > t_0$ and sufficiently small $\varepsilon > 0$, such that the absolutely continuous trajectory segment $x(t)$, $t \in (t_0, t_1)$ is entirely contained in the set $S(\varepsilon, x_0) - M_s$ and along which $\dot{V} > 0$. But in view of Lemma (8.252), the assumptions of the proposition imply that $\dot{V} < 0$ along trajectories at all points in $S(\varepsilon, x_0) - M_s$. This is a contradiction. ∎

One distinguishing feature of many variable structure control systems is that trajectories beginning in a vicinity of the sliding surface reach the surface in finite time. This clearly is the case if $\dot{V}$ is bounded below by a negative number. However, such a bound is not necessary as the following proposition illustrates.

Proposition 8.256 *Suppose that the conditions of proposition (8.255) hold and in addition $V(x) = \sigma \|s(x)\|^2$, $\sigma > 0$ on a δ-vicinity of D_s. Then trajectories which reach D_s from a δ-vicinity of D_s do so in finite time.*

Proof: Suppose a trajectory beginning at state x_0 in a δ-vicinity of D_s reaches a point $x_1 \in D_s$. Now, $\|s(x_0)\| \leq \delta$. Since $V(x) = \sigma \|s(x)\|^2$ we have

$$\dot{V} = 2\sigma \|s(x)\| \frac{d \|s(x)\|}{dt} \leq -\rho \|s(x)\|$$

which in view of Lemma (8.252) holds throughout the δ-vicinity of D_s. Thus,

$$\frac{d\,\|s(x)\|}{dt} \leq -\frac{\rho}{2\sigma}$$

which implies that the trajectory reaches D_s in time not greater than $\delta(2\sigma/\rho)$.

∎

8.3   Sliding

In this section we consider the design of sliding surfaces for affine systems of the form

$$\dot{x} = f(x) + G(x)u \tag{8.7}$$

$$y = h(x) \tag{8.8}$$

The procedure begins with the reduction of the given affine system to the *regular form* of

$$\dot{z} = Az + E\left[\alpha(\xi, z) + \rho(\xi, z)u\right] \tag{8.9}$$

$$y = Cz \tag{8.10}$$

as described earlier. Now, we do not feedback linearize as was done in Chapter 6. Instead, we choose a variable structure control law with switching surface, $s(x)$. The variable structure control law is of the form:

$$u_i(x) = \begin{cases} u_i^+(x) & s_i(x) > 0 \\ u_i^-(x) & s_i(x) < 0 \end{cases}$$

Notice that the control is not defined during sliding, i.e., for trajectories completely contained within the surface $s(x) = 0$. We can prove that during sliding the equivalent or effective control is , such that feedback linearized behavior is achieved in the sliding phase (see, [3-6]).

Proposition 8.257 *Let the switching surface $s(x)$ be such that $s(x) = 0$ if and only if $Kz(x) = 0$ for some specified $K \in R^{m \times r}$ and suppose that*

1. *$\rho(x)$ has continuous first derivatives with $\det\{\rho(x)\} \neq 0$ on $M_0 = \{x | z(x) = 0\}$.*

2. *$\partial s(x)/\partial x$ is of maximum rank on the set $M_s = \{x | s(x) = 0\}$.*

Then M_s is a regular $n - m$ dimensional submanifold of R^n which contains M_0. Moreover, if K is structured so that the m columns numbered $r_1, r_1 + r_2, \ldots, r$ compose an identity I_m, then for any trajectory segment $x(t), t \in T$, T an open interval of R, that lies entirely in M_s, the control which obtains on T is

$$u_{eq} = -\rho^{-1}(x)KAz - \rho^{-1}(x)\alpha(x) \tag{8.11}$$

and every such trajectory with boundary condition $x(t_0) = x_0 \in M_s$, $t_0 \in T$ *satisfies*

$$\dot{x} = f(x) - G(x)\rho^{-1}(x)\left[\alpha(x) + KAz(x)\right], \quad Kz(x(t_0)) = 0 \qquad (8.12)$$

Proof: The maximum rank condition insures that M_s is a regular manifold of dimension $n - m$. M_0 is a submanifold of M_s in view of the definition of $s(x)$. Motion constrained by $s(x(t)) = 0$ must satisfy the sliding condition $\dot{s} = 0$, equivalently, $K\dot{z}(x) = 0$. Direct computation leads to (8.11) and (8.12). ■

In this case observe that the manifold M_s is invariant with respect to the dynamics (8.12). The flow defined by (8.12) on M_s is called the *sliding dynamics* and the control defined by (8.11) is the *equivalent control*. Note that the equivalent control behaves as a linearizing feedback control. The partial state dynamics in sliding is obtained from (8.9) and (8.11):

$$\dot{z} = [I - EK]Az, \quad Kz(t_0) = 0 \qquad (8.13)$$

Proposition 8.258 *Suppose the conditions of proposition (8.257) apply. Then M_0 is an invariant manifold of the sliding dynamics (8.12). Moreover, if K is specified as*

$$K = \mathrm{diag}(k_1, \ldots, k_m), \quad k_i = [a_{i1}, \ldots, a_{ir_i-1}, 1] \qquad (8.14)$$

where the m ordered sets of coefficients $\{a_{i1}, \ldots, a_{ir_i-1}\}$, $i = 1, \ldots, m$ each constitute a set of coefficients of a Hurwitz polynomial. Then every trajectory of (8.12) not beginning in M_0 approaches M_0 exponentially.

Proof: Notice that (8.13) implies that the only trajectory of (8.12) with boundary condition $z(t_0) = 0$ is $z(t) = 0$ for all t and hence M_0 is an invariant set.

Note that $\mathrm{Im}[E] \oplus \ker[K] = R^r$ so that the motion of (8.13) can be conveniently divided into a motion in $\mathrm{Im}[E]$ and a motion in $\ker[K]$ and the latter has eigenvalues which coincide with the transmission zeros of the triple (K, A, E), Young et al [7]. To prove that trajectories of (8.12) approach M_0 exponentially we need only show that all trajectories of (8.13) in $\ker[K]$ approach the origin asymptotically. Let the matrix N be chosen so that its columns form a basis for $\ker[K]$ and introduce the coordinate vectors $w \in R^{r-m}$ and $v \in R^m$, and write

$$z = Nw + Ev \qquad (8.15)$$

The inverse of (8.15) may be written

$$\begin{bmatrix} w \\ v \end{bmatrix} = \begin{bmatrix} M \\ K \end{bmatrix} z \qquad (8.16)$$

Direct calculation verifies that (8.13) is replaced by

$$\frac{d}{dt}\begin{bmatrix} w \\ v \end{bmatrix} = \begin{bmatrix} MAN & MAE \\ 0 & 0 \end{bmatrix}\begin{bmatrix} w \\ v \end{bmatrix}, \quad v(0) = 0 \qquad (8.17)$$

The result obtains if $\mathrm{Re}\lambda\{MAN\} < 0$. If the matrix K is chosen in accordance
with (8.14), then the eigenvalues of MAN are precisely the $r - m$ eigenvalues
of the matrices

$$
\begin{bmatrix}
0 & 1 & 0 & . & & 0 \\
0 & 0 & 1 & 0 & & \\
. & . & 0 & 1 & & 0 \\
. & & . & . & & 1 \\
-a_{i1} & -a_{i2} & . & . & -a_{i(r_i-1)}
\end{bmatrix}, \quad i = 1,\ldots,m \tag{8.18}
$$

which are lie in the open left half plane by assumption.   ∎

8.4   Reaching

The second step in VS control system design is the specification of the control
functions $u_i^{\pm}$ such that the manifold $s(x) = 0$ contains a stable submanifold
which insures that sliding occurs. Thus we seek to choose a control that drives
trajectories into $s(x) = 0$, or equivalently, $Kz(x) = 0$. There are many ways
of approaching the reaching design problem, Utkin [1]. We consider only one.
Define a positive definite quadratic form in $\eta = Kz$

$$
V(x) = \eta^T Q\eta, \quad Q > 0 \tag{8.19}
$$

Consider the set of states that satisfy $\eta(x) = 0$. A subset of this set is attractive
if it lies in a region of the state on which the time rate of change V along
trajectories is negative. Upon differentiation we obtain

$$
\frac{d}{dt}V = 2\dot{\eta}^T Q\eta = 2\left[KAz + \alpha\right]^T QKz + 2u^T \rho^T QKz \tag{8.20}
$$

8.4.1   Bounded Controls

If the controls are bounded, $0 > U_{\min,i} \le u_i \le U_{\max,i} > 0$, then, obviously, to
minimize the time rate of change of V, we should choose

$$
u_i = \begin{cases} U_{\min,i} & s_i(x) > 0 \\ U_{\max,i} & s_i(x) < 0 \end{cases} \quad i = 1,\ldots,m \tag{8.21}
$$

$$
s(x) = \rho^T(x)QKz(x) \tag{8.22}
$$

Clearly, $s(x) = 0 \Leftrightarrow Kz(x) = 0$. Notice that if $U_{\min,i} = -U_{\max,i}$, the control
reduces to

$$
u_i = -U_{\max,i}\mathrm{sgn}(s_i)
$$

In this case it follows that $\dot{V}$ is negative (for $s \ne 0$) provided

$$
\left|U_{\max}^T\rho^T QKz\right| > \left|[KAz + \alpha]^T QKz\right| \tag{8.23}
$$

A useful sufficient condition is that

$$|(\rho(x)U_{\max})_i| > |(KAz(x) + \alpha(x))_i| \tag{8.24}$$

Conditions (8.23) or (8.24) may be used to insure that the control bounds are of sufficient magnitude to guarantee sliding and to provide adequate reaching dynamics. This rather simple approach to reaching design is satisfactory when a "bang-bang" control is acceptable.

8.4.2 Unconstrained Controls

Suppose the controls are not constrained to fixed bounds and there exists a continuous bound on the function $\alpha(x)$, i.e.,

$$\|\alpha(x)\| < \sigma_\alpha(x) \tag{8.25}$$

for some continuous function $\sigma_\alpha(x)$. In this case choose u_i and $\sigma(x)$ such that

$$u_i = -\sigma(x)\mathrm{sgn}(s_i(x)), \quad \sigma(x)\|\rho(x)\| > \bar{\sigma}(KA)\|z(x)\| + \sigma_\alpha(x) \tag{8.26}$$

Now, we compute

$$\dot{V} \le \left(\bar{\sigma}(KA)\|z(x)\| + \sigma_\alpha(x) - \sigma(x)\sum_{i=1}^{m}\{|\mathrm{sgn}(s_i(x))|\}\right)\|QKz(x)\| \tag{8.27}$$

Thus $\dot{V}$ is negative when $s \ne 0$ and the sliding manifold is attractive.

8.4.3 A Variation for Unconstrained Controls

Suppose $\alpha(x)$ and $\rho(x)$ are smooth and known with reasonable certainty. A sometimes useful variation of the controller (8.26) is

$$u(x) = u_0(x) + v(x) \tag{8.28}$$

composed of the smooth part

$$u_0(x) = -\rho^{-1}(x)\alpha(x) \tag{8.29}$$

and discontinuous part

$$v_i = -\sigma(x)\mathrm{sgn}(s_i(x)), \quad \sigma(x)\|\rho(x)\| > \bar{\sigma}(KA)\|z(x)\| \tag{8.30}$$

Notice that the required magnitude of the discontinuous part is reduced. We easily compute from (8.20)

$$\dot{V} \le \left(\bar{\sigma}(KA)\|z(x)\| - \sigma(x)\sum_{i=1}^{m}\{|\mathrm{sgn}(s_i(x))|\}\right)\|QKz(x)\| \tag{8.31}$$

8.4.4   Closed Loop Stability

$\mathcal{A} \subset M_0$ is a *stable attractor* of the zero dynamics if it is a closed invariant set and if for every neighborhood U of $\mathcal{A}$ in M_0 there is a neighborhood V of $\mathcal{A}$ in M_0 such that every trajectory of (8.13) beginning in V remains in U and tends to $\mathcal{A}$ as $t \to \infty$. The following proposition establishes conditions under which the variable structure controller applied to (8.7) stabilizes $\mathcal{A}$ in R^n.

Proposition 8.259 *Suppose that the conditions of propositions (8.257) and (8.258) apply; $\mathcal{D}$ is an open region in R^n in which (8.23) (or (8.25)) is satisfied; $\mathcal{D}_s = \mathcal{D} \cap M_s$ is nonempty; and $\mathcal{A} \subset M_0$ is a bounded, stable attractor of the zero dynamics which is contained in $\mathcal{D}_s \cap M_0$. Then $\mathcal{A}$ is a stable attractor of the feedback system composed of (8.7) with feedback control law (8.21) (or (8.26), or (8.28)).*

Proof: Since $\mathcal{D}$ is an open region in R^n in which (8.23) is satisfied, a sliding mode exists in $\mathcal{D}_s = \mathcal{D} \cap M_s$ which is nonempty. In fact, $\mathcal{D}_0 = \mathcal{D}_s \cap M_0$ is also nonempty and it contains a bounded, stable attractor $\mathcal{A}$ of the zero dynamics. Proposition (8.258) implies that $\mathcal{A}$ is also a stable attractor of the sliding dynamics (8.12). Thus, for any neighborhood $\widetilde{U}$ of $\mathcal{A}$ in M_s there is a neighborhood $\widetilde{V}$ of $\mathcal{A}$ in M_s such that trajectories of (8.12) beginning in $\widetilde{V}$ remain in $\widetilde{U}$ and tend to $\mathcal{A}$ with increasing time. We must show that a similar property applies for neighborhoods of $\mathcal{A}$ in R^n with respect to the dynamics defined by (8.7) and (8.21). Let

$$\kappa_{min} = \inf_{\mathcal{D}} \{ U_{max}^T \rho^T Q K z - [KAz + \alpha]^T Q K z \} > 0 \qquad (8.32)$$

which exists by virtue of (8.23), and

$$\kappa_{max} = \sup_{\mathcal{D}} \left\{ \left\| f(x) - \sum_{i=1}^{m} g_i(x) U_{max,i} \mathrm{sign}(s_i) \right\|^2 \right\} < \infty \qquad (8.33)$$

which exists because f and G are continuous and $\mathcal{D}$ is bounded, and where $\|\bullet\|$ denotes the Euclidean norm. Let $S(r, x_0)$ denote the open sphere in R^n of radius r and centered at x_0 and define the set

$$S(r) := \bigcup_{a \in \mathcal{A}} S(r, a) \qquad (8.34)$$

Note that any element of $S(r)$ is at most a distance r from M_s and hence any trajectory starting in $S(r)$ will reach M_s in a finite time not greater than $t_r = r/\sqrt{\kappa_{min}}$. Thus, any trajectory segment of the of the closed loop system beginning in $S(r)$ and terminating upon reaching M_s is entirely contained in the set $S(R)$ where

$$R = r \left\{ 1 + \sqrt{\frac{\kappa_{max}}{\kappa_{min}}} \right\} \qquad (8.35)$$

Now, let $\widehat{U}$ be any neighborhood of $\mathcal{A}$ in R^n. Define $\widetilde{U} = \widehat{U} \cap M_s$, so that $\widetilde{U}$ is a neighborhood of $\mathcal{A}$ in M_s. Then there exists a neighborhood $\widetilde{V}$ of $\mathcal{A}$ in M_s such that trajectories beginning in $\widetilde{V}$ remain in $\widetilde{U}$ and tend to $\mathcal{A}$ with increasing time. In view of (8.35), we can always choose r sufficiently small so that $S(R) \cap M_s \subset \widetilde{V} \cap \mathcal{D}_s$. Then we identify $\widehat{V} = S(r)$. It follows that trajectories of (8.7), (8.21) beginning in $\widehat{V}$ remain in $\widehat{U}$ and approach $\mathcal{A}$ as $t \to \infty$. ∎

Denote $M_h = \{x \mid h(x) = 0\}$ and we assume that M_h is a regular submanifold of R^n of dimension $n - m$. Note that M_0 is a submanifold of both M_h and M_s so that M_0 lies in the intersection of M_h and M_s. The relationships between these manifolds are illustrated in Figure (8.1).

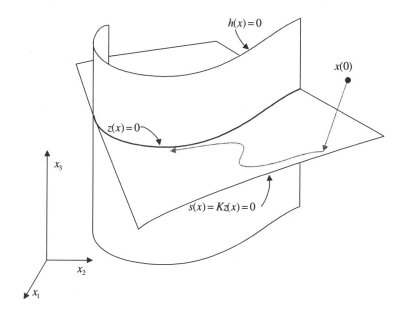

Figure 8.1: The relationship between the output constraint manifold, the sliding manifold and the zero dynamics manifold is illustrated in a three dimensional state space.

Our results imply that the closed loop system behaves as follows. If the initial state is sufficiently close to $\mathcal{D}_s$, the trajectory will eventually reach $\mathcal{D}_s$ and will thereafter approximate ideal sliding. Ideal sliding is characterized by (8.12) and sliding trajectories which remain in $\mathcal{D}_s$ approach $\mathcal{D}_0$ and eventually $\mathcal{A}$. That $\mathcal{A}$ is a stable attractor of (8.12) is obvious. However, this only implies that trajectories of (8.12) beginning sufficiently close to $\mathcal{A}$ approach $\mathcal{A}$.

8.5   Robustness With Respect to Matched Uncertainties

Variable sructure control systems are especially interesting because they exhibit certain robustness properties with respect to model uncertainty. Suppose we have an uncertain system for which the nominal part (f, G, h) has well defined vector relative degree $\{r_1, \ldots, r_m\}$. Then we can proceed as above to design a variable structure control system for the nominal system. The key question is; How does this controller perform when applied to the actual system? When the uncertainty is matched and has a known bound it is possible to design the control to insure that the desired sliding manifold is attractive for any actual plant within the admissible class of systems. Moreover, the sliding behavior is identical to that of the nominal system. If the system has matched uncertainty we can take as our starting point the system:

$$
\begin{aligned}
\dot{\xi} &= F(\xi, z, u) \\
\dot{z} &= Az + E[\alpha(x(\xi, z)) + \Delta(\xi, z, t) + \rho(x(\xi, z))u]
\end{aligned}
\tag{8.36}
$$

where Δ is a function that represents uncertainties and/or disturbances. We assume that $\Delta(\xi, z, t)$ is bounded by a continuous function $\sigma_\Delta(\xi, z) > 0$:

$$
\|\Delta(\xi, z, t)\| < \sigma_\Delta(\xi, z), \quad \forall t
\tag{8.37}
$$

The following proposition establishes the basic robustness result for variable structure controls applied to systems with matched uncertainty.

Proposition 8.260 *Consider a class of* admissible *systems of the form (8.36) satisfying the following conditions*

1. *There is a known and continuous uncertainty bound $\sigma_\Delta(\xi, z) > 0$ such that (8.37) is satisfied.*

2. *There is a continuous bounding function $\sigma_\alpha(x) > 0$ such that*

$$
|\alpha(x)| < \sigma_\alpha(x)
$$

Then there exists a variable structure controller such that for all admissible systems the switching surface $s(x) = 0$ is a sliding manifold and the sliding behavior is identical to the nominal system sliding behavior.

Furthermore, the control is given by:

$$
u_i = -\sigma(x)\mathrm{sgn}(s_i(x))
\tag{8.38}
$$

with

$$
\sigma(x)\,\|\rho(x)\| > \bar{\sigma}(KA)\,\|z(x)\| + \sigma_\alpha(x) + \sigma_\Delta(x)
\tag{8.39}
$$

and

$$
s(x) = \rho^T(x)QKz(x)
\tag{8.40}
$$

where K chosen in accordance with Proposition (8.258).

Proof: First, assume that sliding does occur in the surface $s(x) = 0 \Rightarrow Kz(x) = 0 \Rightarrow K\dot{z} = 0$. Then we have u_{eq} defined by

$$KAz + \alpha(x(\xi, z)) + \Delta(\xi, z, t) + \rho(x(\xi, z))u_{eq} = 0$$

and the sliding dynamics reduce, once again, to (8.13). Thus, the actual sliding dynamics are indeed identical to the nominal system sliding dynamics.

Now, we need to show that it is possible to design control functions $u^{\pm}(x)$ such that sliding occurs in $s(x) = 0$ for all admissible systems. The proposition assumes that both α and Δ are bounded by a continuous functions σ_α and σ_Δ. Consider the positive definite quadratic form in $\eta = Kz$

$$V(x) = \eta^T Q \eta$$

A sliding mode exists on a subset of $\eta(x) = 0$, equivalently $s(x) = 0$, that lies in a region of the state space on which the time rate of change V is negative. Upon differentiation we obtain

$$\frac{d}{dt}V = 2\dot{\eta}^T Q\eta = 2\left[KAz + \alpha + \Delta\right]^T QKz + 2u^T \rho^T QKz$$

Now, choose the control u in accordance with Equations (8.38), (8.39) and (8.40) so that

$$\dot{V} \leq \left(\bar{\sigma}(KA)\,\|z(x)\| + \sigma_\alpha(x) + \sigma_\Delta(x) - \sigma(x)\sum_{i=1}^{m}\{|\mathrm{sgn}(s_i(x))|\}\right)\|QKz(x)\|$$

(8.41)

It follows that $\dot{V}$ is negative wherever it is defined (everywhere but on the sliding manifold), so the sliding manifold is indeed attractive as required. ∎

8.6 Chattering Reduction

The state trajectories of ideal sliding motions are continuous functions of time contained entirely within the sliding manifold. These trajectories correspond to the equivalent control $u_{eq}(t)$. However, the actual control signal, $u(t)$ – definable only for nonideal trajectories – is discontinuous as a consequence of the switching mechanism which generates it. Persistent switching is undesirable in most applications. Several techniques have been proposed to reduce or eliminate it. These include: 'regularization' of the switch by replacing it with a continuous approximation; 'extension' of the dynamics by using additional integrators to separate an applied discontinuous pseudo-control from the actual plant inputs; and 'moderation' of the reaching control magnitude as errors become small.

Switch regularization entails replacing the ideal switching function, $\mathrm{sgn}(s(x))$ with a continuous function such as

$$\mathrm{sat}\left(\frac{1}{\varepsilon}s(x)\right) \quad \text{or} \quad \frac{s(x)}{\varepsilon + |s(x)|} \quad \text{or} \quad \tanh\left(\frac{s(x)}{\varepsilon}\right)$$

This intuitive approach is employed by Young and Kwatny [8] and Slotine and Sastry [9, 10] and there are probably historical precedents. Regularization induces a boundary layer around the switching manifold whose size is $O(\epsilon)$. The reaching behavior is altered significantly because the approach to the manifold is now exponential and the manifold is not reached in finite time as is the case with ideal switching. On the other hand within the boundary layer the trajectories are $O(\epsilon)$ approximations to the sliding trajectories as established by Young et al [7] for linear dynamics with linear switching surfaces. Some of those results have been extended to single input–single output nonlinear systems by Marino [5]. Switch regularization for nonlinear systems has been extensively discussed by Slotine and coworkers, e.g., [9, 10]. With nonlinear systems there are subtleties and regularization can result in an unstable system. However, we can state the following result.

Suppose that each ideal switch is replaced by a smooth version of a switch . Specifically, $\text{sgn}(s) \rightarrow \tanh(s/\varepsilon)$, $\varepsilon > 0$ so that

$$u_i = -\sigma(x)\text{sgn}(s_i(x)) \rightarrow -\sigma(x)\tanh(s_i(x)/\varepsilon)$$

Then $\dot{V}$ is not necessarily negative for $\|s\|$ small. However, for any given $\delta > 0$ there exists a sufficiently small $\varepsilon > 0$ such that $\dot{V} < 0$, for $\|s\| > \delta$ so that all trajectories enter the strip $\|s(x)\| < \delta$. We wish to establish more than that. Namely, we will show that the smoothed control steers the state into a neighborhood of $z = 0$ the size of which shrinks with the design (smoothing) parameter ε.

Proposition 8.261 *Consider the system*

$$\dot{z} = Az + E[\alpha(x(\xi, z)) + \Delta(x(\xi, z), t) + \rho(x(\xi, z))u]$$

Suppose that

1. *there exists a continuous bound on α, $\|\alpha(x)\| < \sigma_\alpha(x)$*

2. *and a continuous bound on Δ, $\|\Delta(x, t)\| < \sigma_\Delta(x)$, $\forall t$*

3. *K is chosen in accordance with Proposition (8.258)*

4. *$u_i = -\sigma(x)\tanh(s_i(x)/\varepsilon)$, where*

$$\sigma(x) > (\bar{\sigma}(KA)\|z(x)\| + \sigma_\alpha(x) + \sigma_\Delta(x))\|QKz(x)\|$$

and $s^*(x) = \rho^T(x)QKz(x)$

Then for any $\delta > 0$ there exists a sufficiently small $\varepsilon > 0$ such that all trajectories enter the ball $\|z\| < \delta$ in finite time.

Proof: Since $KE = I$, we can divide the state space into $\text{Im}E \oplus \ker K$. Thus, we define a transformation (recall the proof of Proposition (8.258):

$$z = [\, E \quad N\,] \begin{bmatrix} \zeta_1 \\ \zeta_2 \end{bmatrix}$$

where the columns of N span $\ker K$. Notice that we can choose a matrix M such that

$$\begin{bmatrix} K \\ M \end{bmatrix} [\, E \quad N\,] = I$$

which implies $K\,[I - EK] = 0$, and $M\,[I - EK] = M$. In these new coordinates the evolution equations are

$$\begin{bmatrix} \dot\zeta_1 \\ \dot\zeta_2 \end{bmatrix} = \begin{bmatrix} KAE & KAN \\ MAE & MAN \end{bmatrix} \begin{bmatrix} \zeta_1 \\ \zeta_2 \end{bmatrix} + \begin{bmatrix} I \\ 0 \end{bmatrix} (\alpha(x) + \Delta(x,t) - \rho(x)\sigma(x)\tanh(s^*(x)/\varepsilon))$$

where

$$\tanh(s^*(x)/\varepsilon) = \begin{bmatrix} \tanh(s_1^*(x)/\varepsilon) \\ \vdots \\ \tanh(s_m^*(x)/\varepsilon) \end{bmatrix}$$

In addition, $s = Kz = \zeta_1$. Furthermore, $\text{Re}\lambda(MAN) < 0$ by design ($MAN \sim A_s$). Hence, there exist matrices, $Q_0 \geq 0, R \geq 0$ such that

1. $z^T Q_0 z = 0$ for $z \in \text{Im}E$ and $z^T Q_0 z > 0$ otherwise.

2. $d(z^T Q_0 z)/dt = -z^T R z \leq -\lambda_{\min} \|\zeta 2\|^2$, where $\lambda_{\min}$ is the smallest nonzero eigenvalue of R.

Now, consider the Liapunov function

$$V(z) = z^T Q_0 z + (Kz)^T QKz > 0, \quad \|z\| \neq 0$$

and compute

$$\frac{d}{dt}V = 2\dot z Q_0 z + 2\dot s^T Q s$$
$$= 2\left\{ Az + b[\alpha + \Delta + \rho u] \right\}^T Q_0 z + 2\,[KAz + \alpha + \Delta]^T QKz + 2u^T \rho^T QKz$$

$$\frac{d}{dt}V = 2\left\{ Az \right\}^T Q_0 z + 2\,[KAz + \alpha + \Delta]^T QKz + 2u^T \rho^T QKz$$

Now, we have

$$[KAz + \alpha + \Delta]^T QKz + u^T \rho^T QKx \leq$$
$$(\bar\sigma(KA)\|z(x)\| + \sigma_\alpha(x) + \sigma_\Delta(x))\,\|QKz(x)\| - \sigma(x)\sum_{i=1}^{m} |\tanh(s_i^*(x)/\varepsilon)|$$

and

$$2\left\{ Az \right\}^T Q_0 z \leq -\lambda_{\min} \|\zeta_2\|^2$$

so that

$$\frac{d}{dt}V \leq -\lambda_{\min} \|\zeta_2\|^2 + 2 \left[\hat{\sigma} - \sigma \sum_{i=1}^{m} |\tanh(s_i^*/\varepsilon)| \right]$$

where

$$\hat{\sigma} = (\bar{\sigma}(KA) \|z(x)\| + \sigma_\alpha(x) + \sigma_\Delta(x)) \|QKz(x)\|$$

Thus, since $\sigma > \hat{\sigma}$ by assumption, for any specified δ there is an ε such that $\dot{V} \leq -c < 0$. Consequently, we have all trajectories entering the strip $\|s\| < \delta(\varepsilon)$ in finite time. In fact, for any given $\delta > 0$ there exists a corresponding sufficiently small $\varepsilon > 0$. Now, since $s = \zeta_1$, it follows that $\|s\| < \delta \Rightarrow \|\zeta_1\| < \delta$. Consequently, from the evolution equations and since MAN is asymptotically (exponentially) stable we can conclude that all trajectories enter a ball with radius proportional to δ in finite time.   ■

Dynamic extension is another effective approach to control input smoothing, Emelyanov et al [11]. A sliding mode is said to be of p-th order relative to an output y if the time derivatives $\dot{y}, \ddot{y}, \ldots, y^{(p-1)}$ are continuous in t but y^p is not. The following observation is a straightforward consequence of the regular form proposition: Suppose the system (8.7) and (8.8) is input-output linearizable with vector relative degree $(r_1, \ldots, r_m)$. Then the sliding mode corresponding to the control law (8.22) is of order $p = \min(r_1, \ldots, r_m)$ relative to the output y. We may modify the relative degree by augmenting the system with input dynamics as described. Hence, we can directly control the smoothness of the output vector y.

When parasitic dynamics of sufficiently high order are present a form of persistent switching can arise that is not removed by the above smoothing strategies. This form of switching can be associated with a (series of) bifurcation(s) in the fast dynamics. It is commonly referred to as chattering. Control moderation can be effective in eliminating chattering. Control moderation involves design of the reaching control functions $u_i(x)$ such that the effective gain is reduced when errors are small, i.e., $|u_i(x)| \to$ small as $|e(x)| \to 0$. For example,

$$u_i(x) = |e(x)|\text{sgn}\,(s_i(x))$$

Control moderation was used by Young and Kwatny [8] and the significance of this approach for chattering reduction in the presence of parasitic dynamics was discussed by Kwatny and Siu [12].

8.7   Computing Tools

We need to be able to reduce the system to normal form, compute an appropriate switching surface, assemble the switching control and insert smoothing and/or moderating functions as desired. Functions to do this are implemented in *ProPac* .

8.7.1   Sliding Surface Computations

There are several methods for determining the sliding surface, $s(x) = Kz(x)$, once the system has been reduced to normal form. We have included a function SlidingSurface that implements two alternatives depending on the arguments provided. The function may be called via

```
{rho,s}=SlidingSurface[f,g,h,x,lam]
```

or

```
s=SlidingSurface[rho,vro,z,lam]
```

In the first case the data provided is the nonlinear system definition f, g, h, x and an m-vector *lam* which contains a list of desired exponential decay rates, one for each input channel. The function returns the decoupling matrix *rho* and the switching surfaces s as functions of the state x. The matrix K is obtained by solving the appropriate Ricatti equation.

The second use of the function assumes that the input-output linearization has already been performed so that the decoupling matrix *rho*, the vector relative degree *vro*, and the normal coordinate (partial) transformation $z(x)$ are known. In this case the dimension of each of the m switching surfaces is known so that it is possible to specify a complete set of eigenvalues for each surface. Thus, *lam* is a list of m-sublists containing the specified eigenvalues, grouped according to the vector relative degree. Only the switching surfaces are returned. In this case K is obtained via pole placement.

8.7.2   Switching Control

The function SwitchingControl[rho,s,bounds,Q,opts] where *rho* is the decoupling matrix, s is the vector of switching surfaces, *bounds* is a list of controller bounds each in the form {lower bound, upper bound}, Q is an $m \times m$ positive definite matrix (a design parameter that can be used, for example, to weight switching surfaces, see Utkin [1]), and *opts* are options which allow the inclusion of smoothing and/or moderating functions in the control. The bounds may be functions of the state. The alternative syntax

```
        SwitchingControl[alpha,rho,s,bounds,Q,opts]
```

returns the control in the form of (8.28), i.e., it contains a smooth feedback linearizing part plus the discontinuous stabilizing part.

Smoothing functions are specified by a rule of the form

```
SmoothingFunctions[x_]->{function1[x],...,functionm[x]}
```

Where m is the number of controls. Moderating functions are similarly specified by a rule

```
ModeratingFunctions->{function1[z],...,functionm[z]}
```

The smoothing function option replaces the pure switch sgn(s) by a smooth switch as specified. The moderating function option multiplies the switch by a specified function. We give an example below.

Example 8.262 (Variable structure control) *We apply some of the above computations in the single input – single output, third order example shown below. First, we display the moderating* $(|x|/(.002 + |x|))$ *and smoothing* $(1 - e^{-|x|/0.1})$ *functions that will be employed.*

In[263]:= $Plot[Abs[x]/(0.002 + Abs[x]), \{x, -3, 3\}]$

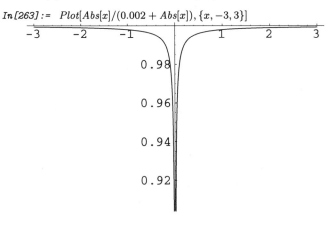

In[264]:= $Plot[Sign[x] (1 - exp[-Abs[x/0.1]]), \{x, -3, 3\}, PlotRange-> All]$

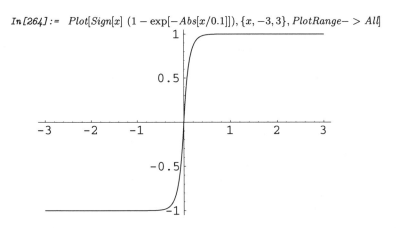

Now we apply input-output linearization. Since the relative degree is 2, there are one-dimensional zero dynamics. They are checked for stability before the control is designed.

In[265]:= $x = \{x1, x2, x3\};$
$f = \{x2, x3, -x1 - x2^\wedge 2 - \sin[x3]\};$
$g = \{0, 1 + x1^\wedge 3, 1\};$
$h = \{x1\};$
$\{\rho, \alpha, vro, control\} = IOLinearize[f, g, h, \{x1, x2, x3\}]$

Out[265]= $\left\{\{\{1 + x1^3\}\}, \{x3\}, \{2\}, \left\{\dfrac{v1 - x3}{1 + x1^3}\right\}\right\}$

In[266]:= $z = NormalCoordinates[f, g, h, \{x1, x2, x3\}, vro];$
$u0 = control/.v1- > 0;$
$LocalZeroDynamics[f, g, h, x, u0, z]$

Out[266]= $\left\{-2\ w1 + \dfrac{w1^3}{6}\right\}$

Since we have a stable equilibrium point we proceed to design a sliding surface. We have already computed the normal coordinates, so we can specify poles at -2,-3 and compute the sliding surface.

In[267]:= $s = SlidingSurface[\rho, vro, z, \{\{-2, -3\}\}];$
$SwitchingControl[\rho, s, \{\{-1, 1\}\}, \{\{1\}\}]$
Out[267]= $\{-Sign[(1 + x1^3)\ (6\ x1 + 5\ x2)]\}$

Now, we compute the switching control using various combinations of smoothing and moderating functions. The particular functions chosen for this example are shown below in Figure 6. Results can change significantly when other functions are used or when the parameters of the functions are varied. We specify the control bounds as ± 1 and $Q = 1$. The following computation yields the controls.

In[268]:= $SwitchingControl[\rho, s, \{\{-1, 1\}\}, \{\{1\}\},$
$SmoothingFunctions[xx\_]- > \{(1 - \exp[-Abs[xx/10]])\}]$
Out[268]= $\Big\{-Sign[(1 + x1^3)\ (6\ x1 + 5\ x2)]+$

$e^{-\frac{1}{10}\ Abs\left[\left(1+x1^3\right)\ (6\ x1 + 5\ x2)\right]}\ Sign[(1 + x1^3)\ (6\ x1 + 5\ x2)]\Big\}$

In[269]:= $SwitchingControl[\rho, s, \{\{-1, 1\}\}, \{\{1\}\},$
$ModeratingFunctions- > \{Abs[z[[1]]]/(0.005 + Abs[z[[1]]])\}]$
Out[269]= $\left\{-\dfrac{Abs[x1]\ Sign[(1 + x1^3)\ (6\ x1 + 5\ x2)]}{0.005 + Abs[x1]}\right\}$

8.8 Backstepping Design of VS Controls

We will describe a backstepping procedure for SISO variable structure control system design in the presence of uncertain, possibly nonsmooth, nonlinearities. The method differs from the backstepping techniques described in the previous chapter in the following ways: (1) the states are grouped in accordance with the appearance of the uncertainty in the system, and (2) the control designed at each step is a variable structure control. We do not assume that uncertainty enters in every state equation. Thus, the number of steps can be reduced. This is important when nonsmooth uncertainties are present as will be evident in the examples below.

Consider a SISO nonlinear system in the (multi-state backstepping) form:

$$\begin{aligned}
x_i^{(n_i)} &= x_{i+1} + \Delta_i(x,t), \quad i = 1, \ldots, p-1 \\
x_p^{(n_p)} &= \alpha(x) + \rho(x)u + \Delta_p(x,t) \\
y &= x_1
\end{aligned} \tag{8.42}$$

We assume that the (possibly nonsmooth) uncertainties $\Delta_i(x,t)$ are bounded by smooth, non-negative functions $\sigma_i(x)$, i.e.,

$$0 \le |\Delta_i(x,t)| \le \sigma_i(x), \quad \forall t \tag{8.43}$$

As noted before, such a model might arise by reduction of a smooth nominal system to regular from and applying the transformation to the (possibly nonsmooth) uncertain system.

The basic idea is very simple. At each of $p-1$ stages we design a 'pseudo-control' v_k, at the kth step (with $v_0 = 0$), using the system

$$\begin{aligned}
x_i^{(n_i)} &= v_i + \Delta_i(x,t), \quad i = 1, \ldots, k < p \\
y_k &= x_k - v_{k-1}(x_1, \ldots, x_{k-1}^{n_{k-1}})
\end{aligned}$$

and at the last (p^{th}) stage we design the actual control, u, using the system

$$\begin{aligned}
x_i^{(n_i)} &= v_i + \Delta_i(x,t), \quad i = 1, \ldots, p-1 \\
x_p^{(n_p)} &= \alpha(x) + \rho(x)u + \Delta_p(x,t) \\
y_p &= x_p - v_{p-1}(x_1, \ldots, x_{p-1}^{n_{p-1}})
\end{aligned}$$

To design the control v_k we first reduce the kth system to normal form by successive differentiation in the usual way. Thus, we identify the new coordinate $y_k, \ldots, y_k^{(n_k-1)}$ that will replace $x_k, \ldots, x_k^{(n_k-1)}$. The transformed evolution equation is

$$y_k^{(n_k)} = L_{f_k}^{n_k} h_k + L_{g_k} L f_k^{n_k-1} v_k = L_{f_k}^{n_k} h_k + v_k = \alpha_k + v_k \tag{8.44}$$

For analysis purposes it is convenient to carry the equations along in the transformed coordinates as the process proceeds. Doing so explicitly, the kth control is obtained by designing a stabilizing smoothed VS controller for a ('nominal') system in the form:

1. $k = 1$

$$\begin{aligned}
x_1^{(n_1)} &= v_1 \\
y_1 &= x_1
\end{aligned} \tag{8.45}$$

2. $k = 2$

$$y_1^{(n_1)} = x_2$$
$$x_2^{(n_2)} = v_2 \qquad (8.46)$$
$$y_2 = x_2 - v_1$$

3. $k = 3, \ldots, p - 1$

$$y_i^{(n_i)} = y_{i+1} + \alpha_i + v_i, \quad i = 1, \ldots, k - 2$$
$$y_{k-1}^{(n_{k-1})} = \alpha_{k-1} + x_k,$$
$$x_k^{(n_k)} = v_k \qquad (8.47)$$
$$y_k = x_k - v_{k-1}$$

4. $k = p$

$$y_i^{(n_i)} = y_{i+1} + \alpha_i + v_i, \quad i = 1, \ldots, p - 2$$
$$y_{p-1}^{(n_{p-1})} = \alpha_{p-1} + x_p,$$
$$x_p^{(n_p)} = \alpha + \rho v_p \qquad (8.48)$$
$$y_p = x_p - v_{p-1}$$

Notice that the zero dynamics of the kth system (8.47) are

$$y_i^{(n_i)} = y_{i+1} + \alpha_i + v_i, \quad i = 1, \ldots, k - 2$$
$$y_{k-1}^{(n_{k-1})} = \alpha_{k-1} + v_{k-1} \qquad (8.49)$$

Now, we design a VS stabilizing controller, $v_k(y_k, \ldots, k_k^{(n_i)})$ such that $y_k(t) \to 0$ as $t \to \infty$. For each $k < p$ we smooth the controller so that the process can be continued. Working in this way through the p stages, and redefining the states $(x \to y)$ at each stage we arrive at the final set of dynamical equations.

$$y_i^{(n_i)} = y_{i+1} + \alpha_i + v_i(y_i, \ldots y_i^{(n_i-1)}) \quad i = 1, \ldots, p - 1$$
$$y_p^{(n_p)} = \alpha + \alpha_p + \rho u(y_p, \ldots y_p^{(n_p-1)}) \qquad (8.50)$$

Notice the triangular structure of the transformed nominal system (8.50). It is illustrated in Figure (8.2).

Now, let us define the procedure in detail.

Algorithm 8.263 (Variable Structure Backstepping Algorithm) *The state transformation and control are constructed sequentially as follows:*

1. $k = 1$ define the vector fields $\widehat{f}_1$, g_1 and the scalar function $\widehat{h}_1$:

$$\widehat{f}_1 = \begin{bmatrix} \dot{x}_1 \\ \vdots \\ x_1^{(n_1-1)} \\ 0 \end{bmatrix}, \quad g_1 = \begin{bmatrix} 0 \\ \vdots \\ 0 \\ 1 \end{bmatrix}$$

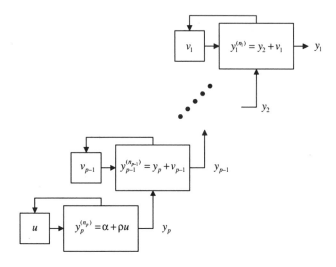

Figure 8.2: The triangular structure of the closed loop (nominal) dynamics achieved with the multistate backstep control design.

$$y_1 = \widehat{h}_1(x_1) = x_1$$

Now define the new state variables:

$$z_1^1 = y_1 = \widehat{h}_1$$
$$z_2^1 = \dot{y}_1 = L_{\widehat{f}_1}\widehat{h}_1$$
$$\vdots$$
$$z_{n_1}^1 = y_1^{(n_1-1)} = L_{\widehat{f}_1}^{n_1-1}\widehat{h}_1$$

which leads to the state space description

$$\dot{Z}^1 = f_1(Z^1) + g_1 v_1 = \begin{bmatrix} z_2^1 \\ \vdots \\ z_{n_1}^1 \\ L_{\widehat{f}_1}^{n_1}\widehat{h}_1 \end{bmatrix} + \begin{bmatrix} 0 \\ \vdots \\ 0 \\ 1 \end{bmatrix} v_1$$

$$y_1 = h_1(Z^1) = z_1^1$$

where $Z^1 = \begin{bmatrix} z_1^1, \ldots, z_{n_1}^1 \end{bmatrix}^T$. This is the state space equivalent to Equation (8.45). Now, design the smoothed variable structure controller v_1.

2. $k = 2, \ldots, p - 1$ Define $\widehat{f}_k$, g_k, and $\widehat{h}_k$

$$\widehat{f}_k = \begin{bmatrix} f_{k-1}(Z^{k-1}) + g_{k-1}v_{k-2}(z^{k-2}) \\ \dot{x}_k \\ \vdots \\ x_k^{(n_k-1)} \\ 0 \end{bmatrix}, \quad g_k = \begin{bmatrix} 0 \\ \vdots \\ \vdots \\ 0 \\ 1 \end{bmatrix}$$

$$y_k = \widehat{h}_k(Z^{k-1}, x_k) = x_k - v_{k-1}(z^{k-1})$$

where

$$Z^{k-1} = \left[(Z^{k-2})^T, z_1^{k-1}, \ldots, z_{n_{k-1}}^{k-1} \right]^T$$

Define the next group of new states

$$z_1^k = y_k = \widehat{h}_k$$
$$z_2^k = \dot{y}_k = L_{\widehat{f}_k} \widehat{h}_k$$
$$\vdots$$
$$z_{n_k}^k = y_k^{(n_k-1)} = L_{\widehat{f}_k}^{n_k-1} \widehat{h}_k$$

Write the state space equivalent to (8.47).

$$\dot{Z}^k = f_k(Z^k) + g_k v_k = \begin{bmatrix} f_{k-1}(Z^{k-1}) + g_{k-1}v_{k-2}(z^{k-2}) \\ z_2^k \\ \vdots \\ z_{n_k}^k \\ L_{\widehat{f}}^{n_k} \widehat{h}_k \end{bmatrix} + \begin{bmatrix} 0 \\ \vdots \\ \vdots \\ 0 \\ 1 \end{bmatrix} v_k$$

$$y_k = h_k(Z^k) = z_1^k - v_{k-1}(z^{k-1})$$

and design the smoothed variable structure control v_k.

3. $k = p$ $\widehat{f}_p$, g_p, and $\widehat{h}_p$ are defined as above for general k. Now introduce the last group of new states

$$z_1^p = y_p = \widehat{h}_p$$
$$z_2^p = \dot{y}_p = L_{\widehat{f}_p} \widehat{h}_p$$
$$\vdots$$
$$z_{n_p}^p = y_p^{(n_p-1)} = L_{\widehat{f}_p}^{n_p-1} \widehat{h}_p$$

to obtain the state space equivalent to (8.48).

$$\dot{Z}^p = f_p(Z^p) + g_p(\alpha + \rho v_p) = \begin{bmatrix} f_{p-1}(Z^{p-1}) + g_{p-1}v_{p-2}(z^{p-2}) \\ z_2^p \\ \vdots \\ z_{n_p}^p \\ L_{\widehat{f}_p}^{n_p} \widehat{h}_p \end{bmatrix} + \begin{bmatrix} 0 \\ \vdots \\ \vdots \\ 0 \\ 1 \end{bmatrix} (\alpha + \rho v_p)$$

$$Z^p = \begin{bmatrix} Z^{p-1} \\ z^p \end{bmatrix} \in R^{n_1 + \cdots + n_p}$$

$$y_p = h_p(Z^p) = z_1^p - v_{p-1}(z^{p-1})$$

Finally, design the variable structure controller v_p.

Now we apply this transformation to the actual system (8.42).

Lemma 8.264 *Consider the transformation defined recursively according to Algorithm (8.263). When applied to the actual system (8.42) the transformed evolution equations are*

$$\begin{aligned} y_i^{(n_i)} &= y_{i+1} + \alpha_i + \Delta_i + v_i(y_i, \ldots y_i^{(n_i-1)}) \quad i = 1, \ldots, p-1 \\ y_p^{(n_p)} &= \alpha + \alpha_p + \Delta_p + \rho u(y_p, \ldots y_p^{(n_p-1)}) \end{aligned} \tag{8.51}$$

Proof: Notice that at each stage of Algorithm (8.263), for $k = 1, \ldots, p-1$, n_k new state variables are defined and n_k first order equations are added to the system. The first $n_k - 1$ equations come from the state definitions, i.e. the defining equations

$$\begin{aligned} z_1^k &= y_k = \widehat{h}_k \\ z_2^k &= \dot{y}_k = L_{\widehat{f}_k} \widehat{h}_k \\ &\vdots \\ z_{n_k}^k &= y_k^{(n_k-1)} = L_{\widehat{f}_k}^{n_k-1} \widehat{h}_k \end{aligned}$$

imply

$$\begin{aligned} \dot{y}_k &= \dot{z}_1^k = z_2^k \\ &\vdots \\ y_k^{(n_k-1)} &= \dot{z}_{n_k-1}^k = z_{n_k}^k \end{aligned}$$

The final equation is obtained by differentiating the last definition and using the evolution equation $x_k^{(n_k)} = v_k$ in the nominal case and $x_k^{(n_k)} = \Delta_k + v_k$ in the actual case, leading to

$$y_k^{(n_k)} = \dot{z}_{n_k}^k = L_{\widehat{f}_k}^{n_k} \widehat{h}_k + L_{\widehat{g}_k} L_{\widehat{f}_k}^{n_k-1} \widehat{h}_k v_k = L_{\widehat{f}_k}^{n_k} \widehat{h}_k + v_k$$

in the nominal case, and

$$y_k^{(n_k)} = \dot{z}_{n_k}^k = L_{\widehat{f}_k}^{n_k} \widehat{h}_k + L_{\widehat{g}_k} L_{\widehat{f}_k}^{n_k-1} \widehat{h}_k (\Delta_k + v_k) = L_{\widehat{f}_k}^{n_k} \widehat{h}_k + \Delta_k + v_k$$

in the actual case.

The case $k = p$ is similar except that $\alpha + \rho v_p$ is replaced by $\alpha + \Delta_p + \rho v_p$. ∎

The idea for establishing stability is roughly as follows. A VS controller is designed for system p, (8.50), via methods described above. The system is stable if and only if the zero dynamics,

$$y_i^{(n_i)} = y_{i+1} + \alpha_i + v_i(y_i, \ldots y_i^{(n_i-1)}) \quad i = 1, \ldots, p-2$$
$$y_{p-1}^{(n_{p-1})} = \alpha_{p-1} + v_{p-1}(y_{p-1}, \ldots y_{p-1}^{(n_{p-1}-1)})$$

(8.52)

are stable. But, v_{p-1} is itself a (smoothed) VS control so that (8.52) is stable if its zero dynamics:

$$y_i^{(n_i)} = y_{i+1} + \alpha_i + v_i(y_i, \ldots y_i^{(n_i-1)}) \quad i = 1, \ldots, p-3$$
$$y_{p-2}^{(n_{p-2})} = \alpha_{p-2} + v_{p-2}(y_{p-2}, \ldots y_{p-2}^{(n_{p-2}-1)})$$

(8.53)

are stable. The argument proceeds in this way.

Proposition 8.265 *Consider the system (8.42) and suppose the uncertainties Δ_i satisfy the inequality (8.43) with continuous bounding functions ε_i, and α also has a continuous bounding function σ_α. Suppose that a controller is designed via the backstepping procedure of Algorithm (8.263) and each control v_k, $k = 1, \ldots, p$ is a smoothed variable structure controller designed in accordance with the assumptions of Proposition (8.261). Then for any given $\delta > 0$ there is a sufficiently small smoothing parameter $\varepsilon > 0$ such that all trajectories enter the ball $\|y\| < \delta$.*

Proof: The p-th system

$$y_p^{(n_p)} = \alpha + \alpha_p + \Delta_p + \rho v_p(y_p, \ldots y_p^{(n_p-1)})$$

(8.54)

satisfies the conditions of Proposition (8.261) with $z_i = y_p^{(i-1)}$, $i = 1, \ldots, n_p$. Hence, we conclude that y_p (and its $n_p - 1$ derivatives) will be driven, in finite time, into a δ-neighborhood of the origin with a suitably small smoothing parameter. Now, the $p - 1$ system is

$$y_{p-1}^{(n_{p-1})} = y_p(t) + \alpha_{p-1} + \Delta_{p-1} + v_{p-1}(y_{p-1}, \ldots y_{p-1}^{(n_{p-1}-1)})$$

(8.55)

and $|y_p(t)| \le \delta$, $\forall t > t^* < \infty$. Thus, we can incorporate $y_p(t)$ into $\Delta_{p-1}(x,t)$. It follows that (8.55) satisfies the conditions of Proposition (8.261) for $t > t^*$, $z_i = y_{p-1}^{(i-1)}$, $i = 1, \ldots, n_{p-1}$, so that y_{p-1} (and its $n_{p-1} - 1$ derivatives) will be driven, in finite time, into a δ-neighborhood of the origin with a suitably small smoothing parameter. We continue in this way for systems $k = p - 2, \ldots, 1$ to establish the conclusion of the theorem. ∎

Example 8.266 (Nonsmooth, Uncertain Friction) *Consider a simple system with a nonlinear, nonsmooth friction:*

$$\dot{\theta} = \omega$$
$$\dot{\omega} = -\phi_{fr}(\omega) + \mu$$
$$\dot{\mu} = u$$

with

$$\phi_{fr} = \mathrm{sgn}(\omega)$$

Suppose the friction to be composed of a smooth nominal part and a nonsmooth uncertain part, i.e.,

$$\phi_{fr} = \tanh(\omega/0.02) + \Delta(\omega), \quad \Delta(\omega) = \mathrm{sgn}(\omega) - \tanh(\omega/\varepsilon), \quad \varepsilon > 0$$

Thus, the nominal system is

$$\dot{\theta} = \omega$$
$$\dot{\omega} = -\tanh(\omega/0.02) + \mu$$
$$\dot{\mu} = u$$

and the uncertainty is bounded by $\sigma_\Delta = const. > 1$.

Now, we complete step 1 and compute the smoothed variable structure (psuedo-) control, vs1:

In[270]:= $f1 = \{\omega, -\tanh[\omega/0.02]\};$
$\quad\quad g1 = \{0, 1\};$
$\quad\quad h1 = \{\theta\};$
$\quad\quad \{rho1, s1\} = SlidingSurface[f1, g1, h1, \{\theta, \omega\}, \{2\}]$

$\quad\quad ctrlbnds = \{\{-5, 5\}\};$
$\quad\quad Q = \{\{1\}\};$
$\quad\quad vsc1 = SwitchingControl[rho1, s1, ctrlbnds, Q, S$
$\quad\quad\quad SmoothingFunctions[x\_] - > \{\tanh[x/0.01]\}]$
Out[270]= $\{-5 \ \mathrm{Tanh}[100. \ \omega + 423.607 \ \theta]\}$

In step 2 we compute the actual control vs2. It is designed without smoothing or moderation.

In[271]:= $f = \{\omega, -\tanh[\omega/0.02] + uu, 0\};$
$\quad\quad g = \{0, 0, 1\};$
$\quad\quad h = \{uu - vsc1[[1]]\};$
$\quad\quad \{rho2, s2\} = SlidingSurface[f, g, h, \{\theta, \omega, uu\}, \{20\}]$

$\quad\quad ctrlbnds = \{\{-5, 5\}\};$
$\quad\quad Q = \{\{1\}\};$
$\quad\quad vsc2 = SwitchingControl[rho2, s2, ctrlbnds, Q]$
Out[271]= $\{5 \ Sign[-uu - 5 \ \mathrm{Tanh}[100. \ \omega + 423.607 \ \theta]]\}$

Now, we set up the equations for numerical computation. Notice that the actual plant friction function is taken to be $\mathrm{sgn}(\omega)$ which corresponds to taking $\varepsilon = 0.02$.

In[272]:= $ReplacementRules =$
$\quad\quad Inner[Rule, \{\theta, \omega, uu\}, \{\theta[t], \omega[t], uu[t]\}, List];$

In[273]:= $\{Surf\} = s2/.ReplacementRules;$
$\quad\quad Surf$

Out[273]= 5 Tanh[100. ω[t] + 423.607 θ[t]] + *uu*[t]

In[274]:= { *VSControl*} = *vsc2*/.*ReplacementRules*;
   VSControl

Out[274]= 5 *Sign*[−5 Tanh[100. ω[t] + 423.607 θ[t]] − *uu*[t]]

In[275]:= *VSsols* =
   NDSolve[{$\partial_t theta$[t] == ω[t], $\partial_t \omega$[t] == −*Sign*[ω[t]] + *uu*[t],
   $\partial_t uu$[t] == *VSControl*, ω[0] == 0.2, θ[0] == 0, *uu*[0] == 0},
   {θ[t], ω[t], *uu*[t]}, {t, 0, 10}, *AccuracyGoal* → 2,
   PrecisionGoal− > 1, *MaxStepSize*− > 10/60000, *MaxSteps* → 60000];

Here are some selected results.

In[276]:= *Plot*[*Evaluate*[{θ[t]} /. *VSsols*],
   {t, 0, 9}, *PlotRange*− > *All*, *AxesLabel* → {t, θ}];
   Plot[*Evaluate*[{ω[t]} /. *VSsols*],
   {t, 0, 9}, *PlotRange*− > *All*, *AxesLabel* → {t, ω}];
   Plot[*Evaluate*[{ *VSControl*}/. *VSsols*],
   {t, 0, 9}, *PlotRange*− > *All*, *AxesLabel* → {t, u}];
   ParametricPlot[*Evaluate*[{θ[t], ω[t]}/. *VSsols*], {t, 0, 9},
   PlotRange− > *All*, *AxesLabel* → {θ, ω}]

Here is the output θ as a function of time. Stability is clearly evident.

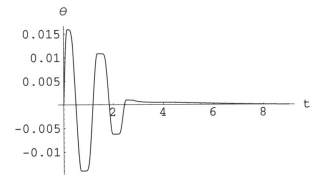

The following plot of angular velocity ω shows the 'stiction' effect of the discontinuous nonlinear friction.

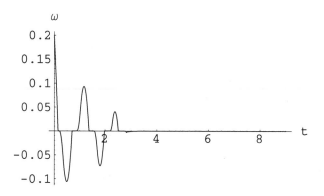

The switching control is shown in the following figure.

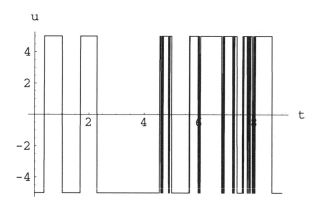

Finally, a phase plot of ω versus θ indicates that theta does not go to the origin. Of course, the ultimate error is controlled by choice of smoothing parameter in step 1.

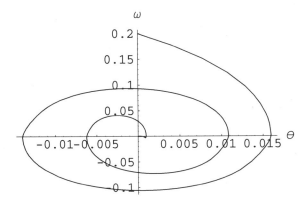

These results clearly illustrate the anticipated properties.

Example 8.267 (Motor-Load System with Nonsmooth friction) *Consider a motor-load system illustrated in Figure (8.3) and described by Equation (8.56).*

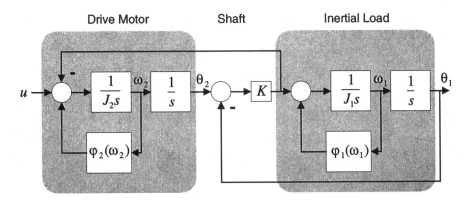

Figure 8.3: A typical drive system consisting of a motor and an inertial load. The nonlinear friction functions, φ_1 and φ_2 contain uncertain discontinuous components.

$$\dot{x} = \begin{bmatrix} \omega_1 \\ \theta_2 - \theta_1 - \omega_1/10 \\ \omega_2 \\ \theta_1 - \theta_2 - \omega_2/2 \end{bmatrix} + \begin{bmatrix} 0 \\ -\frac{1}{10}\left(1 + \frac{1}{10}\exp^{-(\omega_1/0.02)^2}\right)\mathrm{sgn}\omega_1 \\ 0 \\ -\frac{1}{10}\mathrm{sgn}\omega_2 \end{bmatrix} + \begin{bmatrix} 0 \\ 0 \\ 0 \\ 1 \end{bmatrix} u \tag{8.56}$$

We begin by reducing the nominal system to normal form.

In[277]:= $f =$
  $\{omega1, theta2 - theta1 - omega1/10, omega2, theta1 - theta2 - omega2/2\};$
  $g = \{0, 0, 0, 1\};$
  $h = theta1;$

In[278]:= $Df = \{0, -(1 + \exp[-(omega1/0.02)\verb|^|2]/10)\ Sign[omega1]/10, 0,$
  $-Sign[omega2]/10\};$

In[279]:= $\{T1, T2\} =$
  $Chop[SISONormalFormTrans[f, g, h, \{theta1, omega1, theta2, omega2\}]];$

In[280]:= $InvTrans = InverseTransformation[\{theta1, omega1, theta2, omega2\},$
  $\{x1, x2, x3, x4\}, T1];$

InverseTransformation : $\{theta1, omega1, theta2, omega2\} = \{x1, x2, \dfrac{1}{10}\ (10\ x1+x2+10\ x3),$
$\dfrac{1}{10}\ (10\ x2 + x3 + 10\ x4)\}$

The following calculation applies the transformation to the actual (perturbed) system.

In[281]:= $\{fnew, gnew, hnew\} = Chop[TransformSystem[f, g, h,$
  $\{theta1, omega1, theta2, omega2\}, \{x1, x2, x3, x4\}, T1, InvTrans]];$

In[282]:= $\{ff, gg, hh\} = Chop[TransformSystem[f + Df, g, h,$
  $\{theta1, omega1, theta2, omega2\}, \{x1, x2, x3, x4\}, T1, InvTrans]];$

In[283]:= $N[ff]$

Out[283]= $\Big\{x2, x3 + 0.01\ \Big(-10. - 1.\ 2.71828^{-2500.\ \times2^2}\Big)\ Sign[x2],$
  $x4 + 0.001\ \Big(10. + 2.71828^{-2500.\ \times2^2}\Big)\ Sign[x2],$
  $0.0099\ \Big(10. + 2.71828^{-2500.\ \times2^2}\Big)\ Sign[x2]+$
  $0.05\ (-12.\ x2 - 41.\ x3 - 12.\ x4 - 2.\ Sign[x2 + 0.1\ x3 + x4])\Big\}$

Now, the backstepping procedure is applied. Observe the structure of the actual system in normal form. Because uncertainties enter the right hand sides of the second, third and fourth equations, three steps will be required.

1st Step

In[284]:= $f1 = \{x2, 0\};$
  $g1\ =\ \{0, 1\};$
  $h1 = \{x1\};$
  $\{rho1, s1\}\ =\ SlidingSurface[f1, g1, h1, \{x1, x2\}, \{2\}];$

In[285]:= $ctrlbnds\ =\ \{\{-1, 1\}\};$
  $Q\ =\ \{\{1\}\};$
  $vsc1 = SwitchingControl[rho1, s1, ctrlbnds, Q,$
  $SmoothingFunctions[x\_]-> \{\tanh[x/0.1]\}];$

2nd Step

$In[286]:=$ $f2 = \{x2, x3, 0\};$
$g2 = \{0, 0, 1\};$
$h2 = \{x3 - vsc1[[1]]\};$
$\{rho2, s2\} = SlidingSurface[f2, g2, h2, \{x1, x2, x3\}, \{5\}];$

$In[287]:=$ $ctrlbnds = \{\{-5, 5\}\};$
$Q = \{\{1\}\};$
$vsc2 = SwitchingControl[rho2, s2, ctrlbnds, Q,$
$SmoothingFunctions[x\_] - > \{\tanh[x/0.1]\}];$

$In[288]:=$ $\{T1, T2\} = Chop[SISONormalFormTrans[f2, g2, h2, \{x1, x2, x3\}]];$

3rd Step

$In[289]:=$ $f3 = fnew;$
$g3 = gnew;$
$h3 = Chop[SetAccuracy[\{x4 - vsc2[[1]]\}, 4], 10^{\wedge}(-4)];$

It is important to get an estimate of bounds on α in order to set appropriate control bounds.

$In[290]:=$ $\{\rho, \alpha, ro, control\} = IOLinearize[f3, g3, h3, \{x1, x2, x3, x4\}];$

$In[291]:=$ $Coefficient[Truncate[\alpha[[1]], \{x1, x2, x3, x4\}, 1], \{x1, x2, x3, x4\}]$
$Out[291]=$ $\{0, 2117., 498., 49.4\}$

$In[292]:=$ $b = 5 (1 + 10 \, Abs[x4] + 100 \, Abs[x3] + 500 \, Abs[x2]);$

As it turns out, these bounds are fairly tight. Reducing them significantly rsults in very degraded performance – even of the nominal system.

$In[293]:=$ $\{rho3, s3\} = SlidingSurface[f3, g3, h3, \{x1, x2, z3, x4\}, \{20\}];$

$In[294]:=$ $ctrlbnds = \{\{-b, b\}\};$
$Q = \{\{1\}\};$
$vsc3 = SwitchingControl[rho3, s3, ctrlbnds, Q,$
$SmoothingFunctions[x\_] - > \{\tanh[x/0.04]\}];$

Simulation of the Actual Plant

$In[295]:=$ $Eqns = MakeODEs[\{x1, x2, x3, x4\}, ff + gg \, vsc3[[1]], t];$
$InitialConds = \{x1[0] == 0, x2[0] == 0.2, x3[0] == 0, x4[0] == 0\};$

$In[296]:=$ $VSsols = NDSolve[Join[Eqns, InitialConds],$
$\{x1[t], x2[t], x3[t], x4[t]\}, \{t, 0, 4\}, AccuracyGoal \rightarrow 2,$
$PrecisionGoal- > 1, MaxStepSize- > 4/60000, MaxSteps \rightarrow 60000];$

In[297]:= *Plot*[*Evaluate*[{*x1*[*t*]} /. *VSsols*],
         {*t*, 0, 2}, *PlotRange−* > *All*, *AxesLabel* → {*t*, *θ1*}];
       Plot[*Evaluate*[{*x2*[*t*]} /. *VSsols*],
         {*t*, 0, 2}, *PlotRange−* > *All*, *AxesLabel* → {*t*, *ω1*}];
       Plot[*Evaluate*[{*x3*[*t*]} /. *VSsols*],
         {*t*, 0, 2}, *PlotRange−* > *All*, *AxesLabel* → {*t*, *x3*}];
       Plot[*Evaluate*[{*x4*[*t*]} /. *VSsols*], {*t*, 0, 2}, *PlotRange−* > *All*,
         AxesLabel → {*t*, *x4*}];

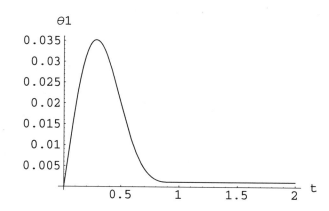

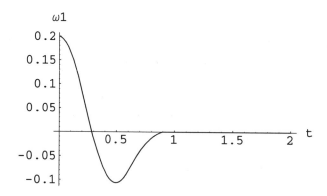

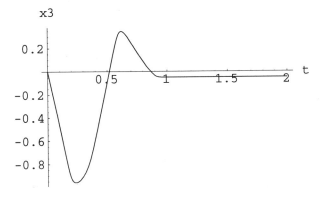

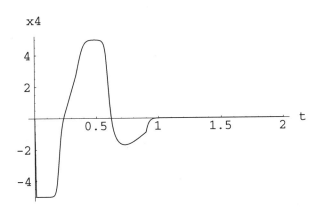

$In[298] :=$ $Control = vsc3[[1]] /. ReplacementRules;$
$Plot[Evaluate[\{Control\} /. VSsols], \{t, 0, 2\}, AxesLabel \to \{t, u\}];$

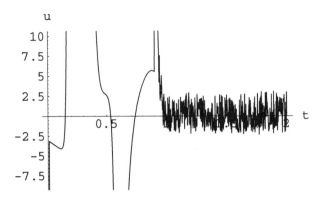

Notice that the position error does not reduce to zero. This is as expected, because of the smoothing of the controllers. By decreasing the smoothing parameter, of course, the error is reduce. On the other hand the (peak) control effort increases. As it is, control effort is quite substantial.

Simulation with Different Initial Conditions

The following computations illustrate response from a different set of initial conditions.

In[299]:=   InitialConds = {x1[0] == 0.2, x2[0] == 0, x3[0] == 0, x4[0] == 0};

In[300]:=   VSsols = NDSolve[Join[Eqns, InitialConds],
            {x1[t], x2[t], x3[t], x4[t]}, {t, 0, 10}, AccuracyGoal → 2,
            PrecisionGoal− > 1, MaxStepSize− > 10/60000, MaxSteps → 60000];

In[301]:=   Plot[Evaluate[{x1[t]} /. VSsols],
            {t, 0, 4}, PlotRange− > All, AxesLabel → {t, θ1}];
            Plot[Evaluate[{x2[t]} /. VSsols],
            {t, 0, 4}, PlotRange− > All, AxesLabel → {t, ω1}];
            Plot[Evaluate[{x3[t]} /. VSsols],
            {t, 0, 4}, PlotRange− > All, AxesLabel → {t, x3}];
            Plot[Evaluate[{x4[t]} /. VSsols], {t, 0, 4}, PlotRange− > All,
            AxesLabel → {t, x4}];

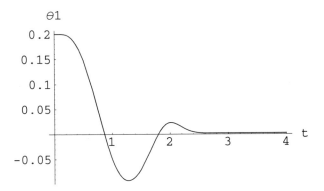

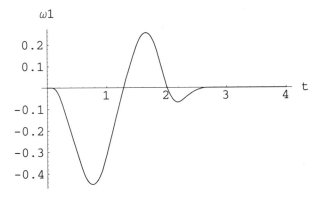

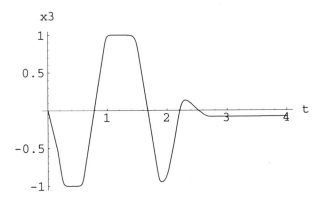

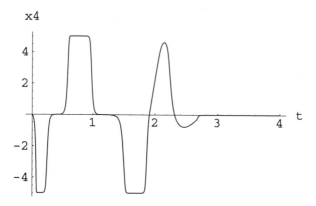

$In[302] := Plot[Evaluate[\{Control\}/.VSsols], \{t, 0, 4\}, AxesLabel \rightarrow \{t, u\}];$

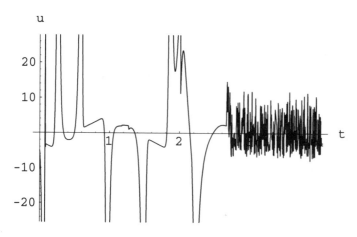

It appears that from these initial conditions the ultimate error is quite small, but it is not zero. Notice also the stiction effects. The control plots give a clear indication of where sliding begins.

8.9 Problems

Problem 8.268 *Consider the magnetic suspension system shown in Figure (8.4) Consider the voltage $v(t)$ to be the control input. The attracting force*

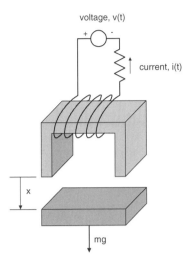

Figure 8.4: Magnetic suspension system.

suspending the mass is

$$F = \frac{k_1 i^2}{(x + k_2)^2}$$

Suppose the circuit resistence is R and the combined inductance of the coil and suspended mass is L.

(a) *Develop Lagrange's equations for the system.*

(b) *Design a variable structure control system. Assume all states are available for measurement. Assume that the mass, m, and the constants k_1 and k_2 are uncertain and can vary ±20% from their nominal values.*

(c) *Develop a simulation of the control system designed in (b).*

(d) *Assume that only the current and gap width can be measured and incorporate an observer in the control design. Compare the state feedback and observer based designs via simulation.*

(e) *Design a state feedback adaptive controller and compare its performance with the controller of (b) by simulation.*

Problem 8.269 *Repeat Problem (7.249) using a variable structure controller.*

Problem 8.270 *Repeat Problem (7.248) using a variable structure controller.*

Problem 8.271 (Synchronous motor, revisited) *Consider the synchronous motor described in Problem (5.154) Suppose that the load torque T_L can be*

measured. Design a variable structure control using the four control variables
v_d, v_q, v_0, v_f. *Assume that there is a power supply that provides a dc voltage of*
$\pm V_S$ *for the stator and* $\pm V_f$ *for the field. The outputs to be regulated are defined*
as follows:

1. *Speed regulation. Define*

$$y_1 = \dot{\omega} + c(\omega - \omega_0) = \left(-\sqrt{\tfrac{3}{2}}L_5 i_f i_q - T_L\right)/J + c(\omega - \omega_0), \; c > 0$$

   where ω_0 *is the desired speed.*

2. *Balanced motor operation. Normally, a 3-phase machine is driven with*
   line voltages v_1, v_2, v_3 *that are sinusoids of the same magnetude and fre-*
   quency and 120 degrees out of phase. Thus, they sum to zero. Some
   deviation from this balance will be allowed, but to regulate it introduce the
   new state

$$\chi = \int_0^t v_0 \, dt$$

   (recall, $v_0 = (v_1 + v_2 + v_3)/\sqrt{3}$*), and define the output*

$$y_2 = \chi$$

3. *Constant d-axis current. Define*

$$y_3 = i_d - i_{d0}$$

   where i_{d0} *is assumed given.*

4. *Constant field current. Define*

$$y_4 = i_f - i_{f0}$$

   where i_{f0} *is also given.*

Some motivation for choosing this set of regulated output follows from the ob-
servation that the electrical torque is $T_E = \sqrt{\tfrac{3}{2}} L_5 i_f i_q$. *Thus, we regulate to*
an equilibrium point in which i_d *and* i_f *assume specified constant values and* i_q
takes a value that insures $T_E = T_L$. *Since the stator current magnitude is* $I_s = \sqrt{i_d^2 + i_q^2}$, *it is not dificult to show that, in steady-state,* $T_E = \sqrt{\tfrac{3}{2}} L_5 i_{f0} I_s \sin\phi$
where ϕ *is the usual power angle, i.e., the angle between the stator current and*
voltage phasors.

Are the zero dynamics stable?

8.10   References

1. Utkin, V.I., Sliding Modes and Their Application. 1974 (in Russian) 1978 (in English), Moscow: MIR.

2. Filippov, A.F., A Differential Equation with Discontinuous Right Hand Side. Matemat: Cheskii Sbornik, 1960. 51(1): p. 99128.

3. Kwatny, H.G. and J. Berg, Variable Structure Regulation of Power Plant Drum Level, in Systems and Control Theory for Power Systems, J. Chow, R.J. Thomas, and P.V. Kokotovic, Editors. 1995, SpringerVerlag: New York. p. 205-234.

4. Kwatny, H.G. and H. Kim, Variable Structure Regulation of Partially Linearizable Dynamics. Systems and Control Letters, 1990. 15: p. 6780.

5. Marino, R., High Gain Feedback NonLinear Control Systems. International Joural of Control, 1985. 42(6): p. 13691385.

6. Kwatny, H.G., Variable Structure Control of AC Drives, in Variable Structure Control for Robotics and Aerospace Applications, K.D. Young, Editor. 1993, Elsevier: Amsterdam.

7. Young, K.D., P.V. Kokotovic, and V.I. Utkin, Singular Perturbation Analysis of High Gain Feedback Systems. IEEE Transactions on Automatic Control, 1977. AC22(6): p. 931938.

8. Young, K.D. and H.G. Kwatny, Variable Structure Servomechanism and its Application to Overspeed Protection Control. Automatica, 1982. 18(4): p. 385-400.

9. Slotine, J.J. and S.S. Sastry, Tracking Control of NonLinear Systems Using Sliding Surfaces, With Application to Robot Manipulators. International Journal of Control, 1983. 38(2): p. 465492.

10. Slotine, J.J.E., Sliding Controller Design for NonLinear Control Systems. International Journal of Control, 1984. 40(2): p. 421434.

11. Emelyanov, S.V., S.K. Korovin, and L.V. Levantovsky, A Drift Algorithm in Control of Uncertain Processes. Problems of Control and Information Theory, 1986. 15(6): p. 425438.

12. Kwatny, H.G. and T.L. Siu, Chattering in Variable Structure Feedback Systems, Proceedings 10th IFAC World Congress, 1987, Munich.

Appendix A

ProPac

A.1 Getting Help

The *Mathematica* package *ProPac* is an integral part of this book. *ProPac* contains subpackages for multibody dynamics, linear control, and nonlinear control. Once it is installed, as described in Chapter 1, appropriate packages will be loaded automatically as they are required. However, individual packages can be manually loaded by simply entering GeometricTools, Dynamics, LinearControl, NonlinearControl, or MEXTools as desired. Once a package is loaded, enter ?GeometricTools, ?Dynamics, ?LinearControl, ?NonlinearControl, or ?MEXTools, respectively, to obtain a complete list of available functions. Then enter ?FunctionName to obtain usage information for the function `FunctionName`. After *ProPac* is installed, the *Mathematica* Help index should be rebuilt as described in Chapter 1. When this is done, help will also be available in the Help Browser under Add-ons.

The CD that accompanies this book includes several *Mathematica* notebooks that illustrate the use of *ProPac*. The notebooks, Dynamics.nb and Controls.nb are intended to give an overview of the available functions.

Of course, all standard *Mathematica* functions and packages are available and *ProPac* is compatible with the *Mathematica* package Control Systems Profesional, available from Wolfram Research.

A.2 Quick Reference Tables

The following tables provide a summary of the available functions. They are not all inclusive. A complete list of available functions can be obtained as described in the previous paragraph.

Function Name	Operation
Bode	produces a Bode plot of the transfer function of a (scalar) continuous time system
RootLocus	generates the root locus plot for a given transfer function
Nyquist	generates the Nyquist plot for a given transfer function
ColorNyquist	generates a color version of the Nyquist plot
PhasePortrait	computes a family of state space trajectories for a vector field on R^2 and returns a list of graphics objects

Table A.1: Graphics Functions

Function Name	Operation
ControllablePair	test for controllability
ObservablePair	test for observability
ControllabilityMatrix	returns the controllability matrix
ObservabilityMatrix	returns the observability matrix
PolePlace	state feedback pole placement based on Ackermann's formula with options
DecouplingConrol	state feedback and coordinate transformation that decouples input-output map
RelativeDegree	computes the vector relative degree
LyapunovEquation	computes the solution, P, of $A^T + PA = -Q$
AlgebraicRiccatiEquation	computes the positive solution of the algebraic Riccati equation
LQR, LQE	compute optimal quadratic regulator and estimator parameters

Table A.2: Linear Systems: Time Domain

Function Name	Operation
LeastCommonDenominator	finds the least common denominator of the elements of a proper, rational G(s)
Poles	finds the roots of the least common denominator
LaurentSeries	computes the Laurent series up to specified order
AssociatedHankelMatrix	computes the Hankel matrix associated with Laurent expansion of G(s)
McMillanDegree	computes the degree of the minimal realization of G(s)
RelativeDegree	computes the relative degree of a linear system
ControllableRealization	computes the controllable realization of a transfer function
ObservableRealization	computes the observable realization of a transfer function
KalmanDecomposition	returns a Kalman decomposition of a linear system

Table A.3: Linear Systems: Frequency Domain

Function Name	Operation
LieBracket	computes the Lie bracket of a given pair of vector fields
Ad	computes the iterated Lie bracket of specified order of a pair of vector fields
Involutive	tests a set of vector fields to determine if it is involutive
Span	generates a set of basis vector fields for a given set of vector fields
TriangularDecomposition	computes the transformation that trangularizes a vector field from a given involutive distribution, invariant with respect to the vector field
SmallestInvariantDistribution	Computes the smallest distribution containing a given distribution and invariant with respect to a set of vector fields
LargestInvariantDistribution	Computes the largets distribution contained in the annihilator of an exact codistribution and invariant with respect to a set of vector fields

Table A.4: Geometry Tools

Function Name	Operation
Joints	returns all of the kinematic quantities corresponding to a list of joint definitions
TreeInertia	computes the inertia matrix of a multibody system in a tree structure containing flexible and rigid bodies
EndEffector	returns the Euclidean Configuration Matrix of a body fixed frame at a specified node
NodeVelocity	returns the (6 dim) spatial velocity vector of a body fixed frame at a specified node
GeneralizedForce	computes the generalized force at specified node in terms of generalized coordinates
RelativeConfiguration	computes the relative configuration of body fixed frames at specified nodes
KinematicReplacements	sets up temporary replacement rules for repeated groups of expressions to simplify kinematic quantities

Table A.5: Kinematics

Function Name	Operation
TreeInertia	generates the spatial inertia of a tree structure
LeafPotential	returns the elastic potential energy associated with leaf absolute position in terms of the system generalized coordinates
BacklashPotential	Returns the Hertz impact potential associated with a specified material potential
JointFriction	assembles a dissipation function of Lur'e type for a joint that involves viscous, Coulomb and Stribeck effects
CreateModel	builds the kinematic and dynamic equations for tree structures
DifferentialConstraints	adds differential constraints to a tree configuration
AlgebraicConstraints	adds algebraic constraints to a tree configuration
MakeODEs	assembles differential equations in a form that can be integrated in Mathematica
MakeLagrangeEquations	assembles Lagrange's equations in a form that can be integrated in Mathematica

Table A.6: Dynamics

Function Name	Operation
ControlDistribution	computes the controllability distribution of a nonlinear (affine) system
Controllablity	test for controllability of a nonlinear (affine) system
ObservabilityCodistribution	computes the observability codistribution of a nonlinear (affine) system
Observability	test for observability of a nonlinear (affine) system
LocalDecomposition	computes a transformation that puts a nonlinear (affine) system into Kalman-partitioned form

Table A.7: Nonlinear Controllability and Observability

Function Name	Operation
SISONormalFormTrans	Computes the transformation taking an IO linearizable SISO system to its normal form
VectorRelativeOrder	computes the relative degree vector
DecouplingMatrix	computes the decoupling matrix, $\rho(x)$
IOLinearize	computes the linearizing control, $u = \rho^{-1}\left\{-\alpha(x) + v\right\}$
NormalCoordinates	computes the partial state transformation, $z(x)$
LocalZeroDynamics	computes the local form of the zero dynamics $F(\xi, 0)$, near x_0
DynamicExtension	implements dynamic extension process to produce a nonsingular decoupling matrix when posible
StructureAlgorithm	implements the Hirschorn-Singh structure algorithm for assembling a dynamic inverse

Table A.8: Feedback Linearizing Functions

Function Name	Operation
AdaptiveRegulator	generates an adaptive regulator for a class of linearizable systems
AdaptiveBackstepRegulator	computes an adaptive regulator by backstepping for SISO systems in PSFF form
AdaptiveTracking	computes an adaptive tracking controller
PSFFCond	tests a system to determine if it is reducible to PSFF form
PSFFSolve	transforms a system to PSFF form if possible

Table A.9: Adaptive Control

Function Name	Operation
SlidingSurface	generates the sliding (switching) surface for feedback linearizable nonlinear systems
SwitchingControl	computes the switching functions - allows the inclusion of smoothing and moderating functions

Table A.10: Variable Structure Control

Index

RECEIVED

DEC 2001

Mission College
Library